GUIDELINES FOR
PROCESS SAFETY DOCUMENTATION

Publications Available from the
CENTER FOR CHEMICAL PROCESS SAFETY
of the
AMERICAN INSTITUTE OF CHEMICAL ENGINEERS

Guidelines for Process Safety Documentation
Guidelines for Safe Process Operations and Maintenance
Guidelines for Process Safety Fundamantals in General Plant Operations
Guidelines for Chemical Reactivity Evaluation and Application to Process Design
Tools for Making Acute Risk Decisions with Chemical Process Safety Applications
Guidelines for Preventing Human Error in Process Safety
Guidelines for Evaluating the Characteristics of Vapor Cloud Explosions, Flash Fires, and BLEVEs
Guidelines for Implementing Process Safety Management Systems
Guidelines for Safe Automation of Chemical Processes
Guidelines for Engineering Design for Process Safety
Guidelines for Auditing Process Safety Management Systems
Guidelines for Investigating Chemical Process Incidents
Guidelines for Hazard Evaluation Procedures, Second Edition with Worked Examples
Plant Guidelines for Technical Management of Chemical Process Safety, Rev. Ed.
Guidelines for Technical Management of Chemical Process Safety
Guidelines for Chemical Process Quantitative Risk Analysis
Guidelines for Process Equipment Reliability Data, with Data Tables
Guidelines for Vapor Release Mitigation
Guidelines for Safe Storage and Handling of High Toxic Hazard Materials
Guidelines for Use of Vapor Cloud Dispersion Models
Safety, Health, and Loss Prevention in Chemical Processes: Problems for Undergraduate Engineering Curricula
Safety, Health, and Loss Prevention in Chemical Processes: Problems for Undergraduate Engineering Curricula—Instructor's Guide
Workbook of Test Cases for Vapor Cloud Source Dispersion Models
Proceedings of the International Symposium and Workshop on Safe Chemical Process Automation, 1994
Proceedings of the International Process Safety Management Conference and Workshop, 1993
Proceedings of the International Conference on Hazard Identification and Risk Analysis, Human Factors, and Human Reliability in Process Safety, 1992
Proceedings of the International Conference/Workshop on Modeling and Mitigating the Consequences of Accidental Releases of Hazardous Materials, 1991.
Proceedings of the International Symposium on Runaway Reactions, 1989
CCPS/AIChE Directory of Chemical Process Safety Services

GUIDELINES FOR
PROCESS SAFETY DOCUMENTATION

CENTER FOR CHEMICAL PROCESS SAFETY

of the

AMERICAN INSTITUTE OF CHEMICAL ENGINEERS

345 East 47th Street • New York, NY 10017

Library of Congress Cataloging-in Publication Data
Guidelines for process safety documentation.
p. cm.
Includes bibliographic references and index.
ISBN 0-8169-0625-4 : $120.00
1. Chemical industry—Safety measures. I. American Institute of Chemical Engineers. Center for Chemical Process Safety.
TP149.G8365 1994
660'. 2804—dc20 94–22854
CIP

PRINTED IN THE UNITED STATES OF AMERICA
5 4 3 2 1 1 2 3 4 5

It is sincerely hoped that the information presented in this document will lead to an even more impressive safety record for the entire industry; however, the American Institute of Chemical Engineers, its consultants, CCPS subcommittee members, their employers, their employers' officers and directors, and John Brown, Stamford, Connecticut disclaim making or giving any warranties or representations, express or implied, including with respect to fitness, intended purpose, use or merchantability and/or correctness or accuracy of the content of the information presented in this document. As between (1) the American Institute of Chemical Engineers, its consultants, CCPS subcommittee members, their employers, their employers' officers and directors, and John Brown, Stamford, Connecticut and (2) the user of this document, the user accepts any legal liability or responsibility whatsoever for the consequence of its use or misuse.

Contents

Preface

The American Institute of Chemical Engineers (AIChE) has been involved with process safety issues in the chemical and allied industries for many years. AIChE has fostered continuous improvement of the process industry's high safety standards through its strong relationship with designers, constructors, operators, safety professionals, and members of academia.

The Center for Chemical Process Safety (CCPS) was established by the AIChE in 1985 to develop and disseminate technical information for use in the prevention of major chemical accidents. CCPS publications have become major resources for those dedicated to understanding the causes of accidents and developing better means of preventing their occurrence and mitigating their consequences.

From its beginning, CCPS recognized that enhancements in chemical process technologies alone would not be sufficient to prevent catastrophic incidents. Therefore, a multifaceted program was established to address the need for technical management commitment and technical management systems in industry to better protect employees and the public. The framework of the CCPS process safety management (PSM) model and its twelve elements is described in the *Guidelines for Technical Management of Chemical Process Safety* which was published in 1989. These twelve elements are reprinted as Table 1-2 in this book.

Implementation of these PSM elements generates a vast amount of information, much of which must be documented and protected for future use. These *Guidelines* provide detailed guidance on establishing the types and amount of information to be recorded, various alternatives for developing record management systems, and record retention and retrieval programs to ensure a viable corporate memory for this PSM-relevant information.

It is recognized that documentation needs will vary from facility to facility; therefore, the guidance presented here should not be interpreted to be universally applicable requirements. It is the responsibility of each organization to identify those documents and documentation practices that are pertinent to its situation.

The Chair of the Process Safety Management Documentation Subcommittee was Walt Frank of DuPont. The Subcommittee members were Andy Heman,

Fluor Daniel; Terry Swanson, Monsanto; Carl Brown, Brown & Root; Phil Rasch, Hoechst-Celanese; Wayne Williamsen, International Paper; Mark George, MK Ferguson; Peter Puglionesi, Roy F. Weston; Robert Wade, Amoco; Charles Twardowski, ICI Americas; Mark Edison and Mike Sherrod, Stone & Webster; Jay Giffin, Union Carbide; Prabir Basu, Searle; and Robert Rosen, BASF. Ray E. Witter was the CCPS staff liaison and was responsible for the overall administration of the project.

The American Institute of Chemical Engineers and the Center for Chemical Process Safety thanks all of the members of the Process Safety Management Documentation Subcommittee for their dedicated efforts and technical contributions to the preparation of the *Guidelines*. CCPS also expresses appreciation to the members of the Technical Steering Committee for their advice and support.

The members of the Process Safety Management Documentation Subcommittee also wish to thank their employers for providing time to participate in this project and to the many sponsors whose funding made this project possible.

John Brown, Stamford Connecticut, was the contractor for this project. Colin Harris and Jim Coyle were the principal authors and editors. Contributing authors were Robert Slaughter, Harold Dorbin, Donald Byrer, Judith Papp, George Doyle, Gabriel Cordova, Stephen Smegal and Norman Madoian. Frank H. Sawberger of LaPorte, Texas was the Technical Writer. Frank D'Erasmo and Donald Byrer provided graphics support. Last but not least, Janet Camarro, diligently provided secretarial support and typed the manuscript throughout the project.

CCPS also gratefully acknowledges the comments and the suggestions submitted by the following peer reviewers: John Anderson, DuPont; Deric Crosby, PCR, Inc.; Daniel Crowl, Michigan Technological University; Mike Deak, DuPont; William Early, Myers & Early, Ltd.; Manuel Ehrlich, BASF; Randy Freeman, Monsanto; Don Frikken, Monsanto; Barry Gibson, DuPont; Thomas Gibson, Dow Chemical; Kathleen Haines, Zeneca; Dennis C. Hendershot, Rohm and Haas Company; John Hudson, PCR, Inc.; Henry Hyde, BASF; Thomas Lawrence, Monsanto; Joe Louvar, BASF; Victor Maggioli, Feltronics; D.H. Meek, Arco Chemical; David Rhyne, Fluor Daniel; Bruce Sellars, Fluor Daniel; Leslie Scher, W. R. Grace; James Thompson, DuPont; Douglas Turner, Process Safety Engineering; Kent Underwood, Monsanto; and Les Wittenberg, CCPS.

Their insights, comments, and suggestions helped ensure a balanced perspective for the *Guidelines*.

Acronyms

(An additional list of acronyms of organization names appears in Appendix 16A)

ACGIH	American Conference of Governmental Industrial Hygienists
ACRC	Association of Commercial Records Centers
ADS	Architectural Design Specification
AIChE	American Institute of Chemical Engineers
AIIM	Association for Information and Image Management
ANSI	American National Standards Institute
API	American Petroleum Institute
ASME	American Society of Mechanical Engineers
ASTM	American Society for Testing Materials
ATC	Acute Toxicity Concentration
ATSDR	Agency for Toxic Substances and Disease Registry
BLEVE	Boiling Liquid Expanding Vapor Explosion
BNA	Bureau of National Affairs
BPCS	Basic Process Control System
CAD	Computer Aided Drafting
CADET	Critical Action and Decision Evaluation Techniques
CAS	Chemical Abstracts System
CCA	Cause–Consequence Analysis
CCF	Common Cause Failure
CCOHS	Canadian Center for Occupational Health and Safety
CCPS	Center for Chemical Process Safety
CD	Compact Disk
CD-ROM	Compact Disk—Read-Only Memory
CEI	Chemical Exposure Index
CEO	Chief Executive Officer
CERCLA	Comprehensive Environmental Response, Compensation, and Liability Act (of 1980)
CFR	Code of Federal Regulations
CHEMA	Critical Human Error Mode Assessment

CHEMNET	A mutual aid network of chemical shippers and contractors
CHEMTREC	Chemical Transport Emergency Center (Chemical Manufacturers Association)
CHLOREP	Chlorine Emergency Plan (Chlorine Institute)
CI	Chlorine Institute
CL	Checklist Analysis
CM	Configuration Management
CMA	Chemical Manufacturers Association
CPI	Chemical Process Industry
CPQRA	Chemical Process Quantified Risk Analysis
CRP	Close Range Photogrammetry
DCS	Distributed Control System
DIERS	Design Institute for Emergency Relief Systems (AIChE)
DIPPR	Design Institute for Physical Properties (AIChE)
DOT	Department of Transportation
EBS	Emergency Broadcast System
EFCE	European Federation of Chemical Engineers
EFD	Engineering Flow Diagram
EIM	Electronic Information Management
ELS	Error Likely Situations
EOC	Emergency Operating Center
EPA	Environmental Protection Agency
EPRI	Electrical Power Research Institute
ERPG	Emergency Response Planning Guidelines
ERT	Emergency Response Team
ETA	Event Tree Analysis
F&EI	Fire and Explosion Index
FAR	Fatal Accident Rate
FAT	Factory Acceptance Test
FEMA	Federal Emergency Management Agency
FMEA	Failure Modes and Effects Analysis
FMECA	Failure Modes, Effects, and Criticality Analysis
FR	Federal Register
FTA	Fault Tree Analysis
HAZCOM	Hazard Communication
HAZMAT	Hazardous Material
HAZOP	Hazard and Operability Study
HAZWOPER	Hazardous Waste Operations and Emergency Response
HE	Hazard Evaluation
HEP	Hazard Evaluation Procedures
HF	Human Factors
HFA	Human Factors Analysis
HRA	Human Reliability Analysis
HTA	Hierarchial Task Analysis

HVAC	Heating, Ventilating, and Air Conditioning
IARC	International Agency for Research on Cancer
IChemE	Institution of Chemical Engineers (United Kingdom)
IDA	Influence Diagram Approach
IDLH	Immediately Dangerous to Life and Health
IEEE	Institute of Electrical and Electronic Engineers
IFMA	International Facilities Management Association
IMAS	Influence Modelling and Assessment Systems
IMO	International Maritime Organization
I/O	Input/Output
ISA	Instrument Society of America
ISO	International Standards Organization
IT	Information Technology
LAN	Local Area Network
LCD	Liquid Crystal Display
LC_{50}	Lethal Concentration, 50% mortality
LEPC	Local Emergency Planning Committee
LFL	Lower Flammable Limit
LV	Low Voltage
MACT	Maximum Achievable Control Technology
MCC	Motor Control Center
MCS	Minimal Cut Set
MMI	Man–Machine Interface
MOC	Management of Change
MSDS	Material Safety Data Sheet
NACA	National Agricultural Chemical Association
NACE	National Association of Corrosion Engineers
NBS	National Bureau of Standards
NDT	Non-Destructive Testing
NEC	National Electric Code
NEMA	National Electrical Manufacturers Association
NFPA	National Fire Protection Association
NIOSH	National Institute of Occupational Safety and Health
NISO	National Information Standards Organization
NPCA-HIMS	National Paint and Coating Association—Hazardous Materials Identification System
NPP	National Presentation Program
NRT	National Response Team
NSF	National Sanitation Foundation
NTP	National Toxicology Program
OAET	Operator Action Event Tree
OCR	Optical Character Recognition
OJT	On-the-Job Training
OSC	On-Scene Coordinator

OSD	Operational Sequence Diagrams
OSHA	Occupational Safety and Health Administration
P&ID	Piping and Instrumentation Diagram
PC	Personal Computer
PEL	Permissible Exposure Limit
PES	Programmable Electronic Systems
PFD	Process Flow Diagram
PHA	Process Hazard Analysis
PHEA	Predictive Human Error Analysis
PIF	Performance Influencing Factors
PLC	Programmable Logic Controller
PSF	Performance Shaping Factors
PSI	Process Safety Information
PSM	Process Safety Management
QA	Quality Assurance
R&D	Research and Development
RCRA	Resource Conservation and Recovery Act (of 1976)
RMPP	Risk Management and Prevention Program (of California)
RP	Recommended Practice
RRT	Regional Response Team
SAMA	Scientific Apparatus Makers Association
SARA	New York State Archives and Records Administration
SARA	Superfund Amendments Reauthorization Act
SCCM	Subjective Cause–Consequence Models
SCF	System Change Form
SFG	Signal-Flow Graph
SHI	Substance Hazard Index
SIC	Standard Industry Classification
SIS	Safety Interlock System
SLI	Success Likelihood Index
SLIM	Success Likelihood Index Method
SRD	Safety and Reliability Directorate (of the United Kingdom Atomic Energy Authority)
STEL	Short Term Exposure Limit
STP	Standard Temperature and Pressure
TBMS	Text-Based Management System
TCPA	Toxic Catastrophe Prevention Act (of New Jersey)
TEMA	Tubular Exchanger Manufacturers Association
THERP	Technique for Human Error Rate Predictions
TLV	Threshold Limit Value
TPQ	Threshold Planning Quantity
TSCA	Toxic Substances Control Act
TWA	Time Weighted Average
UFD	Utilities Flow Diagram

UFL	Upper Flammable Limit
UKAEA	United Kingdom Atomic Energy Authority
UL	Underwriters Laboratory
UN	United Nations
UPS	Uninterruptible Power Supply
USCG	United States Coast Guard
VCE	Vapor Cloud Explosion
VDU	Visual Display Unit
WAN	Wide Area Network
WI	What-If Analysis
WICL	What-If/Checklist Analysis
WORM	Write-Once, Read-Many

Glossary

Accountability: The obligation to answer for one's actions that are related to an organization's goals and objectives

Consequence: The direct, undesirable result of an accident sequence usually involving a fire, explosion, or release of toxic material. Consequence descriptions may be qualitative or quantitative estimates of the effects of an accident in terms of factors such as health impacts, economic loss, and environmental damage.

Distributed Control System (DCS): One type of controller in a programmable electronic system. A multi-input, multi-output, computer-based controller.

Error Likely Situations (ELS): A work situation in which the performance influencing (or shaping) factors are not compatible with the capabilities, limitations, or needs of the worker. In such situations, workers are much more likely to make mistakes, particularly under stressful conditions.

F-N Curve: A plot of cumulative frequency versus consequences (expressed as number of fatalities)

Hazard Scenario Identification: The process whereby possible malfunctions are reviewed to permit the identification of realistic hazard scenarios. These can then be modelled.

Human Reliability Analysis (HRA): A method used to evaluate whether necessary human actions, tasks, or jobs will be completed successfully within a required time period. Also used to determine the probability that no extraneous human actions detrimental to the system will be performed.

Individual Risk Contour: A curve drawn on a map which joins all points of equal risk to an individual (also known as an Iso-Risk Contour).

Minimal Cut Set (MCS): A combination of failures necessary and sufficient to cause the occurrence of the Top event in a fault tree.

Performance Influencing Factor: Factors that influence the effectiveness of human performance and, hence, the likelihood of errors. Also known as performance shaping factor.

Probability: The expression for the likelihood of occurrence of an event or an event sequence during an interval of time, or the likelihood of the success or failure of an event on test or on demand. By definition, probability must be expressed as a number ranging from 0 to 1.

Probit: A statistical variable relating the magnitude and duration of an exposure to a hazardous condition to the probability of a stated level of harm as the result of the exposure. Short for probability unit.

Process Safety Management (PSM): The application of management systems to the identifica- tion, understanding, and control of process hazards to prevent process-related incidents and injuries. A program or activity involving the application of management principles and analytical techniques to ensure the safety of process facilities. Sometimes called process hazard management.

Process Hazard Analysis (PHA): The analysis of the significance of hazardous situations associated with a process or activity. Uses qualitative techniques to pinpoint weaknesses in the designed operation of facilities that could lead to accidents.

Programmable Electronic System (PES): A computer-based system connected to sensors and final control elements for the purpose of control, protection, or monitoring.

Programmable Logic Controllers (PLC): A computer, hardened for an industrial environment, for implementing specific functions such as logic sequencing, timing, counting, and control.

Ready Only: An electronic file or system which can be read but not altered.

Responsibility: An obligation to perform an action.

Risk: The combination of the expected frequency (events/year) and consequence (effects/event) of a single accident or a group of accidents. A measure of economic loss or human injury in terms of both the incident likelihood an the magnitude of the loss of injury.

1

Introduction

1.1. Process Safety Management Activities of the Center for Chemical Process Safety

In 1985, the American Institute of Chemical Engineers (AIChE) formed the Center for Chemical Process Safety (CCPS) to focus on engineering practices that could help prevent process-related accidents in the chemical and associated industries. However, it soon became apparent to CCPS that technology alone was not enough to ensure process safety. After further consideration, CCPS concluded that a management approach was necessary to enhance the effectiveness of technical solutions. The features and characteristics of the management approach developed by CCPS are summarized in Table 1-1.

After first publishing an overview brochure, entitled *A Challenge to Commitment*, which introduced the concepts of process safety management (PSM) systems, CCPS began publishing *Guidelines* on various facets of this system. Two books in this series, *Guidelines for Technical Management of Chemical Process Safety* and *Plant Guidelines for Technical Management of Chemical Process Safety* provided additional guidance in understanding and implementing all twelve of the elements that were developed by CCPS and are summarized in Table 1-2. Other books in the series have been specifically devoted to individual elements of PSM, such as *Auditing, Hazard Evaluation* and *Incident Investigation*.

As experience was gained in implementing the various PSM elements, and as external factors such as regulatory requirements exerted increased influence, it became apparent to CCPS that the chemical process industries (CPI) could profit from guidance in the area of documentation of the vast amount of PSM-related information that was being generated. This *Guidelines* book is intended to provide that guidance and is to be used in conjunction with other books in the *Guidelines* series. For example, Chapter 14, *Auditing,* has been designed to supplement the CCPS publication entitled *Guidelines for Auditing Process Safety Management Systems*. As such, this book is intended as a reference and an aid in implementing and maintaining a sound PSM program. Guidance is provided not only on what to document, but also how to do so effectively.

TABLE 1-1
Features and Characteristics of a Management System for Chemical Process Safety

Planning
Explicit Goals and Objectives
Well-defined Scope
Clear-cut Desired Outputs
Consideration of Alternative Achievement Mechanismsm
Well-defined Inputs and Resource Requirements
Identification of Needed Tools and Training
Organizing
Strong Sponsorship
Clear Lines of Authority
Explicit Assignments of Roles and Responsibilities
Formal Procedures
Internal Coordination and Communication
Implementing
Detailed Work Plans
Specific Milestones for Accomplishments
Initiating Mechanisms
Controlling
Performance Standards and Measurement Methods
Checks and Balances
Performance Measurement and Reporting
Internal Reviews
Variance Procedures
Audit Mechanisms
Corrective Action Mechanisms
Procedure Renewal and Reauthorization

1.2. Benefits of Process Safety Management

The underlying premise of this *Guidelines* book is that, while each PSM element is important individually, the ultimate worth of each element comes as a component of an integrated PSM program. The potential benefits of a high quality, integrated PSM program are numerous and affect a number of diverse stakeholders. These affected groups of people and the benefits which they might realize from an effective PSM program at a CPI plant include, but are not limited to:

- plant owners:
  —more profitable operations, as a result of fewer incidents and less down time; and
  —confidence that they are upholding their responsibilities to the other stakeholders.

TABLE 1-2
Elements and Components of Process Safety Management

1. **Accountability: Objectives and Goals**
   - Continuity of Operations
   - Continuity of Systems (resources and funding)
   - Continuity of Organizations
   - Company Expectations (vision or master plan)
   - Quality Process
   - Control of Exceptions
   - Alternative Methods (Performance vs. specification)
   - Management Accountability
   - Communications

2. **Process Knowledge and Documentation**
   - Process Definition and Design Criteria
   - Process and Equipment Design
   - Company Memory (management information)
   - Documentation of Risk Management Decisions
   - Protective Systems
   - Normal and Upset Conditions
   - Chemical and Occupational Health Hazards

3. **Capital Project Review and Design Procedures (for new or existing plants, expansions, and acquisitions)**
   - Appropriation Request Procedures
   - Risk Assessment for Investment Purposes
   - Hazards Review (including worst credible cases)
   - Plot Plan
   - Process Design and Review Procedures
   - Project Management Procedures

4. **Process Risk Management**
   - Hazard Identification
   - Risk Assessment of Existing Operations
   - Reduction of Risk
   - Residual Risk Management (in-plant emergency response and mitigation)

5. **Management of Change**
   - Change of Technology
   - Change of Facility
   - Organizational Changes That May Have an Impact on Process Safety
   - Variance Procedures
   - Temporary Changes
   - Permanent Changes

6. **Process and Equipment Integrity**
   - Reliability Engineering
   - Materials of Construction
   - Fabrication and Inspection Procedures
   - Installation Procedures
   - Preventive Maintenance
   - Process, Hardware, and System Inspections and Testing (pre-start-up safety review)
   - Maintenance Procedures
   - Alarm and Instrument Management
   - Demolition Procedures

7. **Human Factors**
   - Human Error Assessment
   - Operator/Process and Equipment Interface
   - Administrative Controls versus Hardware

8. **Training and Performance**
   - Definition of Skills and Knowledge
   - Training Programs, e.g., new employees, contractors, technical employees
   - Design of Operating and Maintenance Procedures
   - Initial Qualification Assessment
   - Ongoing Performance and Refresher Training
   - Instructor Program
   - Records Management

9. **Incident Investigation**
   - Major Incidents
   - Near-Miss Reporting
   - Follow-up and Resolution
   - Communication
   - Incident Recording
   - Third-party Participation as Needed

10. **Standards, Codes, and Regulations**
    - Internal Standards, Guidelines, and Practices (past history, flexible performance standards, amendments, and upgrade)
    - External Standards, Guidelines, and Practices

11. **Audits and Corrective Actions**
    - Process Safety Audits and Compliance Reviews
    - Resolutions and Close-out Procedures

12. **Enhancement of Process Safety Knowledge**
    - Internal and External Research
    - Improved Predictive Systems
    - Process Safety Reference Library

- plant workers:
  —a safer working environment; and
  —greater assurance of continuing employment.
- the public:
  —a basis for confidence that their health, welfare, and homes are not jeopardized;
  —assurance of the continued benefit to the local economy provided by the facility; and
  —faster and more informed response when incidents do occur.
- stockholders:
  —protection of their investment.
- the CPI in general:
  —an improved safety record that, hopefully, can reflectively enhance the image of the CPI.
- the regulatory community:
  —ability to focus on more critical needs.

1.3. Description of Documentation

To achieve the ultimate worth of a PSM program, results from implementation of each individual element must be documented. This ensures that information is available to be communicated to those persons responsible for implementing other PSM elements. In fact, most PSM elements are dependent on this flow of information to function properly, or at all. For example, important information on how to accomplish a task more safely may be identified in a process hazard analysis (PHA); however, this information would be of little value unless communicated to the appropriate personnel through procedures and training. CCPS's goal in publishing this *Guidelines* book is to increase the awareness of PSM documentation: what it is, why it is important, and how it can be better accomplished in order to ensure accuracy, timeliness, and continued availability of critical PSM information.

This *Guidelines* book stresses documentation of eleven of the twelve PSM elements, discussed in detail in *Guidelines for Technical Management of Chemical Process Safety*. The twelfth, *Enhancement of Process Safety Knowledge,* is a CPI-wide initiative that goes beyond the facility or organizational focus of this *Guidelines* book. In addition, certain other PSM-related topics have been addressed. For example:

- control software (actually part of the *process knowledge* element) was segregated for individual treatment due to its unique nature and growing importance;
- emergency response and *contractor issues* have themes common to a number of PSM elements; and
- a chapter on *records management,* addressing available media, aids, and tools as well as practices, was considered to be pertinent for reader assistance.

The book stresses that there are two types of documentation associated with each element or topic. The first type is the documentation of the policies, standards and procedures, etc., that describe why and how the element or topic is to be implemented; and the second type consists of records that come from the actual implementation of the element or topic. For example, when one thinks of documentation of the *training* element, the records that readily come to mind are those concerning who was trained on what, and the test records showing how well the material was assimilated. However, it is recommended in this *Guidelines* book that the documentation of the *training* element would also include, among other topics:

- what the organization attempts to accomplish through *training*,
- how trainers are qualified;
- who is trained on what topics;
- how often retraining is required;
- how the training effectiveness is evaluated; and
- where and how the records are maintained.

1.4. Organizational and Individual Responsibilities

The magnitude of the documentation task has markedly increased in recent years with the onset of quality programs, such as ISO 9000, and the burgeoning of governmental regulations which began in the early 1980s. The responsibility for responding to this growing documentation challenge is shared among all levels of the organization. Corporate management must establish and demonstrate, by their statements, actions, and continued emphasis on the organization's PSM program, that they support *Values* and *Policies* which clearly communicate that this task is important and that its execution in a quality manner is highly valued. From these *Values* and *Policies*, site management or supervisory level personnel must develop more detailed *Criteria* or *Guidelines* from which *Standards* and *Procedures* are ultimately prepared to implement the program. Finally, those with the day-to-day responsibility for creating, maintaining, and using the documentation must adequately reflect the requirements of the *Standards* and *Procedures* in their daily *Practices* and *Behaviors*.

This shared responsibility is consistent with, and predates, recent regulatory emphasis on *employee involvement*. All employees from the operator or mechanic to the Chief Executive Officer have a role to play. These roles and responsibilities must be clearly defined in such a way that all are aware of and know that they will be held accountable for their performance. Finally, the value of PSM documentation must be clearly communicated so that its performance is not perceived as an onerous, purposeless paper chase.

1.5. Regulatory Considerations

Recent regulations such as 29 CFR 1910.119, the OSHA process safety management regulation, and similar state regulations have placed requirements on the CPI to create and maintain numerous categories of PSM documents. This book is not intended as a regulatory compliance guide but, rather, to be a continuation of CCPS principles on how to instill safety in all CPI operations. While some of the techniques discussed in this book may prove to be helpful in regulatory compliance, readers should consult their own safety and legal staffs or the appropriate regulatory bodies for the actual details of compliance requirements.

1.6. How to Use This Book

Any of the stakeholders listed in Section 1.2 may find the content of this *Guideline* to be of value to them; however, the primary audiences intended for this book are those within the CPI that are responsible for policy making, planning, and executing PSM. This encompasses a broad cross-section of personnel, including: engineers, business managers, production managers, safety managers, consultants, designers, constructors, maintainers, operators, mechanics, and clerks. By defining the documentation requirements pertinent to all phases of PSM activity, it is hoped that this book can help improve communications between all who are involved in implementing PSM.

This *Guidelines* book is intended to be user friendly. The first part, Chapters 1 through 4, offers introductory material that presents an overview and sets the stage for the main body of the book. Accountability and records management are discussed. In the second part, Chapters 5 through 19 address the various PSM elements or PSM-related topics. Brief descriptions of the relevant PSM elements or PSM-related topics and reviews of what they are intended to accomplish introduce the individual chapters. References are made to the appropriate volumes in the *Guidelines* series or, where no previous *Guidelines* book exists, reference is made to other pertinent texts for more detailed descriptions of how the elements or topics are implemented. Where there is no concise, readily available reference for a particular element or topic, more detailed *how to* information is provided in the chapter introduction or appendix. The intent is to ensure that the reader understands what is meant to be accomplished by a particular element or topic before attempting to understand the documentation that is generated when implementing the element or topic.

Chapters 5 through 19 describe the documentation associated with each PSM element or PSM-related topic. Both types of documentation are discussed; i.e., the program documentation describing why and how the element or topic will be implemented, as well as the documentation resulting from the implementation.

Additionally, each chapter addressing a PSM element or PSM-related topic contains unique guidance related to records management issues as well as pertinent suggestions for auditing the documentation practices for that particular element or topic. Finally, case studies or examples are provided throughout the book to illustrate the importance of sound documentation practices.

In the last part of the book are two additional chapters intended to supplement the body of the *Guidelines* book. *Document Life Cycle*, Chapter 20, traces how the documentation issue evolves throughout the life cycle of a facility from conception, through research and development, on to design and construction, operation, and modification, and final shutdown and demolition of the facility. Chapter 20 shows how the nature, rate of generation and integrated amount of information changes through each phase of the facility life cycle, as do the source, principal users and custodians of the information. The Chapter attempts to illustrate how the needs for, and goals of, documentation evolve along with these many factors.

Chapter 21, *Emerging Technologies, Research and Development* suggests where constantly evolving documentation needs, practices, and tools might lead in the future (e.g., what additional information management tools might be required to facilitate PSM documentation, or what role will expert systems play in future years).

Process safety management documentation affects everyone, but everyone need not be an authority on all aspects of the documentation program. The introductory material in Chapters 1 through 4 is suggested reading for all, especially policy makers. Chapters 20 and 21 should also be read by all, to put the documentation tasks in perspective. Chapters 5 through 19, which address specific PSM elements, are targeted more toward those with the responsibility for planning and executing these individual program elements. Reference can be made to these chapters on an as-needed basis to understand the documentation requirements of a particular PSM element or PSM-related topic. Those interested in designing overall PSM document management systems, should consult the entire book in order to better understand the flow of PSM information.

1.7. Summary

While it has been said that *knowledge is power*, this adage might be paraphrased to say *knowledge is wealth*. In implementing the various PSM elements, the CPI generates a tremendous wealth of information that can, in turn, be reinvested in other PSM elements. The purpose of this *Guidelines* book is to assist the reader in the task of amassing, protecting, and reinvesting this valuable PSM *working capital*.

1.8. References

AIChE-CCPS, *Chemical Process Safety Management: A Challenge to Commitment*, 1988
AIChE-CCPS, *Guidelines for Technical Management of Chemical Process Safety,* 1989
AIChE-CCPS, *Plant Guidelines for Technical Management of Chemical Process Safety,* 1991

Dowell, A.M., *Getting from Policy to Practices: The Pyramid Model (Or What Is This Standard Really Trying to Do?)*, 1992 Process Plant Safety Symposium, February 1992.
Occupational Safety and Health Administration, 29 CFR 1910.119, *Process Safety Management of Highly Hazardous Chemicals,* Washington, DC, 1992.

2

Process Safety Documentation Overview

2.1. Introduction

Process safety and loss prevention programs date back to the late nineteenth century and have as their objective the prevention of unwanted incidents or the reduction of the consequences of such occurrences. Examples of such programs were the provision of relief valves on boilers, electrical tagout or lockout, and isolation procedures such as the insertion of blinds in flanges to prevent the release of hazardous material during maintenance operations.

Subsequently, companies took additional proactive steps to improve the management of process safety, including:

- systematic Process Hazard Analysis (PHA);
- improved operator training;
- preventive maintenance programs; and
- increased use of instrumentation for process control.

More recently, the chemical process industry (CPI) has progressively developed process safety management (PSM) programs for plant operation. These programs were developed to reduce or eliminate incidents and their major consequences such as injury, loss of life, damage to property or the environment, or business interruption.

The principal elements and components of a PSM system were identified and discussed in the CCPS publication *Guidelines for Technical Management of Chemical Process Safety*. These elements, summarized in Table 1-2 (page 3), address many aspects of chemical process plants, from process data requirements to detailed design, start-up, operation, and maintenance. The entire process life cycle is covered.

The development of integrated PSM programs has resulted in a greater need for documentation throughout the life cycle of chemical, petrochemical, and other manufacturing facilities. In this book, the PSM elements, as well as other PSM-related topics, are identified and discussed to provide an understanding of the importance of good documentation in successful PSM and to provide tools to aid in establishing and maintaining good documentation practices.

2.1.1. *Examples of Incidents Associated with Inadequate Documentation*

Incidents can still occur even where PSM systems exist and the probabilities of such occurrences are increased if documentation is deficient. Following are a few examples of incidents that might have been avoided if satisfactory PSM documentation had existed and been effectively used.

- A major explosion occurred in Pasadena, Texas, in October 1989 because, during a maintenance operation, an 8-inch isolation valve opened, resulting in the release of high-pressure hydrocarbons. The corporate procedure required the use of double block and bleed valves to safely isolate during repair work. These corporate procedures had not been incorporated in the plant operating and maintenance procedures. Multiple fatalities and major financial loss resulted.

- In a western U.S. plant an underground gas line was ruptured with a backhoe. The construction crew obtained an approved excavation permit, and the utilities were properly marked for location. However, the depth of the piping was not documented and the construction crew assumed it was at the normal depth for that facility. Actually, the pipe was buried only 12 inches below the surface. Fortunately, there were no injuries.

- In a midwestern U.S. plant a runaway reaction and resultant explosion occurred 10 years after a similar incident had been caused by the same fault. The agitator in the reactor failed, and over a period of hours, the liquid phases in the vessel separated causing violent local reactions. In the first incident investigation it was determined that, in the event of agitator failure, the potentially disastrous decomposition could be prevented by simply adding water to the reactor. This technique had been reviewed with all operating personnel after the first incident. However, it was not documented in the operating procedures nor in the training program. During the ensuing years, all of the operators and supervisors were replaced and, without proper documentation, there was no company memory of the appropriate response to agitator failure. These incidents resulted in multi-million dollar losses and, in one incident, there was a loss of life.

- In a U.S. plant in 1989, a large amount of liquefied toxic gas was dumped to the floor from a batch reactor when it was filled with a drain valve open. All of the reactor's operating valves, except the drain valve, were automatically controlled. The drain valve was manually opened at the end of each batch cycle for a short period. However, the day of the accident, the valve was not closed before a new batch was started. The manual operation was not integrated with the control system documentation and was overlooked on this occasion.

2.2. Goals and Benefits of Documentation

The first goal of PSM is to take appropriate steps at the right time to prevent incidents in order to protect life, property, and the environment. Preventive maintenance programs, training programs, and Management Of Change (MOC) have all been developed to help achieve this goal. PHA is conducted not only in the project phase of the facility life cycle, but also in the operating phase to identify and address potential causes of incidents. When incidents do occur, investigations are conducted, corrective actions are proposed, and modifications are made to prevent further occurrences. The likelihood of success of these, and other PSM, efforts is enhanced by, or in some cases is dependent upon, accurate, effective documentation.

It may not be possible to prevent every incident. Thus, the second goal of PSM is to minimize the hazardous consequences of incidents that do occur. For example, during a process upset, the pressure in a reactor may exceed the design pressure; therefore, relief systems are provided to restrict the pressure to a safe level by discharging a portion of the reactor's contents. However, the hazardous discharge must be directed safely to prevent serious injury or damage resulting from the operation of the protective device. The objectives of each of these stages of protection should be documented via design drawings, operating procedures, maintenance procedures and training requirements. This documentation enables the many people who may be involved with the reactor system (e.g., operations, maintenance, process control, or engineering personnel) to be more fully aware of the operational requirements of the equipment and of their responsibilities. This reduces the likelihood that the protections will be compromised during the operating life of the facility.

The third goal of PSM is to enhance compliance with industry initiatives such as the Chemical Manufacturers Association (CMA) Responsible Care™ program or the American Petroleum Institute (API) Recommended Practice (RP) 750, and to satisfy regulatory requirements. Adequate documentation of procedures and audit results enhances the confidence of the public, and of regulatory bodies, that industry is being a good corporate citizen.

Among the many benefits of good and comprehensive process safety documentation are the following:

- an ability to link all elements of the PSM program together (e.g., training is linked to procedures, procedures are linked to PHA, as exemplified in Figure 2-1.);
- historical data are preserved for future use throughout the life cycle of a facility, providing an institutional memory that is independent of personnel;
- documentation of safety reviews and related topics is available for periodic audit. This provides a mechanism for identifying deviations before they result in incidents;

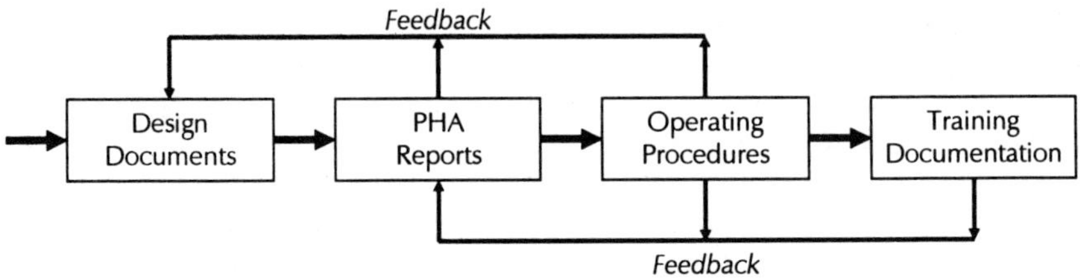

FIGURE 2-1. **Example of PSM Information Flow**

- regulatory compliance is facilitated. Some regulatory requirements are specifically document oriented, and the organized assemblage of PSM documentation greatly assists in providing the necessary information;
- incidents can be reduced through prevention because the increased information that is documented improves the understanding of hazards; and
- downtime can be reduced, resulting in more profitable operations.

2.3. Technological Changes

During the last 50 years, there has been a revolution in information handling technology. For example, many types of documents are now created, revised and stored using computers. The basic information flow relationships illustrated in Figure 2-1 remain the same, but methodologies and tools have changed. The opportunity and challenge today is to make the best use of new technology to more effectively manage the increased volume of documentation. Chapter 21, *Emerging Technologies, Research and Development* takes a look forward to what additional technological changes might enhance the effort in the future.

2.4. Summary

Proper use of good documentation should be part of a quality PSM program that communicates to employees and to the community a commitment to safety. Such documentation serves as a resource for improvement in many areas within and beyond the PSM program. It makes good business sense to have effective documentation; its contribution to improved understanding and communication enhances awareness of hazards in the workplace and results in a better, safer working environment.

This book describes many types of PSM-related documentation and suggests document management programs that can assist in the preparation, revision, storage, protection, retrieval and use of this valuable information resource.

2.5. References

AIChE-CCPS, *Guidelines for Technical Management of Chemical Process Safety*, 1989

AIChE-CCPS, *Plant Guidelines for Technical Management of Chemical Process Safety,* 1991

American Petroleum Institute, *Management of Process Hazards,* RP750, Washington, D.C.

Carson, P. A., and Mumford, C. J., *The Safe Handling of Chemicals in Industry*, 2 Volumes, Longman/Wiley, 1988

Kletz, T. A., *Lessons from Disaster: How Organizations Have No Memory and Accidents Recur,* Institution of Chemical Engineers, 1993

Lees, F. P., *Loss Prevention in the Process Industries*, 2 Volumes, Butterworth, London, England, 1986

National Institute of Safety and Health, *Guide to Chemical Hazards,* 1990

The National Safety Council, *Personnel Safety in Chemical and Allied Industries,* Itasca, IL

Occupational Health and Safety Administration, *Process Safety Management of Highly Hazardous Chemicals,* 29 CFR 1910.119, Washington, D.C., 1992

3

Accountability

3.1. Overview

3.1.1. Introduction and Definition

Accountability is the obligation to answer for one's actions that are related to an organization's goals and objectives. It is an essential part of an effective management system. Because accountability is associated with positive rewards for good performance, as well as penalties for poor performance, it gives teeth to the responsibilities assigned through the management system. Documentation of accountability should be maintained throughout the life cycle of the process through the use of policies and procedures which designate specific individual responsibilities. The accountability for the assigned responsibilities should be stated in a clear and concise manner.

3.1.2. Goals and Benefits

The principal goal of accountability is to ensure the development of a process safety management (PSM) policy and program which will endure throughout the life cycle of a process. Supporting this principal goal are more specific goals to:

- establish a corporate culture in support of the PSM effort;
- provide perspective on the scope of the PSM program; and
- demonstrate management support.

Well-documented accountabilities must be established for each management function; planning, organizing, implementing, and controlling (see Table 1-1). The benefit to be obtained from documented accountabilities is to ensure that the PSM system accomplishes each of these management functions.

3.2. Description of Documentation

In order to effectively manage PSM, there should be a documented program outlining the overall objectives and describing how they will be achieved. Specific objectives which should be considered for documentation include:

- continuity of operations;
- continuity of systems;
- continuity of organization;
- control of exceptions and variances;
- management accessibility;
- employee involvement;
- communications; and
- reduction in incidents, injuries, and property losses.

Specific instructions for how PSM is to be implemented and how these goals are to be achieved are typically described in a system of documents beginning with a general statement of PSM values and policies and concluding with detailed procedures. These increasingly detailed documentation requirements are discussed further in this section.

3.2.1. Statement of Values and Policies

The Chief Executive Officer (CEO) of an organization is typically the *sponsor* or *advocate* of process safety, and leads by example. In this role, the CEO usually issues a statement articulating the organization's values and describing in simple but direct terms what the organization expects to achieve through PSM. The statement should be signed and dated, and should be renewed on a periodic basis and upon replacement of the CEO. This statement should be communicated to all employees. Two examples are included as Figures 3-1 and 3-2.

Note that such statements can also be issued at other levels of the organization, but require the endorsement and commitment of top management at that level.

3.2.2. More Detailed Documentation

Policy statements describe required organizational behavior in the broadest sense and do not specify how to act in particular situations. To achieve an organization's PSM goals, more detailed documentation must be provided. Dowell describes a pyramid model where *criteria* and *guidelines* provide interpretation of *policy statements* and serve, in turn, as the basis for more detailed *standards* and *procedures*. Typical documents promulgated at the corporate level may include:

- corporate safety guidelines;
- corporate engineering standards;

Diamond Shamrock

COMMITMENT TO SAFETY

It is the policy of Diamond Shamrock to:

- Maintain a safe, clean workplace.
- Provide training in safe job performance.
- Ensure that unsafe conditions are recognized, reported, and promptly corrected.
- Comply with applicable safety laws and regulations.

Diamond Shamrock is committed to the safety and well being of our employees and customers. Protecting human life from harm and property from loss is a key responsibility of every employee. We must always THINK SAFETY before and during every task.

Roger

Roger R. Hemminghaus
Chairman and
Chief Executive Officer

Diamond Shamrock P.O. Box 696000, San Antonio, Texas. 78269-6000. Phone: 210 641-6800

FIGURE 3-1. Process Safety Management Policy—Example 1

- process or product specific standards;
- plant design procedures;
- equipment integrity standards;
- emergency response guidelines; and
- PSM audit guidelines.

The DuPont Commitment

Safety, Health and the Environment

We affirm to all our stakeholders, including our employees, customers, shareholders and the public, that we will conduct our business with respect and care for the environment. We will implement those strategies that build successful businesses and achieve the greatest benefit for all our stakeholders without compromising the ability of future generations to meet their needs.

We will continuously improve our practices in light of advances in technology and new understandings in safety, health and environmental science. We will make consistent, measurable progress in implementing this Commitment throughout our worldwide operations. DuPont supports the chemical industry's Responsible Care® and the oil industry's Strategies for Today's Environmental Partnership as key programs to achieve this Commitment.

Highest Standards of Performance, Business Excellence

We will adhere to the highest standards for the safe operation of facilities and the protection of our environment, our employees, our customers and the people of the communities in which we do business.

We will strengthen our businesses by making safety, health and environmental issues an integral part of all business activities and by continuously striving to align our businesses with public expectations.

Goal of Zero Injuries, Illnesses and Incidents

We believe that all injuries and occupational illnesses, as well as safety and environmental incidents, are preventable, and our goal for all of them is zero. We will promote off-the-job safety for our employees.

Continuously Improving Processes, Practices and Products

We will extract, make, use, handle, package, transport and dispose of our materials safely and in an environmentally responsible manner.

We will continuously analyze and improve our practices, processes and products to reduce their risk and impact throughout the product life cycle. We will develop new products and processes that have increasing margins of safety for both human health and the environment.

We will work with our suppliers, carriers, distributors and customers to achieve similar product stewardship, and we will provide information and assistance to support their efforts to do so.

We will assess the environmental impact of each facility we propose to construct and will design, build, operate and maintain all our facilities and transportation equipment so they are safe and acceptable to local communities and protect the environment.

We will be prepared for emergencies and will provide leadership to assist our local communities to improve their emergency preparedness.

Goal of Zero Waste and Emissions

We will drive toward zero waste generation at the source. Materials will be reused and recycled to minimize the need for treatment or disposal and to conserve resources. Where waste is generated, it will be handled and disposed of safely and responsibly.

We will drive toward zero emissions, giving priority to those that may present the greatest potential risk to health or the environment.

Where past practices have created conditions that require correction, we will responsibly correct them.

Conservation of Energy and Natural Resources, Habitat Enhancement

We will excel in the efficient use of coal, oil, natural gas, water, minerals and other natural resources.

We will manage our land to enhance habitats for wildlife.

Open and Public Discussion, Influence on Public Policy

We will promote open discussion with our stakeholders about the materials we make, use and transport and the impacts of our activities on their safety, health and environments.

We will build alliances with governments, policy makers, businesses and advocacy groups to develop sound policies, laws, regulations and practices that improve safety, health and the environment.

Management and Employee Commitment, Accountability

The Board of Directors, including the Chief Executive Officer, will be informed about pertinent safety, health and environmental issues and will ensure that policies are in place and actions taken to achieve this Commitment.

Compliance with this Commitment and applicable laws is the responsibility of every employee and contractor acting on our behalf and a condition of their employment or contract. Management in each business is responsible to educate, train and motivate employees to understand and comply with this Commitment and applicable laws.

We will deploy our resources, including research, development and capital, to meet this Commitment and will do so in a manner that strengthens our businesses.

We will measure and regularly report to the public our global progress in meeting this Commitment.

- Replaces November 1971 Policy
  July 1994

FIGURE 3-2. Process Safety Management Policy—Example 2

Many facilities develop more detailed procedures, conforming to corporate policy and guidelines but addressing the requirements of the particular location. Some examples of such facility documents are listed below.

Safety Manuals—Safety manuals should describe practices and procedures to be followed in operations. These may range from procedures for the safe performance of potentially hazardous operations to practices for personal protective clothing and equipment.

Job Descriptions—The purpose of job descriptions is to provide direction for the employee, assist in training, and provide a paper trail for auditing. The job description should include the responsibilities for specific process safety activities, and should be critiqued to ensure consistency and clarity. Job descriptions may include:

- hazard communications;
- hazards of particular tasks;
- specific operating procedures;
- emergency procedures;
- start-up and shutdown procedures; and
- safe work practices.

Employee Involvement—Facility PSM documentation may include a statement concerning employee involvement. This statement may include the following considerations:

- overall policy;
- employee communication and awareness;
- employee responsibility; and
- employee involvement in accident investigations, process hazard analyses, and operating procedures preparation.

3.2.3. Responsibility and Accountability

The PSM program documentation should describe responsibilities in terms of who, what, when, and how. Accountability for required actions should be addressed throughout the program documentation. The documentation should address:

- who, by job title, will be delegated responsibility and for which sections of the overall program;
- what they will be responsible for (e.g., for approval or modification of procedures, documentation, training or auditing. Different individuals may be responsible for different functions);
- availability of the necessary resources;
- measurement of performance; and

- the precise conditions for authorization of exceptions or variances to the policy.

It is important that the criteria and responsibility for variance approval be clearly documented. In many cases, this responsibility may be reserved to senior management.

3.2.4. Measurement

Many organizations adopt targets for safety performance using measurable indexes. Typically, these include safety and loss prevention statistics and PSM activities, such as:

- reduction of hazardous inventories;
- lost-time accident rate;
- reportable releases; and
- property losses.

The measurements to be used and corresponding goals should be clearly stated in the PSM program documentation.

3.3. Records Management

3.3.1. Policies and Practices

To ensure continued effectiveness of PSM, the accountabilities and responsibilities for PSM program documentation should be documented initially and revised as organizations, processes, and technology change.

Specific guidance for more detailed elements of PSM documentation (e.g., *Operating Procedures*) will be provided in subsequent chapters. It is equally important, however, that responsibilities for maintaining the broader policy and interpretative documentation (e.g., policy statements or corporate PSM guidelines) also be clearly established.

3.3.2. Records Revision and Retention

It is imperative that PSM documentation remain current (e.g., procedures are updated to reflect current requirements and out-dated information is purged from the files). Thus, it is important that PSM documentation be managed in accordance with the overall corporate records revision and retention policy. The policy should specifically address each element of the PSM program, since different elements may require different retention schedules. Accountability and responsibility should be assigned for the maintenance of PSM records for each element. For additional guidance, refer to Chapter 4, *Records Management*.

3.4. Auditing

To ensure continued effectiveness of accountability and responsibility for PSM as organizations, processes, and technology change, the PSM documentation system should be periodically audited. External and internal audits of individual accountability for each of the PSM elements will help ensure that the program objectives are met and any identified deficiencies are corrected. The audit program should address the following issues:

- Is there a documented program for establishing PSM responsibility and accountabilities?
- Does a means exist for updating the program, as required?
- Is the program being followed, as described?
- Are the records from the program implementation being properly managed?

For additional guidance, refer to Chapter 14, *Auditing* and CCPS *Guidelines for Auditing Process Safety Management Systems*.

3.5. References

AIChE-CCPS, *Guidelines for Technical Management of Chemical Process Safety,* 1989.

AIChE-CCPS, *Plant Guidelines for Technical Management of Chemical Process Safety,* 1991.

Dowell, A. M., *Getting from Policy to Practices: The Pyramid Model (Or What Is This Standard Really Trying to Do?)*, 1992 Process Plant Safety Symposium, February 1992.

4

Records Management

4.1. Overview

4.1.1. Introduction

Records management addresses the various tasks involved in compiling, administering, and protecting records that are required as part of the implementation of each element in a process safety management (PSM) program. As was previously pointed out, it is the communication of PSM information between elements that maximizes the effectiveness of the overall PSM program. Sound records management practices help ensure the effectiveness of these communications. Further, by coordinating record keeping for all PSM elements, organizations can minimize the costs of implementation and have a sound basis for future management decisions and actions. In addition to the process safety benefits, there are many side benefits of a comprehensive PSM records system, such as improved operating efficiency, which can lead to enhanced profits.

4.1.2. Goals and Benefits

The goals of *records management* are to:

- identify what records should be maintained in order to ensure availability of the information needed to operate a process facility safely and efficiently;
- collect, categorize, file, disseminate, and protect critical information;
- provide easy, convenient means of entering information into the system;
- ensure storage in proper locations and formats to permit later access and speedy retrieval;
- provide safeguards to allow for access by authorized personnel only; and
- protect records from physical damage, or unauthorized or inadvertent revision.

The principal benefit intended from sound records management is enhanced effectiveness of the PSM program. However, sound records management can additionally provide direct cost savings and other benefits by reducing the time required to find documents vital for normal operations, maintenance, and in emergency situations. For example, losses due to production downtime can be cut by having vendor information and maintenance records readily available. Maintenance workers can then respond to both scheduled and unscheduled needs quickly and safely based on timely, accurate information.

Other indirect benefits can flow from or be contributed to by the existence of an efficient, comprehensive records management program. Some examples are:

- ability to better plan preventive maintenance;
- more efficient and effective execution of turnarounds and modifications;
- more effective use of management time; and
- avoidance of regulatory noncompliance.

4.2. Description of Records Management

Records management encompasses both the procedures and the equipment required to effectively handle the PSM record keeping task. Record management systems should be tailored to the individual needs of the facility. Systems can range from a completely manual, paper-based system with file cabinets and manual indexing to a paperless system where all information is entered and all documents are scanned into a computer. Some combination of these two approaches, addressing the operational and economic needs of the facility, can be effective for many users. System choices can be based on various performance requirements, such as speed of retrieval, indexing or search capabilities, or the data storage capacity required. Whether a computer-based system, a manual system, or some combination is used, the system chosen needs to address the potentially conflicting goals of providing information security while, at the same time, providing for ease of dissemination of information to authorized users.

4.2.1. Manual Systems

Some small organizations and sections of larger organizations may be able to effectively utilize a good manual filing system to handle records management. Such manual systems are often based on a limited number of central storage locations for historical records, data requiring common access by a wide variety of users, or sensitive data requiring special controls. Records with more specific application, or requiring more frequent access, are often stored at satellite locations closer to the point of use. Manual systems incorporate traditional records handling equipment such as:

- flat files for storage of drawing masters;
- file cabinets and/or book shelves for specifications, standards, policies, vendor prints, etc.; and
- manual cataloging or indexing systems indicating document storage locations.

4.2.2. Computerized Document Management Systems

A variety of computerized document management systems is available on the market to handle data and records, to distribute information and to allow communication with all required parties. Smaller facilities may be able to handle their information with one desk top computer. Larger organizations may have a dedicated main frame, minicomputer network, or microcomputer network to handle the increased volume of information and meet required retrieval times. New lower cost microcomputer-based systems are coming into the market which should make electronic document management more widely available. Some examples of computerized document management systems are given in Appendix 4A.

4.2.3. Combination Systems

Many facilities have a combination of manual and computerized data record systems due to the gradual evolution of desk top computers and electronic information management systems. For example, some organizations have centralized maintenance systems which utilize main frame computers. These have been used for many years to track work orders for maintenance, to compile historical data, and for similar uses. More recently, computer aided drafting (CAD) systems have become progressively more useful for engineering drawings. The trend for the future will be to use electronic information systems that integrate graphics such as piping and instrumentation diagrams (P&IDs), and text data such as instrument specifications so that more comprehensive information will be available to the user. See Chapter 20, *Emerging Technologies, Research and Development* for a broader discussion of what future developments might bring to the records management task.

4.2.4. System Selection

The selection of a manual, computerized, or combination system will depend on the requirements established by the organization for its PSM program, such as:

- what types of records are needed, including:
  —drawings and other graphics; and/or
  —text information such as process descriptions or maintenance manuals from equipment manufacturers.
- who needs access to various records, for example:
  —all workers employed in or near a process need all health and hazard information for that process; or
  —only a few maintenance mechanics might need access to the repair manual for an air compressor.
- how quickly must the information be available, for example:
  —within one or two days;
  —during that shift or day;
  —immediately, for emergencies.

4.3. Description of Documentation

4.3.1. Documentation of Records Management Accountability and Responsibility

Organizations implementing records management should include in their PSM program documentation a discussion of the goals of records management, as well as any general procedures required for the implementation of the program. The details of the records management task will be unique to the various PSM elements and the PSM program documentation for the elements should address more specific records management requirements. Such unique guidance is provided in the individuals chapters of this book.

Almost everyone in a facility is responsible for records in some way as either an originator, a user, or a record custodian. All records management responsibilities should be clearly defined and assigned to specific positions and/or individuals at the facility. Any required interfaces with other functions or groups should be well defined. Table 4-1 shows some typical functions involved in PSM records generation, use, and management.

One of the more important elements of records management program documentation is the requirement for a records retention and purge schedule. While it is essential that pertinent records be maintained for their useful lifetime, outdated records unnecessarily consume space and attention in the records management system. Further, there can be valid legal reasons for the timely purging of records that have no further value to the organization. Thus, the PSM program documentation should not only clearly recommend when certain records are to be prepared, or updated, but it should also state when they may be discarded. Some information may be retained indefinitely; for example, it may be necessary to maintain detailed process safety information and much of the original design and equipment information for the life of the facility, with changes in the information documented as they occur. Initial permitting, legal agreements and other legal or regulatory information may also have to be retained for the life of the facility. Where there are legal reasons for long-term storage of records, these

TABLE 4-1
Typical Functions Involved with PSM Records

Plant Management	Auditing
Purchasing	Personnel Safety
Operations	Emergency Response
Training	Industrial Hygiene
Engineering	Incident Investigation
Maintenance	Environmental
Process Safety	Records Management

should be identified. An organization's legal advisor should give advice on what must be kept and for how long. The PSM program should determine how such records are to be identified and appropriately marked to prevent destruction.

4.3.2. *Description of Specific Types of Records*

The list in Table 4-2 covers major types and hierarchies of information typically encountered in the documentation of PSM. It is not considered comprehensive, and is meant only as a guide, since organization size, nature, regulatory requirements, and other factors will affect the levels and number of records needed for various facilities. More detailed treatment of these topics is provided in the individual chapters in this book.

4.4. Records Management

As previously stated, one challenge in administering a PSM records management system is that of simultaneously achieving two potentially conflicting goals; i.e., providing for quick, easy access to information, while protecting it against loss or damage. Such loss or damage my result, for example, from misplacement, physical damage, or inadvertent or unauthorized revision. This section addresses records management practices relevant to both of these goals.

4.4.1. *Where Maintained*

Plant records related to PSM are often maintained in many different places, as governed by the structure and needs of the organization. For example, many types of PSM documentation must be located convenient to the job sites of those persons actually operating, maintaining, and supervising the process. Such records would include operating procedures, maintenance instructions, personnel training records, etc.

Other records may, by their nature, require centralized storage. These may be records having more general utility and requiring access by a broader cross-section of the facility population, such as:

- centralized blueprint files;
- industry standards and codes (e.g., ANSI, API, NFPA);
- governmental regulations (e.g., OSHA and EPA);
- organizational policies and procedures (e.g., PSM program documentation); and
- site standards (e.g., piping specifications).

Additionally, centralized storage offers the potential for greater control of sensitive or valuable records. Key documents may be duplicated for distribution and the original copy stored in a centralized, secure location, protected from

TABLE 4-2
Documentation Categories

Process knowledge:
- hazardous chemicals information:
  —Material Safety Data Sheets (MSDSs) and
  —other sources.
- technology
  —block flow diagrams;
  —process flow diagrams (PFD);
  —process chemistry records;
  —inventory of chemicals used;
  —records of evaluation of consequences of deviation from normal process conditions; and
  —safe upper and lower limits for temperature, pressure, flow composition and other key parameters.
- process equipment:
  —piping and instrumentation diagrams (P&ID's);
  —piping service index with piping service specifications;
  —instrument index and specifications with full description of operating conditions, materials of construction, process fluids;
  —electrical area classification drawings;
  —equipment specifications complete with materials of construction and references to applicable codes;
  —relief system design and design basis;
  —heating, ventilation, and air conditioning (HVAC) systems related to process safety (fume or dust controls);
  —safety systems (e.g., interlocks, detection or suppression systems);
  —design codes and standards used for design; and
  —design basis documents, which refer to compliance with good engineering practice.

Process Hazard Analysis (PHA):
- procedures for each PHA methodology/technique used at the site;
- organization chart of responsible parties complete with job descriptions;
- engineering controls;
- administrative controls;
- records of all past PHAs;
- relevant accidents/incidents;
- facility siting records—complete with basis for selection and a description of all factors considered such as distances, types and quantities of materials, and explosion potential;
- PHA leader and team members—records of qualifications, initial and refresher training;
- recommendation follow-up flow chart along with organization chart and listing of personnel involved and their functions;
- follow-up schedules with records of implementation of past PHA work;
- records of communicating results/changes to all affected employees;
- records of communicating with persons responsible for updating training and operations manuals/procedures;
- records of revalidation of PHA.

Quantitative risk assessment:
- chemical process quantitative risk analysis (CPQRA), policies and practices; and
- records of past CPQRAs.

TABLE 4-2 (*continued*)
Documentation Categories

Process equipment integrity: • records of installation; • equipment lists; • equipment specifications with reference to codes and standards used for design and selection; • manufacturer/supplier documents with all related certifications, mill tests, and similar data that indicate good engineering practice was used; • manufacturer maintenance manuals; • code and standard compliance records; • instrument index; • instrument specifications; • piping systems records; • control systems records (e.g., computers, Programmable Logic Controllers (PLC) and other related); • relief and vent system records; and • maintenance procedures: —preventive maintenance plan; —testing and inspection procedures and records; —fire protection systems maintenance records; —emergency maintenance procedures; —deficiency correction records; • maintenance, training, procedures, and records; • maintenance, testing, inspection personnel records; • outside maintenance testing, inspection contractor qualification records; and • quality assurance program
Human Factors (HF): • employee participation; and • human factors analysis records: —operator/equipment interface; —consequences of operator errors; —work schedules, shifts, hours, breaks, etc.; —control displays; and —location and clarity of signs.
Management of Change: • written procedure: —definition of permanent and temporary; —technical basis requirements; —sequence of events; —authorized personnel identified (organization chart, job description); —implementation requirements; and —follow-up procedure; • flow chart indicating parties to be notified and all affected documents; • records of notification; • follow-up records that process is complete; and • personnel reassignment and replacement procedures.
Operating procedures: • initial start-up records including operator logs; • normal operations; • shutdown; • emergency; and • preparation for maintenance.

Continued on page 30

TABLE 4-2 (*continued*)
Documentation Categories

Training:
- training program records (e.g., names, dates, modules, verification);
- trainer qualifications;
- training procedures; and
- employee and contractor training plan with records of training and verification (written test or other means).

Emergency response:
- emergency action plan;
- escape procedures and routes with drawings;
- site plan drawings which indicate plant areas, muster points, escape routes, wind socks, control rooms, rescue equipment location, fire fighting equipment locations, etc.;
- key personnel organization chart and job descriptions;
- designation of employee routes required for orderly evacuation;
- alarm/escape route/muster point card for issue to employees and contractors;
- alarm systems records of installation, maintenance, testing;
- response plan for small releases;
- emergency lighting plan, records, inspection;
- outside agency coordination plan; and
- community plot plan which indicates location of all support agencies with contacts information.

Auditing:
- audit polices and procedures;
- auditing schedules; and
- audit records.

Incident investigation:
- investigation procedures;
- investigation team members (records of training and qualifications);
- investigation records;
- recommendation records; and
- follow-up records.

Standards, codes, and regulations:
- corporate standards;
- industry codes and standards; and
- local, state, or federal regulations.

Contractor issues:
- evaluation procedures; and
- safety training and performance records.

Safe work practices:
- routine work permits (e.g., cold work);
- hot work permit and procedures;
- lockout/tagout procedures;
- confined space entry permits;
- line breaking procedures;
- drainage and diking procedures; and
- heavy lift procedures.

Control software:
- interlock descriptions; and
- functional specifications for computer control systems.

damage and unauthorized access. For example, the original copies of operating procedures, bearing the signatures authorizing them for use, may be stored in a central location where they can be protected from unauthorized modification. Copies of these original could then be issued for use in the field. A large facility may have several such centralized storage locations, or records centers, focused on the maintenance of records for particular PSM elements.

Other records may be of significant historical value, but require much less frequent access; e.g., original purchase records for facility equipment. Alternatively, some records may require more secure storage than that afforded by central on-site facilities or may require protection from major incidents that might occur on-site. In such instances, long-term archival storage in secure repositories, perhaps off-site, should be considered.

Many PSM documents require considerable effort to create and many would be difficult or impossible to re-create, if lost. An organization should consider the nature and value of each PSM record, define the appropriate degree of control and protection, and reflect this in its records management practices. In certain circumstances, it may be necessary to exercise some degree of control over copies of PSM documentation that are distributed to individuals or locations in the field. This is discussed further below.

4.4.2. Document Control

Some PSM records may require duplication and distribution to allow their use by a broad cross-section of the facility population. This distribution may be accomplished either with or without controls, depending on the nature of the documentation and the intent of the organization.

For example, some organizations maintain their own engineering standards, which are often issued in the form of binders containing a copy of many individual standards. As various standards are created, revised, or superseded, it is important that each set of standards be kept current so that up-to-date information and requirements are available to all users. Some organizations achieve this goal by controlled distribution of standards.

Documents with a controlled distribution are often marked with a serial number and, where appropriate, a revision number. A distribution list of individuals with custodial responsibility for each set of standards is maintained and revisions are directed to these individuals. If a greater degree of control is desired, to ensure that all revised standards are actually filed appropriately, the organization may require that the superseded pages be returned to the point of issue for inventory and controlled destruction.

Such a system of controls may be applied to other types of PSM documentation. For example, some organizations may issue PSM implementation guidelines or operating procedures in this same manner. In any such situation, the organization should weigh the advantages of this degree of document control against the increased operating costs of such a system.

Other PSM documentation may not be amenable to this degree of control, or it may not be desirable to limit the distribution in this manner. For example, an organization may welcome widespread dissemination of the corporate PSM policy statement and may encourage its duplication and distribution to all employees. Alternatively, it may be necessary to allow employees to access and obtain uncontrolled copies of records such as facility design drawings (e.g., P&IDs). The potential adverse effects that could result from the use of an outdated P&ID are more severe than those due to the use of an outdated PSM policy statement and some means of protection against this eventuality should be considered. Generally, organizations will issue indexes of such periodically revised documents showing the name, date of revision, and revision number (or letter) of each document in the index. Reference to such an index will allow the user to gain confidence that the document in hand, whether controlled or uncontrolled, represents the most recent revision. Obviously, greater responsibility is imposed on the users of uncontrolled documents to ensure the use of current document revisions.

It may sometimes be necessary to issue uncontrolled copies of a normally controlled document (for example, for short-term reference needs, or for consideration of a revision to the document). Such uncontrolled copies should be minimized since they fall outside of the administrative system provided for distribution of revisions and, thus, cannot be updated. Many organizations find it helpful to identify such undocumented copies; e.g., by reproducing them on a distinctive color of paper or by stamping them with an appropriate *caveat* such as "For reference only. Do Not Use."

Other PSM documentation may be retained solely for reference purposes and would not be intended for widespread distribution. Such records would normally be maintained in a central records center or archive. Conceivably, only a single copy of such a document might exist; however, to protect against the loss of such information it is common to maintain one record copy and several distribution copies of such documents. The distribution copies could be loaned to users on a checkout and return basis, while the record copy would be maintained in secure storage.

Responsibilities and accountabilities for managing PSM records should be clearly defined, whether they be controlled copies assigned to an individual or those records maintained in a central records center or archive. For larger facilities, such centralized storage locations often have a permanent staff trained in records management practices.

This discussion has focused primarily on hard copy (paper-based) records systems. The advent of electronic data management systems, as discussed below, introduces alternate means of distributing and controlling PSM documentation.

4.4.3. Choice of Media

Organizations will typically need to be able to deal with a variety of media. Many of the documents received or generated by manufacturing organizations will be in hard copy, primarily paper, form. For the convenience of storage, some organizations may convert certain types of records from the hard copy to either microfilm or microfiche. For example, this is commonly done for blueprint files and equipment maintenance manuals.

Some engineering documents such as process flow diagrams, P&IDs, one-line drawings, loop drawings, and site plans are now commonly generated with CAD systems and are stored on some form of magnetic media such as tape or diskettes. Manufacturers of equipment are increasingly offering such electronic files in addition to, or in lieu of, hard copy. Systems with the ability to scan hard copy into computer files are also available and may enhance an organization's capabilities to handle all types of media.

Finally, with the advent of optical disk technology, digital information for computers can be transferred on a nonmagnetic media at information densities many time greater than previously possible. It is increasingly common to obtain text, graphical, and, even pictorial, information in this form. Many industry standards and governmental regulations are now available via optical disk.

The choice of media for a particular type of record will depend on a number of factors such as needed retrieval speed, the volume of information to be stored, required lifetime for the records and the cost of using the particular media. Additional information provided in this section may assist in the selection of media.

4.4.4. Files

File Indexes—Any form of systematic filing of documents, whether in a hard copy filing cabinet, or as a set of computer files, should include an up-to-date index of the contents. File indexes should include revision numbers and dates as well as titles, and should be available in identified locations.

File System Mechanics—PSM files should be organized simply and logically so that, when necessary, the required information can be retrieved without delay by the appropriate personnel. There are many practical filing systems generally in use. For example, files could be organized by:

- chronological order;
- facility area; or
- PSM element.

File organization will be dictated by the needs of the organization and the type of records involved. More specific guidance is not appropriate here.

4.4.5. Document Management Systems

Document management systems can range from a set of file cabinets with a manual cataloging method to a mainframe computer with sophisticated software. No matter what the hardware, the system chosen should be easily understood so that people know where to find documents and how to gain access to them.

Modern computerized systems generally offer the advantages of large storage capacity, speed of retrieval, and the ability to simultaneously disseminate information to multiple users. Computer networks can be used to make information available in multiple locations (e.g., facility offices, control rooms, guardhouses, and remote locations such as emergency response centers or corporate offices). Additionally, such networks can make use of off-site data storage. For example, OSHA standards are accessible via modem from the Department of Labor in Washington, D.C.

Computer software developers have created products that can be used to store and retrieve many different types of documents. These range from systems that handle only text data to systems that integrate text with a graphical interface to computer generated drawings (e.g., CAD). Large mainframe computers are not necessarily required; some advanced desk top computer systems can handle large quantities and numerous types of documents. Refer to Chapter 21, *Emerging Technologies, Research and Development,* for further information.

For many existing facilities, it may be more effective to use a combination of systems to handle the large quantity of information. The system adopted could consist of conventional hard-copy records management facilities supplemented by:

- desk top computers;
- document scanners;
- document management software;
- CAD or similar drafting software;
- drawing printer/plotters;
- magnetic media storage devices such as tape, disks, and diskettes;
- optical CD-ROMs; and
- microfilm readers and microfilming machine (or access to such local services).

Reliance on computer-based systems requires consideration of back-up capabilities to guard against the effects of a computer or other system component failure.

Finally, the most important part of the system will be the people who make the document management system function properly. As previously noted, roles and responsibilities must be clearly defined.

4.4.6. Fire Protection

Some organizations that have experienced fires involving original documents have recognized too late the need for fire protection. Protection against loss of information can be provided by storing multiple copies of documents in physically separated locations. However, in some instances, it is important to preserve the original copy of a document. Special fireproof safes can be used for records that are not too voluminous. However, for many facilities, special fireproof or fire protected buildings or rooms may be necessary. These require special custom designs. Reference can be made to NFPA 232, *Standard for Protection of Records* and NFPA 232AM, *Manual for Fire Protection for Archives and Records Centers*, for guidance.

4.4.7. Environmental Damage Control

Rainwater, burst pipes, and smoke damage, as well as extremes of temperature and humidity can damage or destroy records. The storage location selected should avoid the potential for damage from these causes. Users should follow the manufacturer's recommended storage conditions for the media used, particularly if long term storage is anticipated. Table 4-3 summarizes storage conditions recommended for various media by the Association of Commercial Records Centers.

While storage conditions are important, the quality of the media itself can contribute to the anticipated lifetime. For example, acid-free paper should be used for long term storage. Where old archives already hold paper that may not be acid-free, users may wish to consider copying or microfilming the originals before they deteriorate further. Old drawings may rapidly become illegible due to older methods of reproduction. These too should be considered for copying to a more stable media. Magnetic tapes have a limited shelf life as well, and periodic replacement is necessary.

Additional guidance is available from the National Information Standards Organization (NISO) on paper quality and storage conditions. One example is ANSI-NISO/Z39.48, *Permanence of Paper for Publications and Documents in Libraries and Archives*. The Association of Commercial Records Centers also issues

TABLE 4-3
Recommended Storage Conditions for Various Records Media

	Medium-Term Storage		*Long-Term Storage*	
	Temperature, °F	*Relative Humidity, %*	*Temperature, °F*	*Relative Humidity, %*
Microfilm and Microfiche	77 (max)	20–50	70 (max)	20–30
Magnetic tapes	N/A	N/A	62–68	40–50
Magnetic disks	N/A	N/A	65–75	40–50
Paper	N/A	N/A	70 (max)	40–50

standards on the design of records storage facilities. Other organizations providing records storage guidance are listed in Appendix 4A, beginning on page 38.

Damaged records, even those soaked with water, can sometimes be restored or salvaged by specialized firms such as those who do work of this type for libraries and government archives. The reader may wish to refer to the paper by Wellen cited in Section 4.6.

4.4.8. Security

PSM records must be protected against loss, unauthorized alteration and improper dissemination. For example, the original copy of a critical equipment test report might be removed for reference and not returned to the files, making it impossible to prove the original mechanical integrity of the equipment at some future date. The master copy of an operating procedure might be modified without proper regard for Management of Change considerations. Finally, the general release of company trade secret information, or information covered by secrecy agreements with contractors and licensors, could result in the loss of a competitive advantage or, perhaps, prompt legal action. Thus, organizations should consider information security systems embodying control measures appropriate to the value and sensitivity of the various PSM records.

In particular, access to archives should be restricted to designated individuals. Their responsibilities concerning access, retrieval, removal, copying etc., should be clearly defined and understood. Many organizations find it helpful to maintain a list of those who are allowed access to such records. Access to hard copy files can be controlled through the use of combination or key-locked repositories. Access to computerized records can be controlled with passwords or similar authorization devices. These security devices should not be provided in a form, or left exposed in such a manner that others can obtain unauthorized access, and each individual should be held responsible for maintaining their confidentiality. Organizations may wish to consider changing passwords or combinations both on a periodic basis and upon reassignment of key personnel.

As previously noted, many other PSM documents may need to be generally available to employees. Information security can be maintained by controlling original or master copies of the documentation, while distributing copies, as required. As discussed under *Document Control*, these copies may be either be undocumented or documented and controlled, depending on the nature of the information and the needs of the organization.

Computerized records management and distribution systems offer the advantages of being able to widely disseminate information in a controlled manner by providing users *read only* access to files. This guards against unauthorized deletion or modification of the files and prevents the proliferation of unauthorized paper copies of documents.

In the case of computerized systems, periodic back-up of data is required to ensure against the loss of records due to hardware or storage media failure.

Similarly, the security of hard copy data can be enhanced by storing copies of critical records in secure, off-site repositories. This ensures the availability of such information, even in the event of a major fire, flood, or other such on-site event.

4.4.9. Reproduction

It is sometimes necessary to limit the reproduction of documents to prevent their unauthorized use or diversion, or to ensure that out-of-date copies are not in use. Where original copies of documents are maintained in secure storage, this can be accomplished administratively, with some degree of success. However, once copies are circulated, it is difficult to prevent copying of the copies. In the past, some organizations sought to gain an additional measure of control by ensuring that copies of documents could be distinguished from the originals. However, advances in photocopying technology have made it very easy to produce copies of documents that look essentially identical to the originals. In particular, the advent of color copiers make it possible to circumvent previous controls such as the use of contrasting colors of ink on original documents (which, previously, would have appeared uniformly black on the photocopy).

Fortunately, advances have also been made in security technology. Special stock, similar to financial check stock, can be used for originals. When the stock is photocopied a message such as "VOID" appears on the copy. Additionally, it is now possible, via password controls, to limit access to enhanced capability copiers to approved users. More sophisticated approaches include the use of a pattern recognition computer chip within the copier that will not allow copying of documents containing a particular image, such as a corporate logo. This is a more costly option since it requires custom design of the chip.

Similarly, computer files can be copied, and many devices to prevent unauthorized copying can be defeated. A discussion of advances in this area is beyond the scope of this book. Users may wish to seek assistance from internal computer technology resources.

4.4.10. Destruction

One way in which records can be lost, or compromised, is through an inadequate system for controlling their destruction, since once destroyed, a document, file or drawing may be irreplaceable. The documentation program should set out strict controls, not only for ensuring that designated obsolete documents are destroyed, but that current documents are not accidentally destroyed. A detailed retention and purge schedule, previously described, can help remove ambiguity as to the status of a particular document.

Where confidential hard copy is involved, destruction by shredding is recommended. Additionally, the program may include the destruction of carbon paper, printer ribbons, and any other media containing document images existing from when the originals are prepared.

Alternatively, or in addition, incineration can be used, but this should be done internally under close administrative control.

For computer diskettes, simple *deletion* of files usually does not suffice, since only the file name is deleted. Diskettes must be reformatted, or previously used sectors must be overwritten, before the data is actually obliterated.

The methods to be used for destruction of various media should be specified in the program documentation.

4.5. Auditing

The PSM program documentation should establish the requirements for auditing the Records Management system. Audits should seek to determine whether a documented program such as described in Section 4.3.1 (and elaborated on in Section 4.4) exists and whether it is being administered as described. This is intended to be an audit, not of the records, but of the system.

Any deficiencies identified during the audit must be documented and promptly resolved. See Chapter 14, *Auditing*, and the CCPS book *Guidelines for Process Safety Management Systems* for more guidance on auditing.

4.6. References

The following source documents should provide information useful in establishing, operating, and maintaining records management systems for PSM programs.

AIChE-CCPS, *Guidelines for Auditing Process Safety Management Systems,* 1993.
Association of Commercial Records Centers, *Storage of Records, Media Vault Guidelines.*
National Fire Protection Association, *Manual for Fire Protection for Archives and Records Centers,* NFPA 232AM.
National Fire Protection Association, *Standards for Protection of Records,* NFPA 232.
Wellen, T.G., *Protecting Records from Disaster*, 1994 Process Plant Safety Symposium, AIChE South Texas Section

Appendix 4A. Records Management Resources

1. Computerized Document Management System Suppliers

The following sample software selections are taken from a McGraw-Hill Publication: *Datapro Directory of Microcomputer Software*. The directory contains sections on Databases, File Management, Protection and Security, and Information

Systems Planning. It is updated periodically and can be valuable in obtaining low cost solutions for records management.

Aperture

Aperture Technologies
100-3 Summitt Lake Dr.
Valhalla, NY 10595

Visual retrieval of text & CAD files, with reports, relational database, & CAD capabilities.

MicroMSDS

Gulf Publishing Company,
Book Division
P.O. Box 2608
Houston, TX 77252

A flexible system for storing and retrieving information received on different types of material safety data sheets.

Information Master

High Technology Software Products, Inc.
P.O. Box 60406
Oklahoma City, OK 73116

Data management program with built-in report writer.

2. *Industry Organizations*

ACRC Association of Commercial Records Centers
P.O. Box 20518
Raleigh, NC 27619

AIIM Association for Information & Image Management
110 Wayne Ave., Suite 1100
Silver Spring, MD 20910

—— Commission on Preservation & Access
Washington, D.C.

ANSI American National Standards Institute
1430 Broadway
New York, NY 10018

IFMA International Facilities Management Association
One E. Greenway Plaza, 11th Floor
Houston, TX 77046

NBS National Bureau of Standards, Department of Commerce
Superintendent of Documents
Government Printing Office
Washington, DC 20402

NFPA National Fire Protection Association
1 Batterymarch Park
Quincy, MA 02269

N.I.S.O. National Information Standards Organization
P.O. Box 1056
Bethesda, MD 20827

NPP National Preservation Program
Library of Congress
Washington, DC 20540

SARA New York State Archives & Records Administration
10A46 Cultural Education Center
Albany, NY 12230

3. Document Management Specialty Hardware and Software

Note: These selections are from Dialogue, obtained from an on-line search of the key words "document life cycle."

Global View-for-the-PC

Xerox Integrated Systems Operations
Xerox Centre
101 Continental Blvd.
El Segundo, CA 90245

Coprocessor Board plus software tool kit that turns DOS and OS/2 personal computers into multitasking, networked workstations. An icon-based graphical user interface (GUI) makes it easy for nontechnical professionals to perform complex tasks. Purpose is to integrate text, data, and graphics into compound documents.

DocuTeam

Xerox Integrated Systems Operations
Xerox Centre
101 Continental Blvd.
El Segundo, CA 90245

Software that organizes documents into shared "electronic libraries."

DMS Pro

van der Roest Group, Inc.
Santa Ana, CA

Automatically track and monitor document flow activities. Hyperlink system permits linkages of various CAD files, scanned images, word processor files, databases, and spreadsheets.

Worldview and Relational Document Manager

Interleaf, Inc.
Waltham, MA

Total access to indexed and filed documents from any computer.

Folder View

Filenet, Corp.

Users can create folders of numerous related working documents by assembling groups of icons that resemble each document.

5
Process Knowledge

5.1. Overview

5.1.1. *Introduction*

The design, operation, and maintenance of a process facility requires a considerable amount of process knowledge. This information, which encompasses process chemicals, technology and equipment, is the foundation for understanding the potential hazards of a process and for supporting a comprehensive process safety management (PSM) program. Documentation of process knowledge, therefore, can be detailed and extensive. Process knowledge which is incomplete or incorrect increases the potential for an incident.

Many aspects of process knowledge are discussed elsewhere, especially in other CCPS *Guidelines* books, thus the description here is brief. Emphasis in this chapter is placed on how to assemble, manage, and maintain, throughout the life of a facility, the process knowledge essential to a facility's safe operation.

Process knowledge is also used in preparing the documentation for procurement, fabrication, installation, and testing of equipment that will comprise the hardware of the facility. This aspect is discussed in more detail in Chapter 8, *Process Equipment Integrity.*

5.1.2. *Goals and Benefits*

The main goal of the process knowledge element is to comprehensively document the process knowledge for the facility. Another goal is to ensure that this information describing the process and facility equipment is accurate and up-to-date. This documentation enables those who need to know, both now and in the future, to benefit from this knowledge and to operate the facility safely and efficiently. Documentation also provides much of the basic information needed for successful implementation of other elements of a PSM program.

5.2. Description of Process Knowledge

5.2.1. Objectives

The objective of process knowledge documentation is to record relevant process information so that it is possible to retrieve, use, and update information required to support PSM efforts. Some specific objectives are to provide:

- definition of original design;
- justification of design decisions and changes;
- information for training;
- basis for continued operation; and
- information for other PSM elements;

Each of these objectives is discussed herein.

Definition of Original Design—Unless details of the process design are recorded as they are prepared, the original bases may be difficult to identify in later years, leading to many potential problems. For example, it will be difficult to determine, at some future date, the adequacy of relief valve sizing if the original design basis and calculations were not well documented. Also, new information (such as recognition of the potential for two-phase flow in an emergency relief system) may potentially invalidate original design assumptions. The basis for the original design must be adequately documented to permit evaluation of the impact of such new information.

Justification of Design Decisions and Changes—Information concerning the original design basis of equipment and/or the reasons for a particular design feature may be important when subsequent modifications or replacements are to be made. Documentation can provide this information when required years later, when the original designers are no longer available. Timely and comprehensive recording of design changes is important to ensure that the changes are properly identified, and the reason for them is apparent. This record will become a valuable source of information for those who operate and maintain the facility.

Information for Training—Operations and maintenance personnel should be trained and knowledgeable about the potential hazards inherent in their work. This includes both materials and equipment. Documentation of process knowledge and subsequent training provides an effective way of communicating such essential information.

Basis for Continued Operation—Unplanned shutdowns or incidents can lead to injury or loss of output and profit. By thoroughly documenting the process, a better base of knowledge for quality operation and maintenance on a continuing basis is established, and fewer incidents should occur.

Information for Other PSM Elements—Effective PSM is strongly dependent on thorough and ongoing documentation. Most other elements of a PSM program are based on process knowledge and problems can occur if the documentation is missing, out-of-date, or in error.

5.2.2. Sources and Nature of Process Knowledge

A comprehensive compilation of documented information on the process and related safety information enables employers and the employees involved in operating the process to identify, understand and avoid potential hazards. Documentation described in this section includes:

- information about the chemicals or materials used;
- information about the technology of the process including:
  —process chemistry;
  —inventory;
  —safe upper and lower limits for parameters such as temperature, pressure, flows or compositions; and
  —evaluation of the consequences of deviations, including those affecting the safety and health of employees.
- information about the equipment and protective systems in the process, including:
  —materials of construction;
  —piping and instrumentation diagrams (P&IDs);
  —hazardous area classification;
  —alarms and interlocks;
  —relief system design and design basis; and
  —design codes and standards.

Each of these areas of process knowledge will be discussed separately.

Information about the Chemicals or Materials Used—Process knowledge related to chemicals and materials in a process can typically be found among the following sources:

- Material Safety Data Sheets (MSDSs);
- property test data;
- research reports;
- patents;
- pilot plant reports and data;
- basic design packages;
- licensor documents;
- operating reports for similar processes; and
- articles in the technical literature.

All pertinent data should be considered for documentation.

A common source of information is the MSDS. Categories of information on MSDSs are defined in OSHA's 29 CFR 1910.1200, and ANSI Z400.1 has been issued to standardize the format. The sections recommended by this latter document for inclusion in an MSDS are:

Section 1	Chemical Product and Company Identification
Section 2	Composition/Information on Ingredients
Section 3	Hazards Identification
Section 4	First Aid Measures
Section 5	Fire Fighting Measures
Section 6	Accidental Release Measures
Section 7	Handling and Storage
Section 8	Exposure Controls/Personal Protection
Section 9	Physical and Chemical Properties
Section 10	Stability and Reactivity
Section 11	Toxicological Information
Section 12	Ecological Information
Section 13	Disposal Considerations
Section 14	Transport Information
Section 15	Regulatory Information
Section 16	Other Information

Use of a standard format such as that given above should help all those who use MSDSs to find the required information quickly. Missing information can be a serious problem, and those individuals responsible for MSDS preparation should ensure that pertinent information is not omitted. If a section of the MSDS is not applicable for a particular substance (e.g., substance is not flammable and, therefore, Section 5 does not apply) that section should be so marked.

Purchased raw materials should be accompanied by a MSDS issued by the material supplier. The information on manufacturer or vendor supplied MSDSs should be checked for adequacy prior to use. Occasionally, an MSDS is not provided by a material supplier and, despite prompting, is not produced. Should an MSDS be unavailable from a supplier for any raw material, it is necessary for the organization to prepare or obtain an MSDS from an alternate source, prior to receipt of the material.

MSDSs should be prepared for all facility products. Preparation of MSDSs should also be considered for process intermediates, by-products and waste streams. When MSDSs are prepared for the first time, sufficient time should be allocated to complete the appropriate test work. Standard test procedures should be used and referenced in the documentation.

An example of an MSDS is included as Table 5-1, which illustrates the style and type of information contained. This MSDS is provided for illustration purposes only and should not be used as an actual source of information.

TABLE 5-1
Material Safety Data Sheet

This MSDS is presented for illustration purposes only and should not be used as an actual source of information.

CHLORINE

CHEMICAL PRODUCT/COMPANY IDENTIFICATION

Material Identification

CAS Number:	7782-50-5
Formula:	Cl_2
CAS Name:	CHLORINE

Trade Names and Synonyms

Cl_2

Company Identification

Manufacturer/distributor

ABC CHEMICAL COMPANY
ANYWHERE, USA

Phone Numbers

Product Information :	1-800-555-5555
Transport Emergency :	CHEMTREC: 1-800-424-9300
Medical Emergency :	1-800-555-5555

COMPOSITION/INFORMATION ON INGREDIENTS

Components

Material	CAS Number	%
*CHLORINE	7782-50-5	100

* Regulated as a Toxic Chemical under Section 313 of Title III of the Superfund Amendments and Reauthorization Act of 1986 and 40 CFR part 372.

HAZARDS IDENTIFICATION

Potential Health Effects

Liquid chlorine is corrosive to the skin and eyes. Eye damage may be permanent and may include blindness. Chlorine gas is extremely irritating to the nose, throat, and lungs. Gross overexposure may cause death.

HUMAN HEALTH EFFECTS:

Skin contact may cause skin irritation with discomfort or rash. Eye contact may cause eye irritation with discomfort, tearing, or blurring of vision. Eye damage may be permanent and may include blindness. Inhalation may cause irritation of upper respiratory passages; nonspecific discomfort such as nausea, headache, or weakness; or corrosion of teeth.

Higher exposures may cause skin burns or ulceration; eye corrosion with corneal or conjunctival ulceration; temporary lung irritation with cough, discomfort, difficulty breathing, or shortness of breath; followed in hours by severe shortness of breath, requiring prompt medical attention; asthma-like reactions with shortness of breath, wheezing, or cough, possibly occurring on subsequent reexposure to concentrations below established exposure limits; or temporary alteration of the heart's electrical activity with irregular pulse, palpitations, or inadequate circulation. Death may occur from gross overexposure.

Epidemiologic studies demonstrate no significant risk of human cancer from exposure to this compound.

Individuals with preexisting diseases of the eyes, skin, lungs, central nervous system, or cardiovascular system may have increased susceptibility to the toxicity of excessive exposures.

Carcinogenicity Information

None of the components present in this material at concentrations equal to or greater than 0.1% are listed by IARC, NTP, OSHA or ACGIH as a carcinogen.

FIRST AID MEASURES

First Aid

Contact with moisture in air or tissue may produce hydrochlorous and hydrochloric acids.

INHALATION

If inhaled, remove patient to an uncontaminated atmosphere. Call a physician. Check for breathing and pulse. If not breathing, give artificial respiration. If breathing is difficult, give oxygen as soon as possible (6 liters per minute). Check for other injuries. Keep the patient warm and at rest.

SKIN CONTACT

Immediately, within seconds of contact or suspected contact, shower with large quantities of water and completely remove all personal protective equipment, clothing, and shoes while in the shower. Flush the skin thoroughly with water for at least 5 minutes. Call for medical help while flushing the skin. Keep the affected area cool. Avoid freezing affected area. Wash clothing before reuse.

EYE CONTACT

Immediately flush eyes with large quantities of water while holding the eyelids apart. Continue flushing for 5 minutes. Do not try to neutralize the acid. Call a physician immediately. Transfer promptly to a medical facility. Apply cool packs on the eyes while transporting patient. Avoid freezing affected area.

INGESTION

Do not induce vomiting. Give large quantities of water. Call a physician immediately and transfer promptly to a medical facility. Never give anything by mouth to an unconscious person.

page 2 of 6

FIRE FIGHTING MEASURES

Flammable Properties

Will not burn in air. Strong Oxidizer.

Fire and Explosion Hazards:: Contact with combustible materials may cause fire. Dangerous when heated; emits highly toxic fumes. Follow appropriate National Fire Protection Association (NFPA) codes.

Extinguishing Media

As appropriate for combustibles in area.

Fire Fighting Instructions

Wear self-contained breathing apparatus. Wear full protective equipment. Shut off source of fuel, if possible and without risk.

Keep personnel removed and upwind of fire. Do not apply water directly to leak. Cool tank/container with water spray. Run-off from fire control may cause pollution.

ACCIDENTAL RELEASE MEASURES

Safeguards (Personnel)

NOTE: Review FIRE FIGHTING MEASURES and HANDLING (PERSONNEL) sections before proceeding with clean-up. Use appropriate PERSONAL PROTECTIVE EQUIPMENT during clean-up.

Evacuate personnel, thoroughly ventilate area, use self-contained breathing apparatus. Keep upwind of leak—evacuate until gas has dispersed.

Initial Containment

Dissipate vapor with water spray. Prevent material from entering sewers, waterways, or low areas.

Accidental Release Measures

Specially trained personnel should stop the leak if possible, dike spill, and neutralize any water that may be used with caustic. Comply with Federal, State, and local regulations on reporting releases. The CERCLA Reportable Quantity for a spill, leak, or release is 10 lbs.

HANDLING AND STORAGE

Handling (Personnel)

Do not breathe gas. Do not get in eyes, on skin, or on clothing. Wash thoroughly after handling.

Storage

Store in a cool place away from heat, sparks, and flame. Keep containers tightly closed.

page 3 of 6

EXPOSURE CONTROLS/PERSONAL PROTECTION

Engineering Controls

Use sufficient ventilation to keep employee exposure below recommended exposure limits.

Personal Protective Equipment

Have available and wear as appropriate for exposure conditions: chemical splash goggles; safety glasses, (side shields preferred); full-length face shield; gloves, pants, jacket, apron, and footwear or acid suit made of butyl, "Chemfab", "Chloropel", "Neoprene", nitrile, "Saranex" coated "Tyvek", urethane, or "Viton" and NIOSH/MSHA approved respiratory protection.

Exposure Limits CHLORINE

PEL (OSHA):	1 ppm, 3 mg/m3, Ceiling
TLV (ACGIH) :	0.5 ppm, 1.5 mg/m3, 8 Hr. TWA
	STEL 1 ppm, 2.9 mg/m3 TWA

PHYSICAL AND CHEMICAL PROPERTIES

Physical Data

Boiling Point:	–34.6°C (–30.3°F) @ 760 mm Hg
Vapor Pressure:	4,800 mm/Hg @ 20°C (68°F)
Vapor Density:	2.5 (Air = 1)
Melting Point:	–101°C (–150°F)
Evaporation Rate:	(Butyl Acetate = 1) Greater than 1
Solubility, Water:	0.57 wt% @ 30°C (86°F)
Odor:	Acrid
Odor Threshold:	0.2–0.4 ppm
Form:	Gas at STP/Liquid
Color:	Gas: Greenish yellow;
Liquid:	Amber
Specific Gravity :	1.56 @ –35°C (–31°F) Liquid

STABILITY AND REACTIVITY

Chemical Stability

Dry chlorine is stable in steel containers at room temperature.

Decomposition

Hydrochloric and hypochlorous acids are formed with water or steam.

Polymerization

Polymerization will not occur.

Other Hazards

Incompatibility: Incompatible with alkalies, reducing agents, and organic materials. Reacts explosively or forms explosive compounds with acetylene, turpentine, fuel gas, hydrogen, ether, ammonia gas, and finely divided metals. Reacts vigorously with titanium, zinc, and tin. Reaction with aluminum may result in generation of flammable hydrogen gas. May also form explosive mixtures with combustible organic vapors and cause runaway reactions with certain polymers if contacted in confined areas. Combines with carbon monoxide and sulfur dioxide to form toxic and corrosive phosgene and sulfuryl chloride.

TOXICOLOGICAL INFORMATION

Animal Data

Inhalation 1-hour LC_{50}: 293 ppm in rats

The compound is corrosive to eyes and skin. Toxic effects described in animals from short exposures by inhalation include upper and lower respiratory, kidney, liver, and lung effects. Long-term inhalation exposures caused eye irritation and nonspecific effects such as weight loss. By ingestion, the effects included irritation and corrosion of mucosal surfaces; kidney, liver, and lung effects; and nonspecific effects such as decreased weight gain. No significant adverse effects were observed from long-term dietary administration.

Animal testing indicates that the compound does not have carcinogenic or reproductive effects. Tests for embryotoxic activity in animal species have been inconclusive, with positive results in some studies and negative results in others. Tests in bacterial cell cultures demonstrate mutagenic activity.

ECOLOGICAL INFORMATION

Aquatic Toxicity: 96-hour LC_{50}, rainbow trout: 0.132 mg/liter

DISPOSAL CONSIDERATIONS

Waste Disposal: Treatment, storage, transportation and disposal must be in accordance with applicable Federal, State, and local regulations.

TRANSPORTATION INFORMATION

Shipping Information: DOT/IMO

Proper Shipping Name :	CHLORINE
Hazard Class:	2.3
UN No.:	1017
DOT/IMO Label:	POISON GAS
Special Info.:	POISON—INHALATION HAZARD, ZONE B; MARINE POLLUTANT
Reportable Quantity :	10 lb
Shipping Containers :	Tank Cars.

page 5 of 6

REGULATORY INFORMATION

U.S. Federal Regulations

TSCA Inventory Status: Reported/Included.

TITLE III HAZARD CLASSIFICATIONS SECTIONS 311, 312

Acute:	Yes
Chronic:	Yes
Fire:	No
Reactivity:	Yes
Pressure:	Yes

LISTS:

SARA Extremely Hazardous Substance:	Yes
CERCLA Hazardous Material:	Yes
SARA Toxic Chemicals:	Yes

CHLORINE is specifically listed in Appendix A of 29 CFR 1910.119. Use of chlorine may require compliance with 29 CFR 1910.119, Process Safety Management of Highly Hazardous Chemicals.

OTHER INFORMATION

NFPA, NPCA-HMIS

NFPA Rating

Health:	3
Flammability:	0
Reactivity:	0
Oxidizer	

NPCA-HMIS Rating

Health:	3
Flammability:	0
Reactivity:	1

Personal Protection rating to be supplied by user depending on use conditions.

Additional Information

NSF LIMITS: NSF Maximum Drinking Water Use Concentration: 30 mg/L as chlorine.

The data in this Material Safety Data Sheet relate only to the specific material designated herein and does not relate to use in combination with any other material or in any process.

Responsibility for MSDS:	ABC CHEMICALS COMPANY
Address:	ANYWHERE, USA
Telephone:	555-555-5555

Companies handling large numbers of different substances may amass a data base for MSDSs. While MSDSs can be obtained from inhouse development or from suppliers of purchased material, there are also a few sets of published MSDSs commercially available. These publications seldom include all chemicals used at a particular site, but they may be useful to organizations handling the more common materials. Among these are MSDSs published by:

Genium Publishing Corporation
One Genium Plaza
Schenectady, NY 12304-4690
(518) 377-8855
containing nearly 1000 titles.

MSDS Software sources include:

MSDS-CCOHS
Canadian Center for Occupational Health and Safety (over 70,000 titles);
CHEMTOX
Resource Consultants; and
TAPP
Europa Scientific Software Corp., Hollis, NH.

In addition, other physical or chemical characteristics of materials, beyond those typically found in a MSDS (e.g., specific resistance or dust explosibility), may need to be determined and documented.

Information about the Technology of the Process—Technology information should include a block flow diagram or a simplified process flow diagram. The block flow diagram shows each stage of an operation as a block regardless of the number of items of equipment used in that operation. The block flow diagram should indicate:

- the unit operations in the process;
- the ways the different operations are related to each other; and
- the materials fed to, recycled, and discharged from the process.

A typical block flow diagram is shown in Figure 5-1.

A simplified process flow diagram (PFD) depicts a process in more detail by showing the major pieces of equipment. The process flow diagram should include:

- the major equipment used in the process;
- the chemicals or materials used in the process;
- the utilities used in the process; and
- the major control loops.

A PFD can also include material and energy balances for the process. A typical simplified PFD is shown in Figure 5-2.

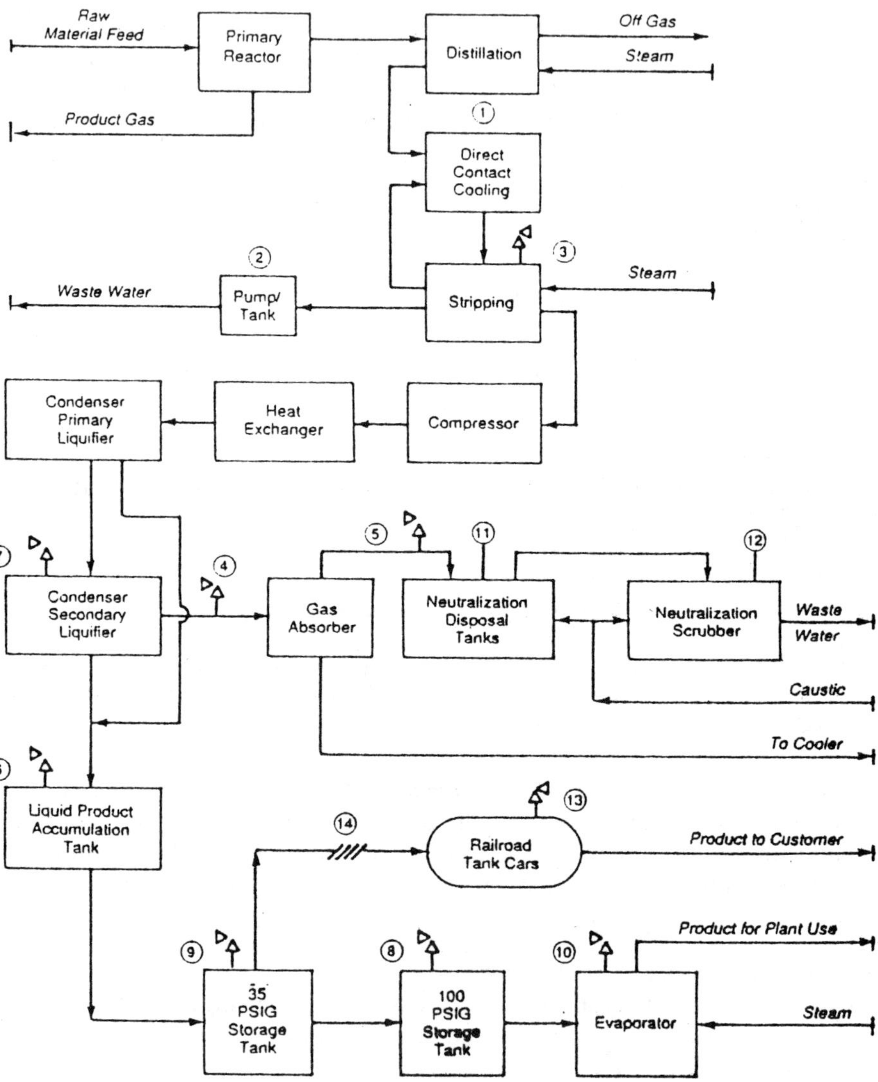

FIGURE 5-1. **Example of Block Flow Diagram.**

Process Chemistry—The documentation of the process chemistry should be comprehensive and include not only the chemistry of the reaction that takes place when the product is made, but also describe the formation and handling of intermediates, by-products and side reactions that might take place in credible abnormal operations. The relative importance of side reactions and the factors which control them should be documented. Critical thermodynamic information such as heat of reaction, dilution, or solution, as well as kinetic data defining rates

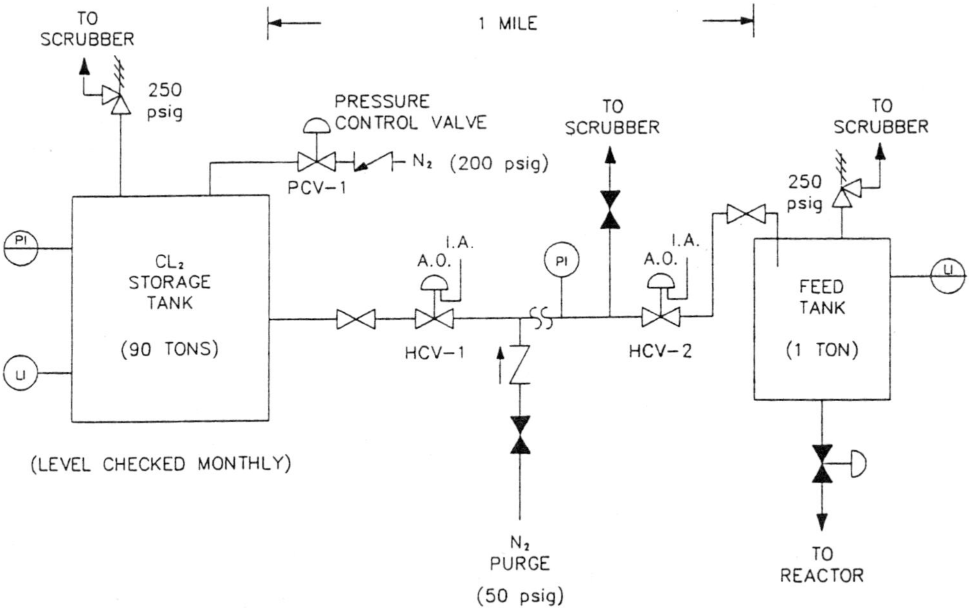

FIGURE 5-2. **Example of Simplified Process Flow Diagram**

of reaction, should also be documented. An example of process chemistry documentation is shown in Table 5-2.

Inventory—Inventory is the quantity of each chemical (raw materials, intermediates, products, by-products, solvents, catalysts, additives, heat transfer fluids, etc.) that is stored and processed in the facility. Inventory includes not only the materials present in the tank farm and warehouse, but those present in pipelines, intermediate feed tanks and operating equipment. The maximum inventory for each material and the basis for this value should be documented. The documentation should also include any notes concerning assumptions which have been made (e.g., size of delivery or shipment).

Safe Upper and Lower Limits for such Parameters as Temperatures, Pressures, Flows or Compositions—Safe operating limits should be considered from several points of view. Each equipment item has a certain upper and lower design temperature and pressure. The equipment documentation should define such limits for safe operation of equipment. Processes have safe operating limits as well, above or below which undesirable or potentially hazardous by-products may form, or run-away reactions may take place. These limits include temperatures, pressures, flows, and concentrations.

TABLE 5-2
Example of Process Chemistry Documentation[a]

	Component	*Abbreviation*	*Formula*
Product	Dimethyl Carbonate	DMC	$CO(OCH_3)_2$
Raw Materials	Methanol		CH_3OH
	Carbon monoxide		CO
Intermediates	Methyl Nitrite		CH_3ONO
	Nitric Oxide		NO
By-products	Dimethyl oxalate	DMO	$(COOCH_3)_2$
Major	Methyl formate		$HCOOCH_3$
Minor	Methyl acetate		CH_3COOCH_3
	Methylal		$CH_2(OCH_3)_2$
	Carbon dioxide		CO_2
	Nitric acid		HNO_3
Purge Gas Contents:	Nitrogen, carbon monoxide, carbon dioxide, nitric oxide, methanol, methyl nitrite and methyl formate		
Wastewater contents:	DMC, DMO, methanol		

Reactions:

$$CO + 2CH_3ONO \rightarrow CO(OCH_3)_2 + 2NO - 45.8 \text{ kcal/gmole (at STP)}$$

$$4NO + O_2 + 4CH_3OH \rightarrow 4CH_3ONO + 2\,H_2O - 59.8 \text{ kcal/gmole}$$

Overall Reaction:

$$2CO + O_2 + 4CH_3OH \rightarrow 2CO(OCH_3)_2 + 2H_2O - 15\ 1.4 \text{ kcal/gmole}$$

By-product, DMO Formation:

$$4CO + O_2 + 4CH_3OH \rightarrow 2(COOCH_3)_2 + 2H_2O - 168.6 \text{ kcal/gmole}$$

[a]Data presented in this table are simulated and are shown only for purpose of illustration of form content.

Evaluation of the Consequences of Deviations, Including those Affecting the Safety and Health of Employees—Significant upper and lower limits of temperature, pressure, flow, viscosity, concentration, etc., outside of which corrective action of some kind is necessary, should be determined and documented, along with the anticipated consequences should corrective action not be taken. For example, the documented record of many PHAs may include significant concerns for potential deviations from normal operation and their expected consequences. These learnings should also be documented in standard operating procedures.

Often, these process operating limits are determined by the equipment design basis or materials of construction. Hence, effective documentation of process knowledge and equipment integrity should consider the interactions between these two elements.

The evaluation of the consequences of deviations is described in Chapter 6, *Process Hazard Analysis* and also in the CCPS book, *Guidelines for Hazard Evaluation Procedures, Second Edition with Worked Examples.*

Information about the Equipment and Protective Systems in the Process—This class of information would typically include:

- materials of construction;
- piping and instrumentation diagrams (P&IDs);
- hazardous area classification drawings;
- alarms and interlocks;
- relief system design and design basis;
- design codes and standards; and
- fire protection system drawings.

Each of these areas is discussed herein.

Materials of Construction (including gaskets, packing, etc.)—The materials of construction of each equipment item should be clearly documented. The knowledge of the materials of construction is especially important when changes in the established technology are considered. See Chapter 8 for the documentation of *Process Equipment Integrity*.

Piping and Instrumentation Diagrams—P&IDs are the major controlling documents for design and construction of a process unit. These diagrams contain basic data pertinent to design, such as:

- vessels and process equipment with size or capacity;
- lines for process and utilities;
- line and nozzle sizes;
- instruments and their interconnections;
- special features (e.g., internals, types of valve, pump, hose, etc.);
- tag or equipment numbers;
- process requirements for venting, draining and purging;
- failure modes of automatic valves;
- set pressures for vacuum and pressure reliefs; and
- other information needed for process design.

Since P&IDs can undergo many changes in the life cycle of a process, it is essential that they be kept up to date with revisions traceable via revision letters and/or numbers and dates. Figure 5-3 shows a section of a typical P&ID.

Hazardous Area Classification—Areas of the plant where inventories of flammable or combustible materials exist are normally classified according to the flammability of the various materials present and the probability of those materials being released from the normal containment system. This analysis is termed

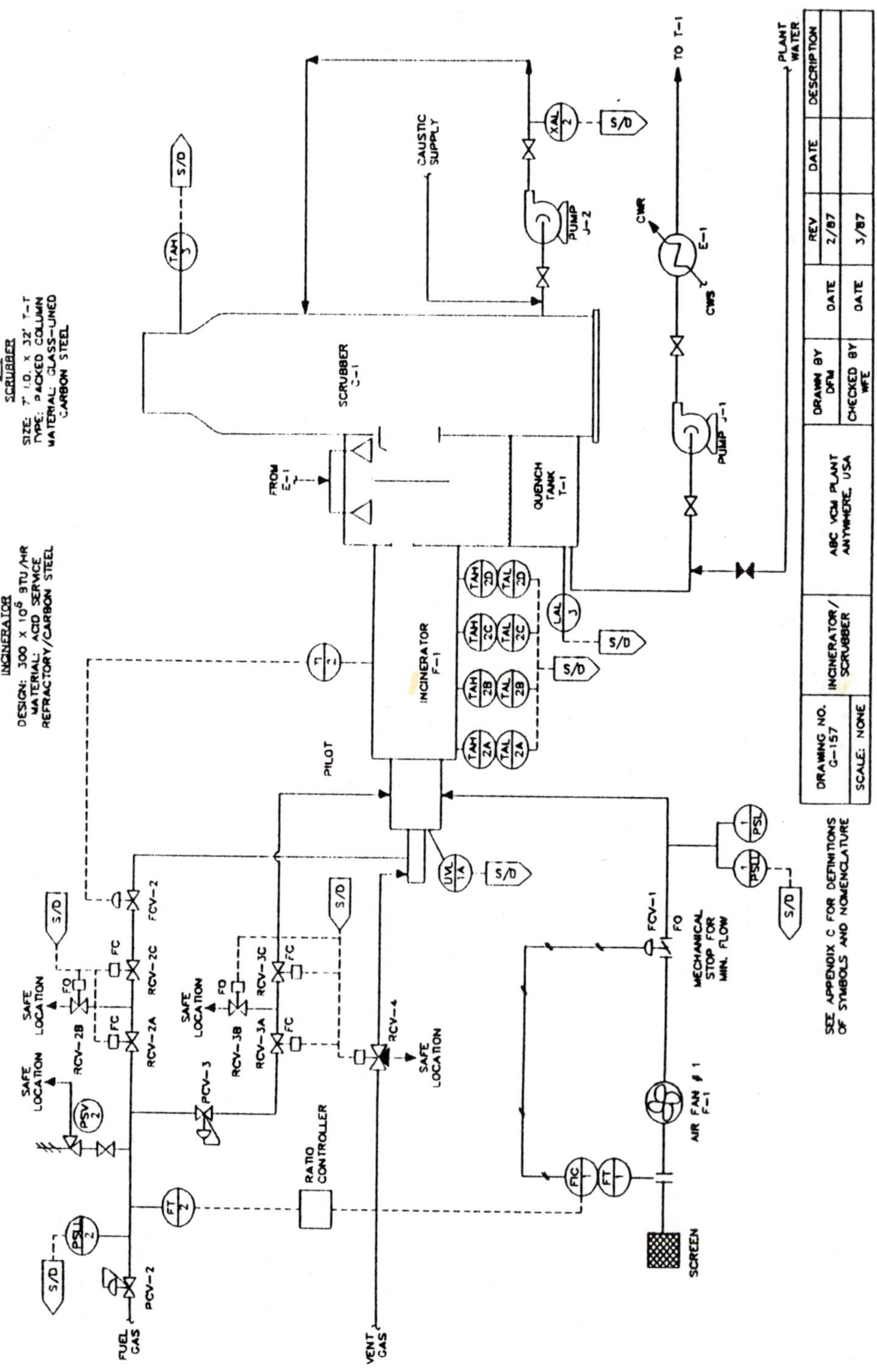

FIGURE 5-3. Example of P&ID

hazardous area classification and is used to determine the degree of protection required to prevent ignition of flammable materials by installed equipment or temporary equipment. Such measures significantly reduce the potential for incidents. Hazardous area classification documentation might include:

- a statement concerning the selection of the code or method used to establish the hazardous area classification;
- a list of flammable or combustible materials with their relevant physical and flammable properties (a typical hazard information data file is shown in Table 5-3.);
- drawings showing the hazard classifications on a plot plan and in detailed equipment arrangements showing both plan and elevation views; the classification for each area should be clearly marked (Figure 5-4 shows typical hazard classification drawings);

TABLE 5-3
Example of Hazard Information Data[a]

Component	*Acetone*	*1,3 Butadiene*	*Tin Metal Powder*
Phase @ 20°C	Liquid	Vapor	Solid Powder
Boiling Point @ atm pressure,°C	56	–4.4	2507
Melting Point/FP°C	–94	–108.9	231.9
Flash Point, °C	–17, Closed Cup	<0, Open Cup	NA
Autoignition Temperature, °C	465	Not found	cloud 630, dust layer 430
LFL, vol %	2.9	2	N/A
LFL, oz/ft^3	N/A	N/A	0.19
UFL, vol %	12.8	11.5	NA
TLV-TWA, ppm	750	10	N/A
TLV-TWA, mg/m^3	N/A	N/A	2
Resistivity, ohm-cm	NA	NA	N/A
Stability	Stable	Without inhibitor hazardous polymerization occurs	Oxidizes in the presence of moisture
Remarks:	NA means not available N/A means not applicable		

[a]Data presented in this table are simulated and are shown only for purpose of illustration of form content.

NOTES:

1. WORK THIS DRAWING WITH THE FOLLOWING HAZARDOUS (CLASSIFIED) LOCATION DESIGN BASIS':

   a.) DESIGN BASIS NO. 11A-LIQUID HANDLING AREA, TANK FARM AND YARD AREAS.

   b.) DESIGN BASIS NO. 11B-REACTOR STRUCTURES.

   c.) DESIGN BASIS NO. 11C-DISTILLATION & EXTRACTION AREAS.

2. FOR DETAILED AREA CLASSIFICATION OF THE REACTOR STRUCTURES, DISTILLATION AREA 1, DISTILLATION AREA 2 AND EXTRACTION BLDG. SEE THE FOLLOWING DRAWINGS:

   REACTOR "R11" STRUCTURE
   #CCPS-7-GROUND FLOOR
   #CCPS-8 PLAN @ EL. 112'-0" & EL. 124'-0"
   #CCPS-9-PLAN @ EL. 139'-1" & EL. 149'-6"

   REACTOR "R33" STRUCTURE & ADJACENT AREAS
   #CCPS-3-GROUND FLOOR
   #CCPS-4-SECOND FLOOR
   #CCPS-5-3RD & 4TH FLOORS
   #CCPS-27-DISTILLATION VENT SCRUBBER AREA

   REACTOR "R44" STRUCTURE
   #CCPS-10-GROUND FLOOR
   #CCPS-11-2ND FLOOR
   #CCPS-12-3RD & 4TH FLOORS

   REACTOR "R55" STRUCTURE
   #CCPS-13-GROUND FLOOR
   #CCPS-14-PLAN @ EL. 112'-0"
   #CCPS-15-3RD FLOOR
   #CCPS-16-4TH & 5TH FLORS

   DISTILLATION AREA 1
   #CCPS-17-PLAN @ GRADE (EL. 100'-0")
   #CCPS-18-PLAN @ EL. 115'-0" & 140'-0"
   #CCPS-19-PLAN @ GRADE (EL. 100'-0")

   DISTILLATION AREA 2
   #CCPS-20-PLAN @ GRADE & UPPER PLATF. LEVELS
   #CCPS-21-PLAN @ GRADE

   EXTRACTION AREA
   #CCPS-22-GROUND FLOOR
   #CCPS-23-2ND LEVEL & ROOF

3. FOR HAZARDOUS (CLASSIFIED) LOCATION DETAILS SEE DWG #CCPS-2 & #CCPS-6.

4. THE ELECTRICAL INSTALLATION IN THE FOLLOWING AREAS SHALL BE SUITABLE FOR THE MATERIAL WITH THE LOWEST IGNITION TEMPERATURE. CLASS I TEMPERATURE MARKINGS SHALL NOT EXCEED THE AUTOIGNITION TEMPERATURE OF THE SPECIFIC MATERIAL LISTED.

PROCESS PLANT AREA	MATERIAL	AUTOIGNITION TEMPERATURE	"T" RATING
TANK FARM	METHANOL	878°F	T1
ANALYZER BLDG. I	REACTION OFF GAS	1075°F	T1
ANALYZER BLDG. II	REACTION OFF GAS	1075°F	T1
ANALYZER BLDG. III	METHANOL	878°F	T1
WASTE ORGANICS STORAGE AREA	CRESOL	1038°F	T1
RAILCAR UNLOADING AREA	METHANOL	878°F	T1
TANK TRUCK UNLOADING AREA	METHANOL	878°F	T1
TANK TRUCK LOADING AREA	TOLUENE	896°F	T1
METHANOL STORAGE TANK AREA	METHANOL	878°F	T1
REACTOR R11,R33,R44 & R55 STRUCTURES	METHANOL	878°F	T1
DISTILLATION AREA	METHANOL	878°F	T1
DISTILLATION 2 AREA	METHANOL	878°F	T1
EXTRACTION BUILDING	METHANOL	878°F	T1

5. LABORATORY SHALL BE CLASSIFIED PER NFPA 45 AND THE GUIDELINES OUTLINE IN DESIGN BASIS 11A PARA. 1.6.13.

AREA CLASSIFICATION LEGEND

- CLASS I, DIV. 1, GROUP B
- CLASS I, DIV. 2, GROUP B
- CLASS I, DIV. 1, GROUP D
- CLASS I, DIV. 2, GROUP D
- NON-HAZARDOUS
- EQUIPMENT VENT TO ATMOSPHERE

REV	DATE	DESCRIPTION	DRWN	CHKD	APPD
1		REVISED NOTES 2 & 4, ADDED PUMPS P-154A/S, CLASSIFIED PIPERACKS. EXTENDED DIV. 2 CLASSIFICATION TO RECOGNIZABLE BOUNDARIES, REVISED BACKGROUND PER CLIENT REVIEW MEETING 08SE93. ISSUED FOR FIELD SURVEY.	GJM	DAF	
0	01SE93	ISSUED FOR CLIENT APPROVAL	GJM	DAF	AA

PROCESS SAFETY MANAGEMENT PROGRAM

JOHN BROWN ENGINEERING & CONSTRUCTION — JOHN BROWN E & C, STAMFORD CT

HAZARDOUS (CLASSIFIED) LOCATIONS - SAMPLE DRAWING PLOT PLAN (NOTES & LEGEND) CHEMICAL PROCESS PLANT

DRAWN F. D'ERASMO	CHKD D. FENYUS	APPR A. ALIPERTI
DATE 02/09/93	DATE 02/12/93	DATE 03/05/93
SCALE: 1" = 40'-0"	CAD: C-ACTITL	
FOR: ABC CHEMICAL COMPANY		
CONTRACT NO. 92398-00	DRAWING NO. CCPS-26	REV. N/A

FIGURE 5-4 Typical Hazard Classification Drawings (Sheet 1 of 4)

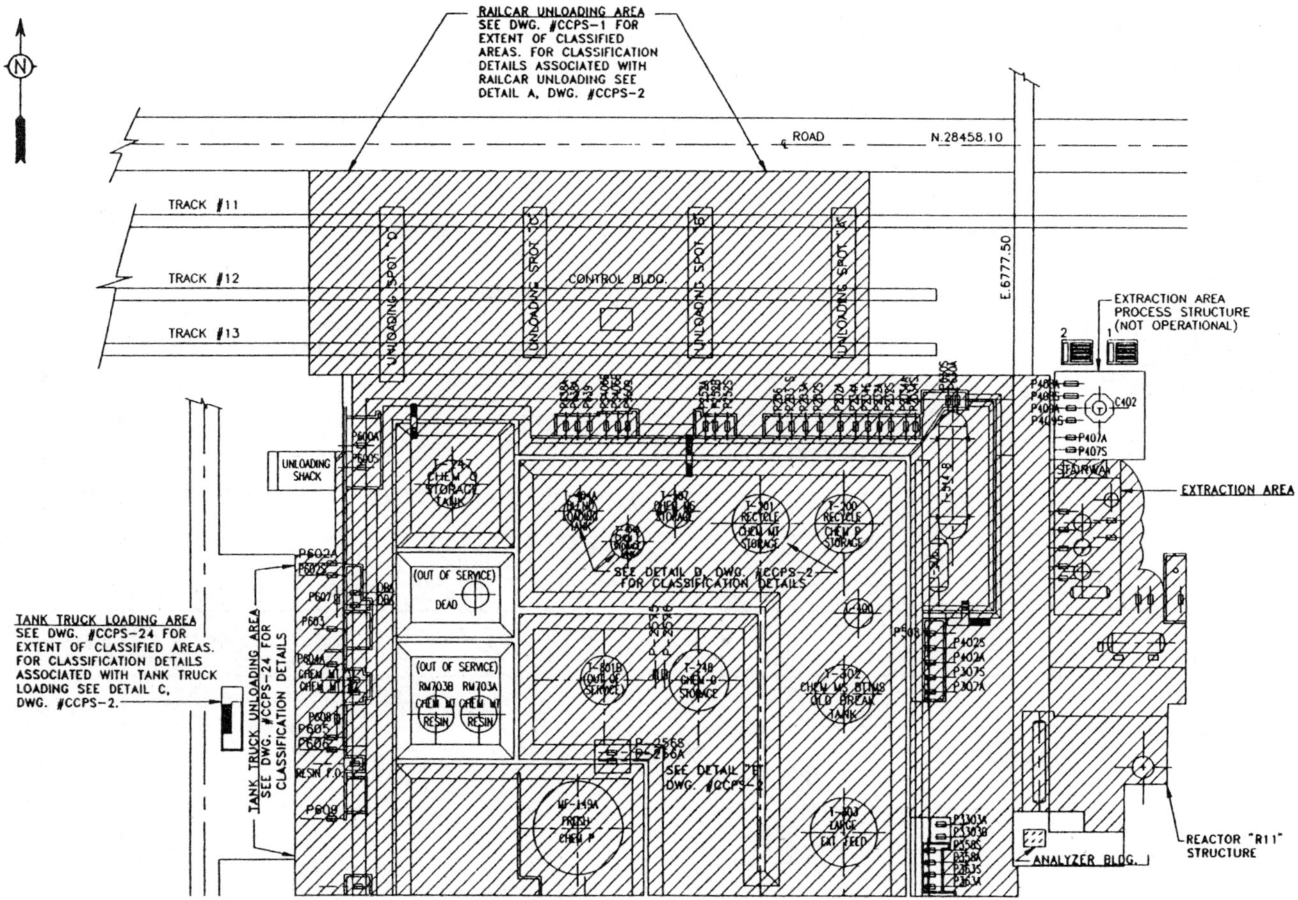

FIGURE 5-4 Typical Hazard Classification Drawings (Sheet 2 of 4)

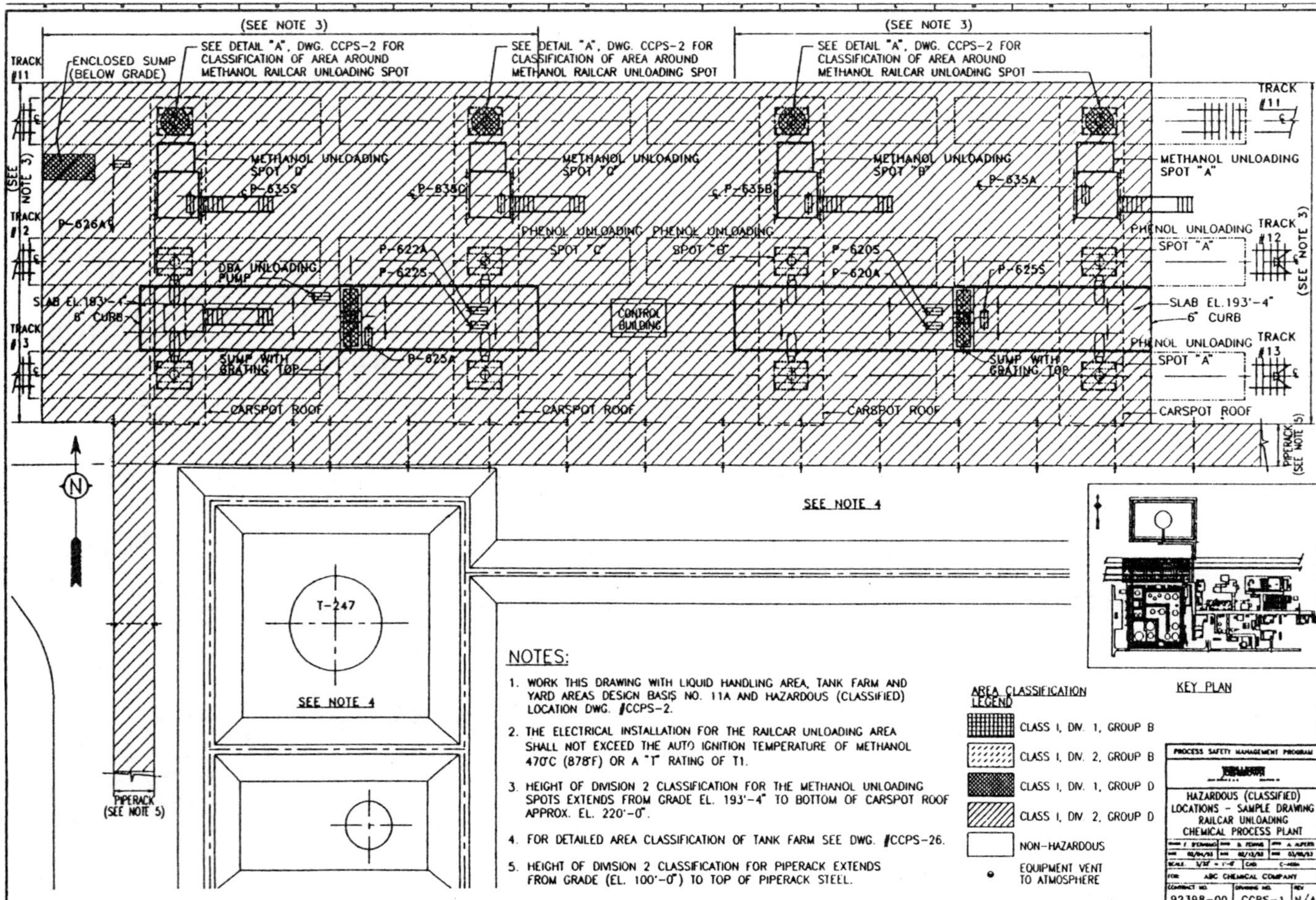

FIGURE 5-4 Typical Hazard Classification Drawings (Sheet 3 of 4)

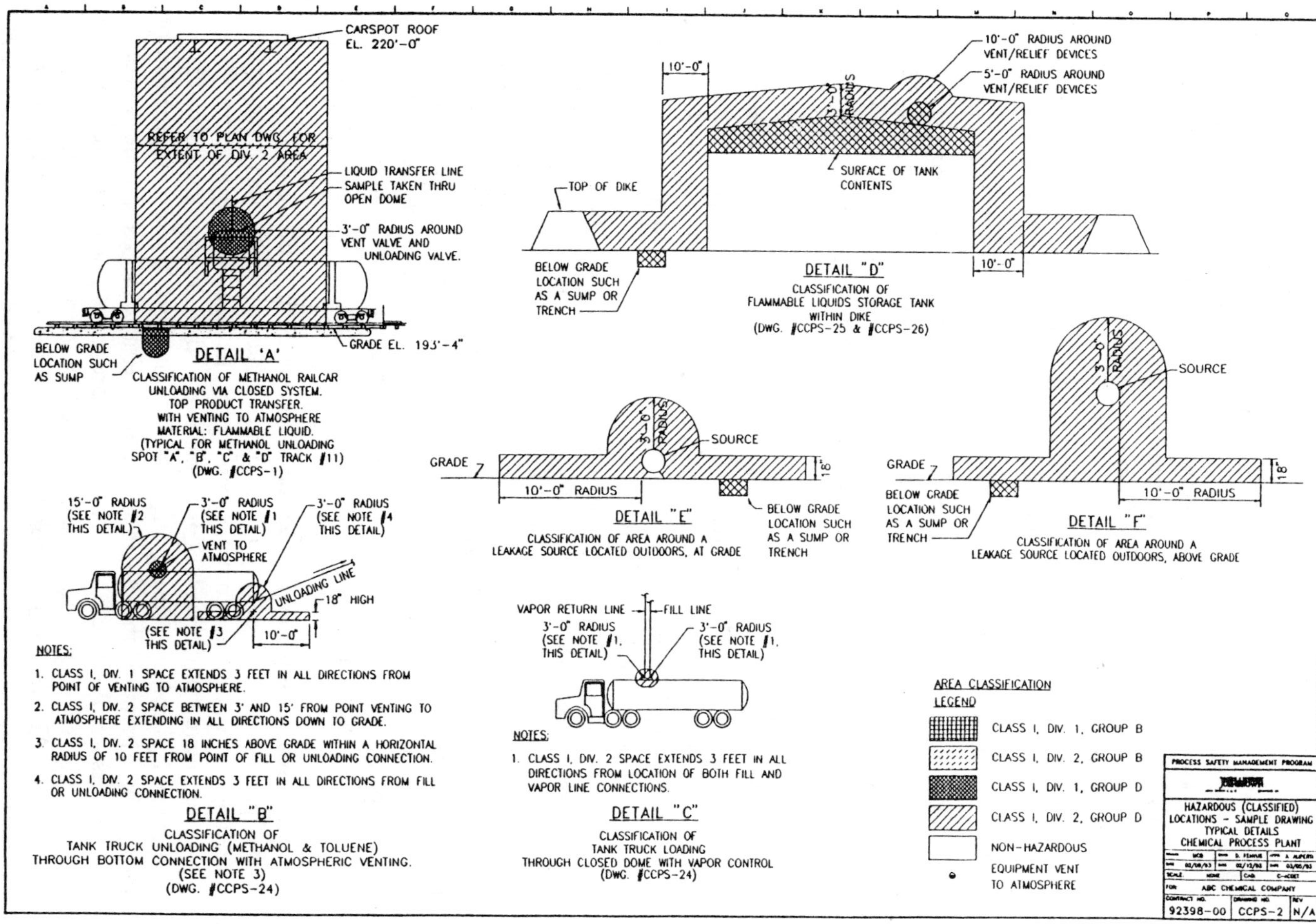

FIGURE 5-4 Typical Hazard Classification Drawings (Sheet 4 of 4)

- any adjustments made to the areas as originally determined to assist in compliance, such as extension to readily identifiable physical boundaries (e.g. fences and drainage ditches, berms);
- the procedures which keep this information accurate and up-to-date;
- provisions which ensure that all electrical equipment, including instrumentation and communication systems, are appropriate for the area in which they are located, and are maintained in good order;
- procedures for control of ignition sources and equipment brought into the area temporarily; and
- procedures, and any necessary equipment, required to handle, collect and safely dispose of flammable material or combustible dusts.

Some standards used for defining hazardous area classifications are:

- American Petroleum Institute (API), *Classification of Locations for Electrical Installations in Petroleum Refineries,* Recommended Practice (RP) 500A, and its referenced documents;
- National Fire Protection Association (NFPA), *National Electrical Code®,* NFPA 70;
- National Fire Protection Association, *Classification of Class I Hazardous (Classified) Locations for Electrical Installations in Chemical Process Areas,* NFPA 497A, and its referenced documents;
- National Fire Protection Association, *Classification of Class II Hazardous (Classified) Locations in Chemical Process Areas,* NFPA 497B, and its referenced documents; and
- American National Standards Institute/Instrument Society of America, *Area Classification in Hazardous (Classified) Dust Locations*, ANSI/ISA S12.10, and its referenced documents.

Alarms and Interlocks—An important aspect of the safe control of a process is the system of alarms and interlocks used to warn of or prevent hazardous conditions from occurring. The reasoning for the selection of each of these is critical, particularly if changes are contemplated, so they should be carefully documented and the information kept up-to-date.

Relief System Design and Design Basis—Relief devices are installed to ensure that a process system, or any of its components, is not subject to pressures exceeding the maximum allowable pressure for the system.

Overpressure is the result of an imbalance or disruption of the normal flows of material and energy, causing one or both of them to build up in some part of the system. Therefore, the system should be carefully analyzed for the possibility of overpressure. Possible causes of overpressure should be considered, and the design of the relieving device should be based on the potentially worst situation. Refer to the AIChE-DIERS book *Emergency Relief System Design Using DIERS Technology* for design aid where potential runaway reactions have been identified.

Comprehensive documentation of the design and design basis should be developed and maintained for each relief system. This documentation should include the following:

- chemicals or materials in the system;
- results of any thermal stability test data to identify runaway reaction potential;
- nature of the materials which may be relieved;
- causes and effects of deviations from normal process conditions;
- design basis for relief case selected;
- specifications of the relief device selected; and
- requirements for safe disposal of the relieved material.

Table 5-4 illustrates a typical design basis for a relief system.

Design Codes and Standards—Chemical process plants, storage and handling areas should be designed in accordance with relevant codes and standards. These should be identified and listed as part of the design records. The listing should indicate the edition used and/or the year of issue. At the design stage, a decision is often made concerning which series of standards will be followed. This decision should be recorded, since it is important that this basic information be known when modifications are made or equipment is replaced. A more extensive discussion on this subject is found in Chapter 16, *Standards, Codes, and Regulations.*

5.3. Process Knowledge Documentation

5.3.1. Process Knowledge Program Documentation

Organizations should develop and document a program defining the goals of the Process Knowledge element and containing the detailed procedures describing how Process Knowledge will be implemented. This documented program should address the following questions:

Who and When?—The task of creating process knowledge documentation for a new facility is relatively easy. Every document is compiled chronologically as research and development, design and construction proceeds. A plan for document preparation, management, and use should be made, especially if the documentation is extensive.

At facility start-up, responsibility for process knowledge is typically transferred to designated members of the facility staff; that is, personnel who will use the information involved. At this stage the ongoing management of the information and any changes to it will be the main focus of the documentation task.

Process information must also be compiled and documented for existing plants. This can be a major task requiring significant research if the original design documents are not available, or are out-of-date. Assigning a knowledgeable person

TABLE 5-4
Example of Relief System Design Basis[a]

Relieving Device ID	PSV-10505	
Equipment or Pipe to Be Relieved	*Name*	*Item Number*
	MC Stripper	C-100-030
	MC Stripper Reboiler	E-110-031
	MC Stripper Condenser	E-110-032
Design Pressure	75 psig/full vacuum	
Set Pressure	75 psig	
Cause of Overpressure	*Considered*	*Basis*
Runaway Reaction	Not Relevant	
Fire	Yes	
Cooling Medium Failure	Yes	MeOH Vapors
Power Failure	Yes	
Utility Failure Excluding Cooling Media and Power	Yes	
Control Valve Failed Open	Yes	
Blocked Discharge	Yes	
Reboiler Tube Failure	Yes	
Special Conditions	Decomposition	
Material Relieving	*Considered*	*Basis*
Liquid	No	
Liquid–Vapor	No	
Vapor	Yes	Vapor
Time of Release	*Considered*	*Basis*
Beginning	Uninterrupted feed assumed.	Normal, continuous
After 50% relieved		
End		
Method of Calculation	Vapor Relief	
Reference	API RP 520 and RP 521	
Basis of Sizing		
Material Relieved	MC with 10 mol % MeOH	
Vapor rate, lbs/hr	4280	
Molecular Weight	70.6	
Maximum Relieving Pressure (MRP), psig	82.5	
Coincidental Temperature, °F	474	
C_p/C_v	1.085	
Heat of Vaporization, BTU/lb	300	
Specific Gravity	Not required	
Viscosity, cp	Not required	

[a]Data presented in this table are simulated and are shown only for the purpose of illustrating form content.

the responsibility for assembling and/or creating such documentation should be considered. See Chapters 4, *Records Management* and 20, *Document Life Cycle* for additional guidance.

What Documents?—The documents may consist of reports, manuals, files, tables, graphs, and drawings, etc. Some of these will be hard copy only, but others may be in electronic or other media such as optical disks. Most documents will be physically included; however, others, such as industry consensus standards, may only be referenced. An index may be provided to identify which documents are included or referenced, and their revision numbers and dates. Appendix 5A shows an example file index.

Where Are the Documents Kept?—Copies of process knowledge documentation may be kept at the most suitable place depending on the stage of the project life cycle and the needs of the users. For operating plants, documents may be kept in several places in order to meet the requirements of document security and accessibility. For example, an operating procedure may be filed in all of the following locations:

- master (or approval) copy in a secure and safe location, for instance in an administrative office;
- working copies in the plant supervisor's office; and
- elsewhere, as appropriate (e.g., in the plant control room).

Documents may be stored and retrieved from electronic data bases or hard copy files. In either case subdivisions of the data base should be clearly marked and kept together for each area of a facility. When using hard copy in areas where considerable handling may occur, such as in control rooms or maintenance shops, consideration should be given to ways to protect the sheets from loss or damage (e.g., lamination or transparent envelopes). Back-up copies may also be desirable. See Chapter 4, *Records Management*.

How Are Files Administered?—The documentation program may also address how the following are to be achieved:

- files are organized for simplicity and timely retrieval;
- revisions to documents are properly authorized and controlled;
- revisions are communicated to the appropriate parties; and
- unauthorized use, removal and copying are prevented.

5.3.2. Records from Implementing the Process Knowledge Element

The records that come from the implementation of this element include those described in Section 5.2.2. As mentioned previously, there is some overlap between this and other PSM elements, such as *Process Hazard Analysis* (Chapter 6) and *Process Equipment Integrity* (Chapter 8). The reader may wish to refer to these other chapters for more detailed document descriptions.

5.4. Records Management

In addition to the description of the Process Knowledge program given in Section 5.3.1, the organization's PSM program documentation should define Process Knowledge records management requirements. There are a number of unique aspects inherent to the records management task for Process Knowledge; for example, the breadth and amount of information involved and the likelihood that this information, and the responsibilities for it, will be distributed among a variety of disciplines. Also, by its nature, Process Knowledge documentation is subject to frequent revisions required to keep information up to date (e.g., P&ID revisions). The program documentation should describe how the records for the comprehensive area of process safety knowledge are to be managed. Since several disciplines will be involved, careful coordination is required, and the person with overall responsibility needs to be aware of not only who looks after individual sections, but also the records management practices pertinent to each section. The records management program should facilitate needed revisions to information, yet provide adequate controls to guard against unauthorized or inadvertent changes. Refer to Chapter 4, *Records Management*, for further guidance.

5.5. Auditing

The PSM program should establish the requirements for auditing Process Knowledge. Audits should seek to determine whether a documented program, such as described in Sections 5.3.1 and 5.4, exists and whether it is being administered as described. The records described in Section 5.3.2 should be audited, not for technical adequacy but, rather, to determine if they are being compiled and managed as required by the program. Such audits may reveal potential improvements to be made in the program itself.

Any deficiencies identified during the audit must be documented and promptly resolved. See Chapter 14 and the CCPS book *Guidelines for Auditing Process Safety Management Systems* for more guidance on auditing.

5.6. Examples

5.6.1. Inadequate Investigation and Documentation of Chemicals Prior to Process Application

In a process involving the mechanical conveying of a unstable organic salt, friction between the barrel and flights of a screw conveyor initiated decomposition of the salt and resulted in a deflagration which caused significant damage to the building and process equipment. Fortunately, no one was injured.

Investigation determined that considerable information on the unstable salt (including ignition temperature, rate of reaction and minimum safe vent area for the deflagration reaction) existed in the technical literature. However, the organization which designed and operated the process was unaware of this information and, thus, could not fully appreciate the hazards posed by the chemical properties of the material. As a consequence, neither the process equipment design nor the maintenance and operating procedures were adequate to prevent the incident.

Had the chemical and physical properties of the feed material been fully investigated and properly documented during the development of the process, a knowledge base would have existed to permit the hazards of the unstable salt to be properly addressed. PHAs could have identified appropriate design features, as well as operating and maintenance procedures required to prevent the incident.

5.6.2. Lack of Documentation of Process Information for Operations Personnel

An explosion and fire occurred in an asphalt tank, ripping open the roof-to-shell seam and sending flames 300 feet in the air. Subsequent investigation determined that asphalt was being pumped into the tank at 500°F when the incident took place. The investigation also revealed that published technical material in the organization's files showed that ignitions and explosions in asphalt tanks, under conditions present in this case, were a matter of record. Such information had not previously been made available to operating personnel of the asphalt facility.

Proper documentation of these known process dangers and dissemination of the information to appropriate operating personnel could have prevented this incident.

5.7. References

AIChE-DIPPR, *Physical Property Data,* various publications available from Design Institute for Physical Properties, 345 E. 47th Street, N.Y., N.Y. 10017

AIChE-CCPS, *Guidelines for Auditing Process Safety Management Systems,* 1993

AIChE-CCPS, *Guidelines for Hazard Evaluation Procedures, Second Edition with Worked Examples,* 1992

AIChE-DIERS, *Emergency Relief System Design using DIERS Technology*, 1992

The American Chemical Society, *The Advances in Chemistry Series,* 1959

American National Standards Institute, *Standard for the Preparation of Material Safety Data Sheets* ANSI Z400.1, 1993

American National Standards Institute/Instrument Society of America, *Area Classification in Hazardous (Classified) Dust Locations,* ANSI/ISA S12.10, New York, NY

American Petroleum Institute, *Classification of Locations for Electrical Installations in Petroleum Refineries*, RP 500A, Washington, D.C.

Bichowsky, F.R. and Rossini, F.D., *The Thermochemistry of Chemical Substances,* Reinhold, 1951

Bretherick, L., *Handbook of Reactive Chemical Hazards,* Butterworths, London (UK)

Carson, P.A., and Mumford, C.J., *The Safe Handling of Chemicals in Industry,* 2 Volumes, Longmans/Wiley, 1988

Gallant, R.W., *Physical Properties of Hydrocarbons Volumes 1 & 2,* Gulf Publishing Company, Houston, 1968

Gmehling, J., Onken, U. and Arlt, W. *Vapor-Liquid Equilibrium, Data Collection, Volumes I through V*

Lewis, R.J., *Hazardous Chemicals—Desk Reference,* Van Nostrand Reinhold, 1993

National Academy of Sciences, *International Critical Tables of Numerical Data, Physics, Chemistry and Technology, Volumes I through VIII,* 1929

National Fire Protection Association, *Classification of Class I (Classified) Hazardous Locations for Electrical Installations in Chemical Process Areas,* NFPA 497A, Quincy, MA

National Fire Protection Association, *Classification of Class II Hazardous (Classified) Locations in Electrical Chemical Process Areas,* NFPA 497B, Quincy, MA

National Institute of Occupational Safety and Health, *Guide to Chemical Hazards*, 1990

National Institute of Occupational Safety and Health, *Register of Toxic Effects of Chemical Substances*

Perry's Chemical Engineers Handbook, McGraw Hill, New York, NY

Occupational Safety and Health Administration, *Hazard Communication,* 29 CFR 1910. 1200(g), Washington, D.C.

Reid, R.C., Prausnitz, J.M. and Poling, B.E., *The Properties of Gases and Liquids,* McGraw Hill, New York, NY, 1987

Sax, N. Irving, *Dangerous Properties of Industrial Materials,* Van Nostrand Reinhold

Simulation Sciences Library, *SimSci Data Bank*

Appendix 5A. Example of Process Knowledge File Index

This is an example only, not intended to be comprehensive, and utilizes a fictional company and process, for purposes of illustration of typical information content.

PLANT: ABC — Chemical Company
PLANT AREA: — Solvent Storage
PLANT AREA LOCATION: — Northeast corner pf plant property, adjacent to railroad tracks

TOPIC	REV. NO.	REV. DATE	RESPONSIBILITY
1. Chemical and occupational health hazard			Industrial Hygiene Coordinator
Material Safety Data Sheets			
Acetone	E	Nov. 1988	
Ethyl Acetate	A	April 1986	
Methyl Alcohol	D	Nov. 1991	
2. Process definition and design criteria			Process Technical Engineer
Acetone storage	1	Oct. 1991	
	2	Dec. 1991	
	3	April 1992	
Ethyl Acetate storage	1	Oct. 1991	
	2	Dec. 1991	
Methyl Alcohol storage	1	Oct. 1991	
	2	Dec. 1991	
3. Process description			Manufacturing Manager
Acetone storage	1	Nov. 1991	
Ethyl Acetate storage	1	Nov. 1991	
Methyl Alcohol storage	1	Nov. 1991	
4. Material and energy balance	2	Dec. 1991	Process Technical Engineer
5. Equipment design			Maintenance Engineer
Storage tanks and their relief devices			
Unloading pumps,			
Feed pumps,			
Filters for all solvents:			
Acetone storage	1	Nov. 1991	
	2	Dec. 1991	
Ethyl Acetate storage	3	April 1992	
	1	Nov. 1991	
	2	Dec. 1991	
Methyl Alcohol storage	1	Nov. 1991	
	2	Dec. 1991	
6. Description of operating procedures and guidelines.	1	April 1992	Manufacturing Manager
	2	June 1992	

7. Historical perspective of process design	Not Applicable		
8. Process hazard identification and process risk management decisions.	1	Nov. 1991	Process Technical Engineer
9. Abnormal operating conditions and responses	1 2	Dec. 1991 June 1992	Manufacturing Manager
10. Safety and protective systems Acetone storage Ethyl Acetate storage Methyl Alcohol storage	 1 2 1 2 1 2	 Dec. 1991 April 1992 Dec. 1991 April 1992 Dec. 1991 April 1992	Process Technical Engineer
11. Predictive and/or preventive maintenance program	1	May 1992	Maintenance Engineer
12. Industrial Hygiene	1	May 1992	Industrial Hygiene Coordinator
13. Training Program	1	April 1992	Manufacturing Manager
14. Emergency Procedure	2	May 1993	Safety Engineer

6

Process Hazard Analysis

6.1. Overview

A Process Hazard Analysis (PHA) is an organized effort to identify and analyze the significance of hazardous scenarios associated with a process or activity. PHAs are used to pinpoint weaknesses in the design and operation of facilities that could lead to accidental chemical releases, fires, or explosions and to provide organizations with information to aid in making decisions for improving safety and managing the risk of operations.

Over the last several decades, the chemical process industry has developed a repertoire of techniques for performing PHA; other techniques have been adapted from the nuclear, aerospace, and electronics industries. This chapter briefly describes the more commonly used PHA techniques, many of which are described in greater detail in the CCPS book, *Guidelines for Hazard Evaluation Procedures, Second Edition with Worked Examples*.

6.1.1. Introduction

The PHA is the cornerstone of an organization's overall process safety management (PSM) program. By identifying the key design features, process conditions, and operating practices required for the safe operation of the facility, PHA serves as the basis for other PSM elements. Therefore, results of the PHA should be concisely and accurately documented in an organized manner that makes this information readily available to those responsible for implementing the other PSM elements.

This chapter briefly describes the results obtained from each technique and provides documentation guidance. Since one of the purposes of PHA is to identify needed changes, there is typically a considerable follow-up effort required to ensure the proper resolution of all proposed changes and recommendations. Follow-up systems intended to track and ensure the resolution of such recommendations are also described.

As with the other PSM elements, there is also a second documentation task required to ensure the consistent, effective implementation of the PHA element, that is, the description of the program defining just how the PHA element will be performed (e.g., who is responsible, how often it will be done, and in what manner). Suggested aspects of the program description that should be considered for documentation are also addressed.

6.1.2. Goals and Benefits

The major goals of performing PHA are to allow implementation and application of potentially hazardous chemical processes and facilities for manufacturing chemical products in a safe and cost effective manner. This is achieved by developing detailed knowledge of the actual hazards, and deriving design, construction, operation, and control approaches to minimize anticipated dangers.

Benefits to be gained include reducing all types of risks: human, property, financial, and legal (e.g., regulatory, litigation stemming from incidents) and providing management with a solid technical basis upon which to plan for long term, profitable business.

6.2. Description of Process Hazard Analysis

6.2.1. Objectives

Five intended objectives in conducting a PHA are to:

- identify the hazards inherent in the process or activity;
- identify the credible human and/or equipment failures likely to lead to accident scenarios;
- evaluate the consequences and likelihood of the various accident/incident scenarios, that is, assess the *risk*;
- propose changes where warranted, to equipment design, process conditions, and operating procedures, to mitigate the risk to a tolerable level; and
- document the PHA study in a sufficiently complete, accurate, and concise manner in order to:
  —provide a historical record of the results of the reviews;
  —establish adequacy and applicability of the PHA;
  —allow the use of the information learned from the review to further enhance the safety of the facility operation; and
  —provide a basis for future safety studies for that facility.

Each PHA technique has unique strengths and weaknesses. The logic underlying the decision of which PHA technique to select is complex and involves many factors, including those listed in Table 6-1. Detailed guidance on technique

TABLE 6-1.
Factors That Could Influence the Selection of Hazard Evaluation Techniques

- motivation for the study;
- type of results needed;
- type of information available to perform the study;
- characteristics of the analysis problem;
- perceived risk associated with the subject process or activity;
- resource availability and analyst/management preference;
- amount of experience available with the technology involved; and
- organization and/or industry accident/incident experience.

selection is provided in the CCPS book, *Guidelines for Hazard Evaluation Procedures, Second Edition with Worked Examples*, previously cited. The results derived from the application of the various techniques will vary somewhat in format and degree of detail, thus affecting documentation methods.

6.2.2. Method 1: Safety Reviews

Description—Safety Reviews appear under a variety of names, such as process safety review, design review, or loss prevention review. The safety review can be applied at any stage of a project or plant life cycle, before or after construction. When performed on existing facilities, the safety review typically involves a walk-through inspection that can vary from an informal routine visual examination to a formal examination, performed by a team, that takes several weeks.

Safety reviews are intended to identify plant conditions or operating procedures that could lead to an accident or incident and result in injuries, significant property damage, or environmental impacts. The review could include the following activities:

- a walk through the plant (or the *design* if the plant is not yet built);
- interviews with key people such as designers, operators, maintenance staff, and management;
- reviews of operating, maintenance, and emergency procedures;
- reviews of equipment test records; and
- reviews of plant design documents (e.g., drawings, and specifications).

Safety review inspections and interviews concentrate on significant process safety issues rather than just facility housekeeping. Safety reviews can be used to identify areas that require more detailed study, areas where some significant change has made study necessary, or areas where some gradual progressive change in operating procedure or philosophy has occurred.

The objectives of the safety review are to:

- note and list safety issues requiring further examination or more detailed review;
- identify apparent departures from the original or current design or operating procedures;
- identify new issues;
- propose a course of action or a recommendation for follow-up; and
- document the results in a comprehensive report. There is no specific style required for this report.

Reference

AIChE–CCPS, *Guidelines for Hazard Evaluation Procedures, Second Edition with Worked Examples,* 1992

6.2.3. Method 2: Checklist Analysis (CL)

Description—In a checklist analysis (CL), the hazard/risk analyst, or team, uses a list of specific items to identify known types of hazards, design deficiencies and potential accident scenarios associated with process equipment and operations. In effect, the traditional checklist represents a set of standardized design or operating practices against which the process or facility can be assessed. Other checklists are more general in scope and prompt the formulation of more original what-if type questions within the review, see Method 5: *What-If Analysis* (WI) and Method 6: *What-If/Checklist Analysis* (WICL). Many checklists of each type exist and have been published in the literature.

Example—Table 6-2 shows the use of a checklist and one method for recording answers; the first column repeats all the questions or prompts in the checklist, and the third column records the basic answer to the question as prompted. If necessary, additional columns can be used to record recommendations, actions completed, dates required for recommendation resolution, and/or responsibilities assigned. Some checklists, as published, do not have adequate space for questions/answers to appear alongside each other. Such checklists may need to be retyped, with the responses shown below or alongside the questions. If such lists are used regularly, they can be converted to word processor or computer files.

A traditional checklist generates a list of responses to the standard checklist questions, usually indicating compliance or noncompliance. The traditional checklist can also be used to develop a list of identified hazards and suggested corrective actions. More general checklists may produce lists of questions that require further evaluation before hazards can be assessed or recommendations proposed. A spreadsheet or a word processing program that permits columnar editing can be used for the record.

TABLE 6-2
Checklist for Hazardous Conditions—Release of Materials[a]

	Equipment	*Protection*	*Action*
F1. Identify all probable sources of spillage, which can be caused by:			
a. Overflow, backing-up, blowback, air lock, vapor lock.	TK-101	LAH/LAHH	OK
	TK-201	Overflow to dike.	OK
	TK-202	Overflow to grade.	Dike
b. Excess pressure, wrong routing, loss of vacuum.	R-101	PAH, PRV	OK
c. Vessel being damaged, tilted, collapsed, vibrated, overstirred.			
d. Overloading of an open flow/conveyor.			
e. Poor isolation, drains or doors open, flanges uncovered.			
f. Failure of control of major services.			
g. Surging, priming, foaming, puking, spraying-over of vessels.			
h. Condensation products present in vapor, spitting.			
i. Change in normal discharge.			

[a]Based on part of table in Wells; *Safety in Process Plant Design*; p. 248)

Specialist commercial software useful for this method includes:

- HAZSEC Plus (DNV-Technica, Temecula, CA);
- PHARA (Midwest Technologies, Inc., Kingsport, TN); and
- SAFEPLAN (DuPont, Westlake Village, CA).

References

AIChE–CCPS, *Guidelines for Hazard Evaluation Procedures, Second Edition with Worked Examples*, 1992.

Wells, G.L., *Safety in Process Plant Design*, Halsted/Wiley, 1980.

6.2.4. *Method 3: Relative Ranking Analysis*

Description—One of the challenges of a PHA program is that of establishing priorities to determine which parts of a large facility most urgently require attention. To address this problem, some chemical companies, such as Dow and ICI, have established review programs to determine the relative ranking of hazards. Other organizations have developed their own internal ranking systems or adapted existing systems to provide more options or to consider different aspects.

The Dow Fire and Explosion Index, the ICI Mond Index, and others are based on assigning quantitative values to various aspects of a plant's design features, operating conditions, and inventory. Using any of these index methods, it is possible to compare one plant with another, or one area of a plant with another. Also, since some general hazard scales are offered as guidance, individual facilities can be ranked in qualitative terms, for example, low, moderate, or severe risk.

Other indexes have been developed more recently, including:

- Dow, Chemical Exposure Index, (CEI);
- EPA, Threshold Planning Quantity, (TPQ); and
- OSHA, Substance Hazard Index, (SHI).

The last two indexes are less comprehensive in that they relate largely to the hazardous properties of individual substances. However, they are useful for general guidance and for regulatory compliance.

All relative ranking methods should result in an ordered list of processes, equipment, operations, or activities. Both the Dow and Mond Indexes incorporate calculation sheets on which ratings are applied to a wide variety of process and operating conditions. Other results, such as index scores, factor scores, graphs, etc., depend upon the particular technique used. Also, lists of recommendations intended to lower individual risk rankings may be produced, along with responsibility assignments and deadlines for recommendation implementation.

It is possible with both the Dow and Mond indexes to use computerized methods—either standard spreadsheets or special software. Available commercial software includes:

- MOND INDEX (ICI Plc, Northwich, UK)
- RISK*RANK (Battelle Columbus Division, Columbus, OH); and
- SPECTRUM/PC–F&E (Battelle Columbus Division, Columbus, OH).

References

AIChE–CCPS, *Guidelines for Hazard Evaluation Procedures, Second Edition with Worked Examples*, 1992.

Dow Chemical Co., *Fire and Explosion Index Hazard Classification Guide*, 7th Edition, AIChE, 1989.

Lewis, D.J., *Mond Fire, Explosion and Toxicity Index*, 13th Annual Loss Prevention Symposium, AIChE, Houston, 1979.

The Mond Index 2nd Edition, ICI Plc, Northwich, UK, 1985

Dow Chemical Company, *Chemical Exposure Index* (CEI), AIChE, 1994

6.2.5. Method 4: Preliminary Hazard Analysis

Description—The preliminary hazard analysis is a technique that was derived from the *Military Standard System Safety Program Requirements.* It can be used early in the design process and is generally followed by more in-depth studies. Facility characteristics considered in the analysis include:

- materials being handled and their properties;
- plant equipment;
- facility layout;
- operating environment;
- operational activities (e.g., testing, maintenance); and
- interfaces among system components.

The study is normally qualitative but can provide information to assist with prioritization of scenarios for further studies, either qualitative or quantitative.

The team assesses the importance of each plant characteristics listed above and identifies hazardous scenarios for each. These scenarios can then be assigned a criticality rating that can be used to address the scenarios in order of importance in later design work or safety studies. The results are usually in text format, but may include some tabular material. Available commercial software useful for this method includes HAZMAN (PLG, Inc., Newport Beach, CA).

References

AIChE–CCPS, *Guidelines for Hazard Evaluation Procedures, Second Edition with Worked Examples,* 1992.

Dept. of Defense, *Military Standard System Safety Program Requirements*, MIL-STD-882, Dept. of Defense, Washington, D.C., 1993.

6.2.6. Method 5: What-If Analysis (WI)

Description—The What-If (WI) analysis technique is a creative, brainstorming approach in which a team of experienced people familiar with the subject process or facility ask questions, voice concerns and develop scenarios for possible undesired events. The purpose is to identify hazardous situations, or specific accident scenarios that could produce an undesired hazardous event.

Results of a WI Analysis are most commonly recorded using a simple tabular form. Information entered on the form can include:

- headings detailing the section of the facility or subsystem being studied, relevant drawings, names of the individuals conducting the analysis, and the date;
- columns setting out the following:
  —the proposed what-if question or concern;
  —the identified scenario;
  —consequence/hazard for each scenario;
  —existing safeguards to prevent or mitigate the consequences;
  —recommendation, required action, including investigation outside the meeting;
  —names of individuals who are to execute action or follow up the question (optional, may be contained in subsequent documents); and
  —required completion dates for action plans (optional, may be contained in subsequent documents).

Such records can either be manually recorded, or recorded during the analysis on a personal computer (PC) using a simple form as described above. Commercially produced software is also available specifically for this type of analysis, although simple alternatives that include the above information are not difficult to create with spreadsheet or word processing software. An example of a WI worksheet is given in Table 6-3.

Some of the commercial software programs useable for WI are:

- HAZSEC Plus (DNV Technica Inc., Temecula, CA);
- HAZPRO (DiGraphics, Inc., Houston, TX);
- PHARA (Midwest Technologies, Inc., Kingsport, TN);
- SAFEPLAN (DuPont, Westlake Village, CA); and
- WHAT IF-PC (PrimaTech, Inc., Columbus, OH).

References

AIChE–CCPS, *Guidelines for Hazard Evaluation Procedures, Second Edition with Worked Examples*, 1992.

Burk, A.F., *What If/Checklist—A Powerful Process Hazards Review Technique,* AIChE National Meeting, Pittsburgh, PA, August, 1991.

Greenberg, H.R., Cramer, J.J., *Risk Assessment and Risk Management for the Chemical Process Industry,* Van Nostrand Reinhold, 1991.

6.2.7. Method 6: What-If/Checklist Analysis (WICL)

Description—The What-If/Checklist (WICL) method is a hybrid of *What-If analysis (WI)* and *Checklist analysis (CL)*. It is used to capitalize on the best attributes of the two methods. The WI analysis is creative, and the CL analysis is systematic and experience based. In combination, the advantages of each method are synergized while the shortcomings are minimized.

The WICL analysis is normally approached in one of two ways. It may begin with a WI team review that allows a creative approach to the hazard analysis, followed by a CL review to ensure completeness. Alternatively, the CL and WI reviews may be reversed, where the CL is reviewed and used as a springboard for WI scenario creation by team members. In either approach, all the advantages of both the WI and CL methods can be realized.

The format for documentation can be in either text or tabular form as described for the individual CL or WI.

Currently available commercial software includes:

- HAZSEC Plus (DNV Technica, Inc., Temecula, CA);
- HAZPRO (DiGraphics, Inc., Houston, TX);
- PHARA (Midwest Technologies, Inc., Kingsport, TN);
- SAFEPLAN (DuPont, Westlake Village CA); and
- WHAT IF-PC (PrimaTech, Inc., Columbus, OH).

TABLE 6-3
What If Worksheet Example (Page 1 of 3)

What-If Analysis Worksheet

Date: 8/15/85
Page: 1

Process and Location: *VCM Plant Concept* **Topic Investigated:** *Safety Hazards*
Team: Mr. Safety (Leader—Anywhere Plant Safety Coordinator), Mr. Chemist (ABC Research & Development), Ms. Engineer (Anywhere Plant Engineer), Mr. Petros (Consultant)

Equipment/Task Intentions: *Direct Chlorination Reactor Feeds*

What if . . .?	Consequence/Hazard	Recommendation	Responsible Individual	Initial and Date when Resolved
1. Ethylene feed is contaminated	1. Typical contaminant in ethylene is oil. Oil will react energetically with chlorine. However, the amount of oil in ethylene is usually small, and the large quantity of ethylene dichloride (EDC) in the reactor should quench any oil/chlorine reaction. Water is also a trace contaminant	1.a Verify availability of high-purity ethylene and reliability of supply 1.b Determine the reaction kinetics for oil/chlorine reactions. Examine the reaction kinetics for chlorine/water reactions	1.a Ethylene expert 1.b Chemist	

TABLE 6-3

What If Worksheet Example (Page 2 of 3)

What-If Analysis Worksheet

Date: 8/15/85
Page: 2

Process and Location: *VCM Plant Concept* **Topic Investigated:** *Safety Hazards*

Team: Mr. Safety (Leader—Anywhere Plant Safety Coordinator), Mr. Chemist (ABC Research & Development), Ms. Engineer (Anywhere Plant Engineer), Mr. Petros (Consultant)

Equipment/Task Intentions: *Direct Chlorination Reactor Feeds*

What if . . .?	Consequence/Hazard	Recommendation	Responsible Individual	Initial and Date when Resolved
2. Cl_2 feed is contaminated	2. Typical contaminant in chlorine is water. Large quantities of water in chlorine would cause equipment damage in the chlorine plant and cause a shutdown well before it made it to the VCM plant. Small quantities of water should be no problem	2.a Verify water content in chlorine supplies is very low	2.a Chlorine expert	

TABLE 6-3

What If Worksheet Example (Page 3 of 3)

What if . . .?	Consequence/Hazard	Recommendation	Responsible Individual	Initial and Date when Resolved
3. A feed pipeline breaks	3. Chlorine — A large release of liquid chlorine would occur, which would create a large chlorine vapor cloud Ethylene — A large liquid ethylene release would occur, resulting in a large ethylene vapor cloud, which is a potential fire and explosion hazard	3.a Consider supplying chlorine vapor to the VCM plant 3.b Evaluate ABC's abilities to handle highly flammable materials. Consider additional fire safety training and protection equipment 3.c Consider remote-controlled feed	3.a Chemist 3.b Plant fire chief, corporate training officer 3.c Engineer	
4. The feed rates are out-of-balance	4. A runaway reaction may be possible. An acceptable operating range is not yet known	4.a Examine the reaction rates at various ethylene/chlorine feed ratios	4.a Chemist (to communicate with research)	

References

AIChE–CCPS, *Guidelines for Hazard Evaluation Procedures, Second Edition with Worked Examples*, 1992.

Burk, A.F., *What If/Checklist—A Powerful Process Hazards Review Technique,* AIChE National Meeting, Pittsburgh, PA, August, 1991.

6.2.8. *Method 7: Hazard and Operability Study (HAZOP)*

Description—The Hazard and Operability Study (HAZOP) is a rigorous, systematic study into the consequences of process deviations from the design intent that may cause undesirable hazard or operability problems. The HAZOP was originally developed by ICI in the United Kingdom in the early 1970s from the application of method study techniques to design problems.

The basic format of the documentation for a HAZOP study is universal, but many variants exist. Whether the study is recorded manually on blank forms, on chart pads, or on a PC using special software, documentation requirements are similar.

The worksheet consists of two parts:

- headings detailing the study team and the precise design intent of the line, vessel, node, subsystem or subsection being studied; and
- a record, usually tabular, of potential deviations from the declared design intent, and the likely expected outcome.

A third section exists, sometimes separately, to record actions and responses for management control. A typical HAZOP worksheet is shown in Table 6-4 (the analysis was conducted on the process illustrated by the P&ID shown on Figure 5.3). The following features of Table 6-4 should be noted:

- *Part 1—Heading*. These items should be recorded:
  —who was on the study team and when (or use an attendance roster in the report);
  —project number/title;
  —area of facility being studied;
  —precise identification of the line or vessel subject to study in the table;
  —the design intent or objective of the line/vessel (e.g., liquid transfer, storage) flow rate, temperature, pressure, normal composition; and
  —other key parameters.
  This information defines the normal or design intent from which the team will identify how deviations can occur.
- *Part 2—Deviations.* This section should list, in sequential order of questioning:
  —deviation from the *design intent* (parameter and guide word);
  —possible causes of deviation;
  —probable consequences of the deviation;

—safeguards or protection;
—required actions/investigations/questions—Note (a);
—response to actions/investigations/questions—Note (b);
—persons responsible for actions/investigations/questions—Note (c); and
—priority rating and/or hazard potential rating—Note (d).

Note (a) *Required Actions/Investigations/Questions* may also include serial numbers for future management reference.

Note (b) *Responses* may be recorded separately, along with the relevant question and reference to the study table, for subsequent follow-up.

Note (c) *Persons Responsible* should identify the individual who is to be responsible for the reply or action. It may also include a required completion date.

Note (d) *Priority Ratings* may include numbers or symbols for ratings for priority or hazard potential.

Storage and Retrieval—The worksheet record of a HAZOP study can extend from a few sheets for a small/simple facility to possibly 1,000 sheets or more for a large/complex installation. During the recording process, a system that provides a unique record of every sheet should be considered. One method is to number each study node, table or section, from 1 upward. Each of these nodes might contain one or more pages, which could be numbered, *page 1 of 4*, *page 2 of 4, etc.* The convention selected for naming computer files should allow for quick identification and retrieval, for example by incorporating the drawing number.

HAZOP studies can be performed with any of a variety of commercially prepared software for use on personal computers. These include:

- CAHAZOP (NUS Corporation, San Diego, CA);
- HAZOP (Lihou Technical & Software Services, Aston, UK);
- HAZOP-PC (PrimaTech, Inc., Columbus, OH);
- HAZOPtimizer (A.D. Little, Cambridge, MA);
- HAZSEC Plus (DNV Technica, Inc., Temecula, CA);
- HAZPRO (DiGraphics, Inc., Houston, TX);
- HAZTEK (Westinghouse Electric Corp, Pittsburgh, PA);
- LEADER (JBF Associates, Knoxville, TN); and
- SAFEPLAN (DuPont, Westlake Village, CA).

In addition, some organizations have adapted word processing or database software to provide their own customized worksheets for HAZOP recording.

References

AIChE–CCPS, *Guidelines for Hazard Evaluation Procedures, Second Edition with Worked Examples* 1992.

Freeman, R.A., "Documentation of Hazard and Operability Studies." Plant/Operations Progress 10:155–158, July 1991.

Knowlton, R. Ellis, *An Introduction to Hazard and Operability Studies—The Guide Word Approach.* Chemetics International Company, Vancouver B.C., Canada, February 1992.

TABLE 6-4
HAZOP Worksheet Example

P&ID No: E-250
Revision: D
Meeting Date: 9/5/90
Team: Mr. Smart, Mr. Associate, Ms. Piper, Mr. Stedman, Mr. Volt (all from the ABC Anywhere Plant)

Item Number	Deviation	Causes[a]	Consequences	Safeguards	Actions
1.0 LINE — AIR SUPPLY LINE TO INCINERATOR (INTENTION: SUPPLY 15,000 SCFM OF AIR TO INCINERATOR AT AMBIENT TEMPERATURE AND 3 IN. WC)					
1.1	No flow	1 — Air fan #1 fails off 2 — FCV-1 fails closed 3 — FT-1 fails — high signal 4 — FT-2 fails — low signal 5 — Loss of electric power 6 — Plugged air screen	A — Incinerator shuts down. Possible release out the scrubber stack. Potential incinerator explosion if shutdown interlocks fail	1 — Redundant fan on standby with autostart A — Low-low air pressure (PSLL-1) shutdown interlock 1,2,3,4,6 — Multiple incinerator shutdown interlocks (temperature, flame) 1,5 — Automatic shutdown upon loss of electric power 6 — Air screen cleaned weekly 5 — FCV-1 fails open on loss of electric power 2,3,4 — Mechanical stop on FCV-1	1 2

1.2	Low flow	1 — FCV-1 fails — partially closed 2 — FT-1 fails — high signal 3 — FT-2 fails — low signal 4 — Plugged air screen	A — Incinerator shuts down. Possible release out the scrubber stack. Potential incinerator explosion if shutdown interlocks fail	1,2,3 — Mechanical stop on FCV-1 4 — Air screen cleaned weekly A — Low-low air pressure (PSLL-1) shutdown interlock A — Multiple incinerator shutdown interlocks (temperature, flame)	
1.3	High flow	1 — FCV-1 fails open 2 — FT-2 fails — high signal 3 — FT-1 fails — low signal 4 — Operator inadvertently starts fan #2 (fan #1 still running)	A — Poor combustion, with potential release of flammable gas out the scrubber stack. Excessively high air flow may blow flame out causing a shutdown	A — Multiple incinerator shutdown interlocks (temperature, flame) 4 — FT-1 closing to reduce air flow	2 3
1.4	Low temperature		No consequences of interest		
1.5	High temperature		No consequences of interest		

[a]The consequences and safeguard numbers correspond to the numbered causes. The letter "A" indicates that the respective consequences and safeguards apply to all of the listed causes.

6.2.9. Method 8: Failure Modes and Effects Analysis (FMEA)

Description—Failure Modes and Effects Analysis (FMEA), sometimes extended to Failure Modes, Effects and Criticality Analysis (FMECA), is used to examine the way each item of equipment can fail and the effect the failure could have on the plant or system. In addition, a criticality rating for each failure can be developed, if required, leading to prioritization of necessary actions.

An FMEA analysis generates an itemized list of equipment pieces and an associated list of failure modes appropriate to each. For each failure mode for each equipment item, the analyst or team identifies the effects (consequences) of the failure, as well as any existing safeguards. Recommendation for further safeguards are proposed, where warranted. Additionally, in a FMECA, a semiquantitative evaluation is proposed for the frequency of the failure and the significance or severity of its consequence.

Text or a multicolumn analysis form may be used to display the results. The heading of the analysis table should adequately identify the study team and date, and the equipment or system being studied. An example of a FMEA worksheet is given in Table 6-5 (the analysis was conducted on the process illustrated by the P&ID shown on Figure 5.3).

Available commercial software includes:

- CARA (DNV Technica, Inc., Temecula, CA);
- FIABEX (Institution of Quality Assurance, London, UK);
- FMEA-PC (PrimaTech, Columbus, OH);
- HAZOPtimizer (A.D. Little, Cambridge, MA);
- HAZSEC Plus (DNV Technica, Inc., Temecula, CA); and
- SAFEPLAN (DuPont, Westlake Village, CA).

References

AIChE–CCPS, *Guidelines for Hazard Evaluation Procedures, Second Edition with Worked Examples,* 1992

Department of Defense, *Procedures for Performing a Failure Mode, Effects and Criticality Analysis,* MIL-STD-1629A, Dept. of Defense, Washington, D.C., November, 1980.

International Organization for Standardization, *Failure Mode and Effects Analysis,* ISO CEI/812: 1985, Geneva, Switzerland

6.2.10. Method 9: Fault Tree Analysis (FTA)

Description—A fault tree is a graphical model that illustrates combinations of failures that will cause one specific failure scenario of interest, called a top event. FTA is a deductive technique that uses Boolean logic symbols—that is, AND gates and OR gates—to break down the causes of a top event into basic events involving equipment failures and human errors. The analyst begins with an accident or undesirable event that is to be avoided and identifies the immediate causes of that event. Each of the immediate causes, called fault events, is further examined in the same manner until the analyst has identified the basic causes of each fault event

TABLE 6-5
FMEA Workshhet Example

P&ID No: E-250
Revision: D
Meeting Date: 9/5/90
Team: Mr. Smart, Mr. Associate, Ms. Piper, Mr. Stedman, Mr. Volt (Anywhere VCM Plant)

Item Number	Component	Failure Mode	Effects	Safeguards	Actions
1	Flame scanner UVL-1B	No signal change	Loss of capability to initiate an incinerator shutdown upon loss of flame. Potential incinerator fire or explosion if flame extinguished	Redundant UVL Multiple incinerator interlocks (temperature, fuel, and air)	
		False flameout signal	Inadvertent incinerator shutdown. Potential incinerator explosion if incinerator fuel not shut off	Shutdown is alarmed. Operators verify shutdown actions Double block and bleed valves in fuel lines Three-way shutoff valve in vent line	Verify the reliability of the UVLs

or reaches the boundary established for the analysis. The resulting fault tree model displays the logical relationships between basic events and the selected top event. The advantage of the schematic nature of FTA is that it clearly shows what the causal events are, which can lead to developing means to control or eliminate those events.

Top events are specific hazard scenarios that are typically identified through the use of other hazard scenario identification techniques, e.g., WICL, HAZOP, FMEA. A fault tree model can be used to generate a list of the combinations of failure modes that can cause the top event of interest. These failure mode combinations are known as cut sets. A minimal cut set (MCS) is a smallest combination of component failures which, if they all occur or exist simultaneously, will cause the top event to occur. These are the *smallest* combinations in that all of the failures in a MCS must occur if the top event is to occur as a result of that particular MCS. A list of minimal cut sets represents the known ways the undesired event can occur, stated in terms of equipment failures, human errors, and associated circumstances.

Special symbols are used, and International Organization for Standards (ISO) symbology should be used in the preparation of an FTA diagram. An example of one of the various types of FTA diagrams is given in Figure 6-1 (the FTA was conducted on the process illustrated by the P&ID shown on Figure 5.3).

There is usually no single ultimate event of interest in a plant, but rather several, perhaps of different types or locations. Hence, no single fault tree normally represents a plant; instead, many fault trees are usually required. It is important that the selection of the scope of the FTA be fully documented.

The FTA diagram can grow, and often single events will themselves develop into smaller fault trees. The recording system chosen should be able to accommodate this evolutionary progress.

The FTA diagram can be created and documented in several ways:

- manual drawings (in this case, the use of a template for FTA symbols can be helpful; see Berol: RapiDesign R-555 Fault Tree);
- CAD drawings or other computer-based drafting systems;
- word processor or spreadsheet graphics (which may have limitations); and
- specialized software written for FTA.

In addition, a summary report may be prepared with drawings and statements identifying sources of data. The report will also typically contain a ranked list of minimal cut sets for the system under study.

Available commercial software includes:

- AFTQM (University of New Mexico, Las Cruces, NM);
- AFTP3 (S.I.A.I, Paris, France);
- BRAVO (JBF Associates Inc., Knoxville, TN);
- CAFTA+/SAIPLOT/SAICUT (SAIC, Los Altos, CA);
- CARA (DNV Technica, Inc., Temecula, CA);

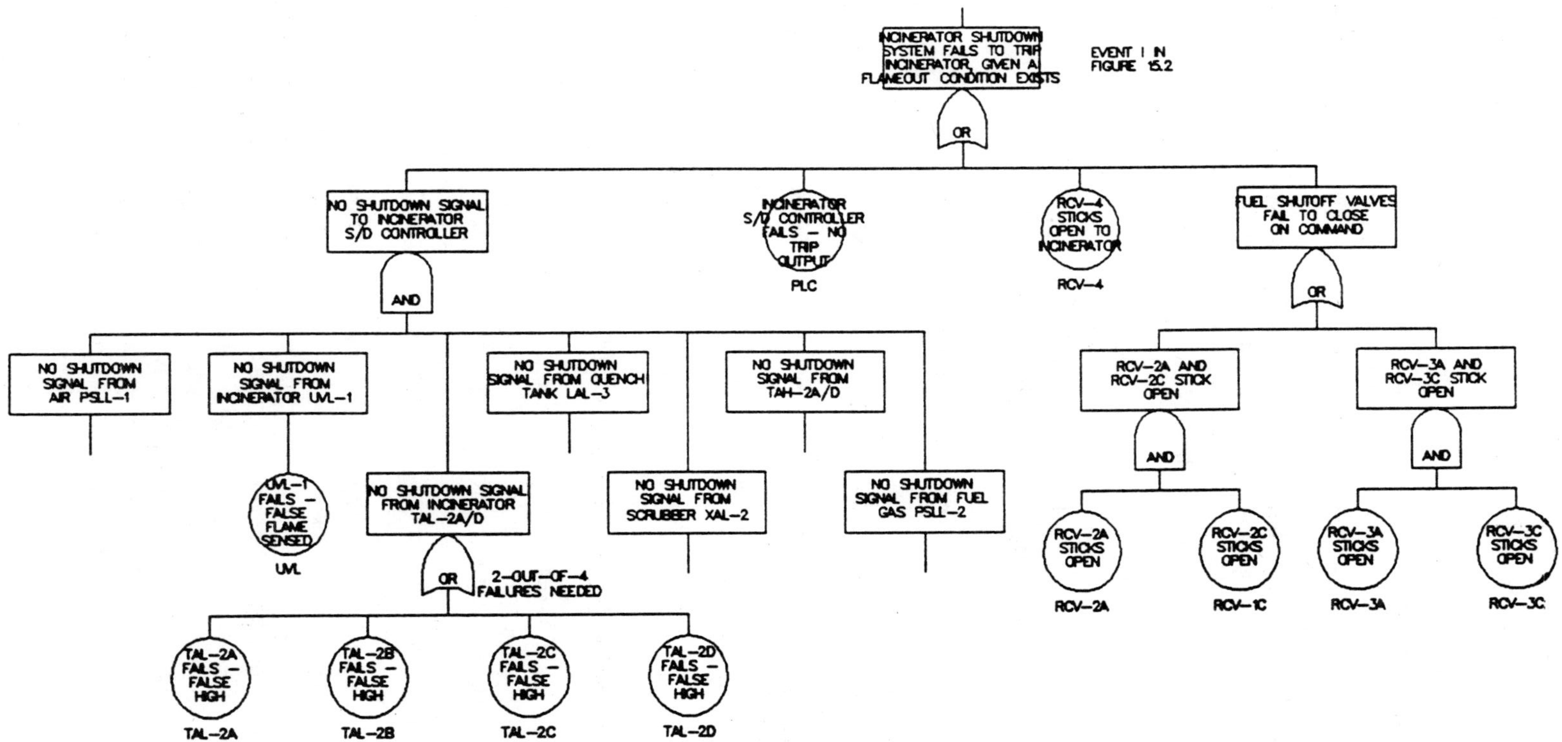

FIGURE 6-1. Example FTA Diagram.

- CAT (EPRI NP-705);
- FaultrEASE (A.D. Little, Cambridge, MA);
- FIABEX (Institution of Quality Assurance, London, UK);
- FTL (a reliability data collection, used in risk and availability assessment), Springer-Verlag, NY;
- IRRAS-PC (Idaho National Engineering Laboratory, Idaho Falls, ID);
- ISOGRAPH (Isograph Ltd, Manchester, UK);
- LOGAN/GRAAFT (RM Consultants, Abingdon, UK);
- MOCUS, ANCR-1156, (Idaho National Engineering Laboratory, Idaho Falls, ID);
- NUSSAR-PC (NUS Corp., San Diego, CA);
- PC-RISA (RISA GmbH, Berlin, Germany);
- PHARA (Midwest Technical, Inc., Kingsport, TN); and
- RISKMAN (PLG Inc., Newport Beach CA).

References

AIChE–CCPS, *Guidelines for Chemical Process Quantitative Risk Analysis,* 1989.

AIChE–CCPS, *Guidelines for Hazard Evaluation Procedures, Second Edition with Worked Examples*, 1992.

US Nuclear Regulatory Commission, *Fault Tree Handbook,* NUREG-0492, Washington, D.C., 1981

International Organization for Standardization, *Symbols,* ISO CEI/IEC 617: Geneva, Switzerland

International Organization for Standardization, *Fault Tree Analysis*, ISO CEI/IEC 1025:1990, Geneva, Switzerland

6.2.11. Method 10: Event Tree Analysis (ETA)

Description—Fault Tree Analysis (see Section 6.2.10) is a deductive reasoning method; after identifying the ultimate or top event, the analyst deduces the sequence of events that can lead to the top event. In contrast, Event Tree Analysis uses inductive reasoning. After identifying an event, the possible outcomes of the event are developed. Sometimes this method is referred to as the *bottom-up* approach, in contrast to FTA's *top-down* approach.

Event trees are important because they can help in developing the sequence of events that can occur following a primary accident or event. They can help in the writing of emergency preparedness plans, since many of the possible secondary outcomes may not be within the experience of facility management, and these may not be anticipated during an actual accident situation.

ETA can be used when examining in detail the performance of plant protective systems. Using the structure of ETA, it is easy to evaluate and, if necessary, quantify the potential failures of equipment and their likely consequences. With the results, it is then possible to evaluate potential improvements to such systems. Basic steps in conducting an ETA are as follows:

- identify the need for ETA, define its scope, assemble the available information, and determine what additional information is required;
- identify the primary or initiating events;
- construct the event tree (diagram);
- describe the resulting accident sequence (text);
- determine the accident minimal cuts sets (MCS); and
- document the results, to include:
  —diagrams of ETAs performed;
  —lists of accident sequence MCSs;
  —a discussion of consequences of various accident sequences; and
  —recommendations for action;

ETA diagrams are normally represented either sequentially from left to right, or sometimes vertically, with the initial event below the sequential event. The diagrams can be drawn manually, or constructed with simple word processing or graphics programs on PCs or with specialty software. An example of an ETA diagram is given in Figure 6-2 (the ETA was conducted on the process illustrated by the P&ID shown on Figure 5.3).

Available commercial software includes:

- CARA (DNV Technica, Inc., Temecula, CA);
- CHEM-RISK (NUS Corp. San Diego, CA);
- ETA-11 (SAIC, Los Altos, CA);
- IRRAS-PC (Idaho National Engineering Laboratory, Idaho Falls, ID);
- METEOR (RM Consultants, Abingdon, UK); and
- RISKMAN (PLG Inc., Newport Beach, CA).

References

AIChE–CCPS, *Guidelines for Chemical Process Quantitative Risk Analysis,* 1989

AIChE–CCPS, *Guidelines for Hazard Evaluation Procedures, Second Edition with Worked Examples*, 1992.

Greenberg, H.R., Cramer, J.J., *Risk Assessment and Risk Management for the Chemical Process Industry,* Van Nostrand Reinhold, 1991.

US Nuclear Regulatory Commission, *Fault Tree Handbook,* NUREG-0492, Washington, D.C., 1981

6.2.12. Method 11: Cause–Consequence Analysis (CCA)

Description—Cause–consequence analysis (CCA) is basically a combination of FTA and ETA. The CCA can benefit from the deductive and inductive reasoning used in the separate application of these methods. Unfortunately, because the diagrams produced can be extensive, there is a limit to their practical application.

As with FTA and ETA, the outlines of the CCA can be sketched on a large sheet of paper, and the completed diagram can be transferred to a manually prepared drawing or to a computer recording system. As with FTA and ETA, the recording system must accommodate the evolutionary progress of the study.

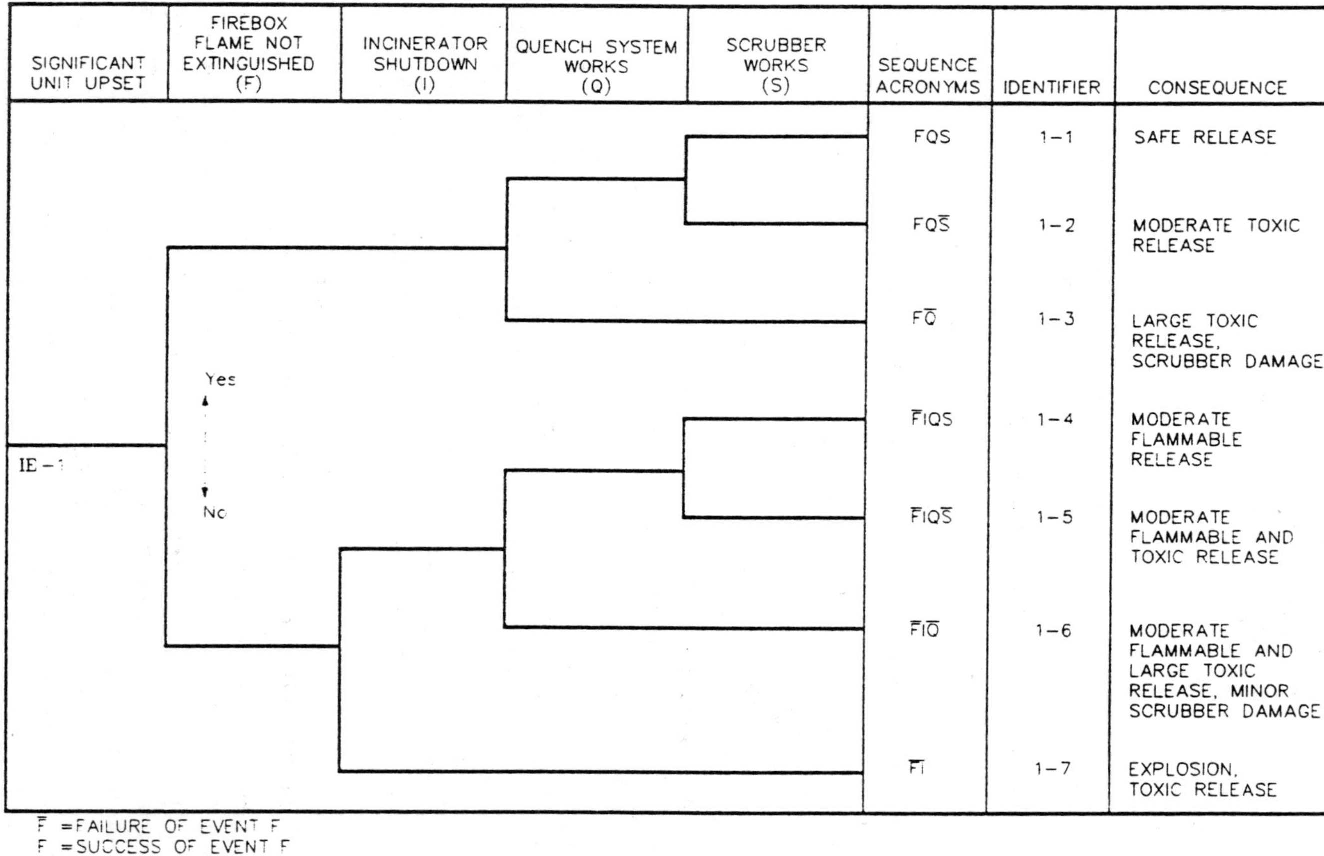

FIGURE 6-2. Example ETA Diagram.

Documentation requirements parallel these for FTA and ETA and use can be made of software that handles and quantifies FTA and ETA (see earlier sections). CCA-specific software includes CARA (DNV Technica, Inc., Temecula, CA).

As with FTA and ETA, the completed graphic documentation is typically accompanied by a text summary report.

References

AIChE–CCPS, *Guidelines for Hazard Evaluation Procedures Second Edition with Worked Examples,* 1992.

Nielsen, D.S., *The Cause/Consequence Diagram Method as a basis for Quantitative Accident Analysis,* RISØ, Atomic Energy Commission Research Establishment, RISØ, Denmark, 1971.

Nielsen, D.S., *Use of Cause–Consequence Charts in Practical Systems Analysis,* in Reliability and Fault Tree Analysis, SIAM, Philadelphia, PA, 1975.

Taylor, J.R., *Cause–Consequence Diagrams,* NATO Advanced Study Institute, 1978.

6.2.13. Method 12: Human Factors Analysis (HFA)

Description—Virtually every aspect of process engineering, plant operations and maintenance is directly associated with the performance of human beings. People are known to be fallible and also to be influenced in their performance by a wide variety of performance influencing factors (PIFs), also known as performance shaping factors (PSFs). Some of these ergonomic factors such as lighting, heating, noise, etc., can be optimized to minimize errors, but others are very personal and include such aspects as emotional state, stress, training level, and experience.

Considerable research has gone into improving human reliability and to the removal of adverse PIFs, but the possibility of error remains. Techniques for Human Factors Analysis (HFA), also known as human reliability analysis (HRA), are applied to the identification of PIFs and ways in which they can be improved.

An HFA should be considered whenever human error is known or expected to be a potentially significant contributor to process risk. HFA may be performed separately after other hazard evaluation procedures have been conducted, or at the same time as any other PHA study method discussed earlier.

The HFA typically follows these steps:

- describe the personnel, their tasks and their work environment;
- evaluate the man/machine interface using HFA;
- perform a task analysis of operator functions;
- perform a human error analysis of operator functions; and
- develop recommendations to reduce human error;
- document the results.

Different levels of human reliability analysis and use of different techniques result in various work products. The HFA typically should produce a list of corrective actions that will reduce the likelihood of specific human errors. The HFA should also produce a detailed description of operator tasks, in list or graph form, which can be used to establish or improve procedures, training or policies. In addition, the analysis may generate a set of combinations of human errors that may be ranked using human error data. HFA event trees, such as illustrated in Figure 6-3, may also be included. Thus, the documentation may include text, tables, charts, and graphic representations of the analysis results.

For a more detailed description of the documentation of the application of Human Factors in PSM, see Chapter 9 of this book. The CCPS book *Guidelines for Preventing Human Error in Chemical Process Safety* provides details of a large number of different methodologies for Human Factors Analysis (HFA), with examples of their application.

Available commercial software includes SHERI (Battelle Columbus Laboratory, Columbus, OH) which can handle the construction of diagrams and their analysis. Help can also be obtained by using FTA and ETA drawing packages and software.

References

AIChE–CCPS, *Guidelines for Chemical Process Quantitative Risk Analysis,* 1989, pp. 237–247

AIChE–CCPS, *Guidelines for Hazard Evaluation Procedures, Second Edition, with Worked Examples,* 1992.

AIChE–CCPS, *Guidelines for Preventing Human Error in Chemical Process Safety*, 1994.

Greenberg, H.R., Cramer, J.J., *Risk Assessment and Risk Management for the Chemical Process Industry,* Van Nostrand Reinhold, 1991.

Lorenzo, D.K., *A Manager's Guide to Reducing Human Errors—Improving Human Performance*, Chemical Manufacturers Association, Washington, D.C., July, 1990.

6.3. Process Hazard Analysis Documentation

In the preceding sections, the more commonly used PHA techniques and the types of results that come from their application were briefly described. Where PHA is appropriate, guidelines should be followed to ensure that PHA is performed and documented in a consistent and effective manner. This Section includes issues relative to documentation of the PHA element itself, and to documentation of the program describing how the organization will implement PHA.

6.3.1. PHA Program Documentation

Organizations implementing PHA should should include in their PSM program documentation a discussion of the goals of PHA as well as any general procedures that outline how PHA will be implemented. The PHA program documentation should set the standard of performance and demonstrate the support of upper management for what can be a resource-consuming endeavor. Without the visible support of upper management, there is a risk of perfunctory efforts from those responsible for the actual implementation of PHA. Some organizations may issue

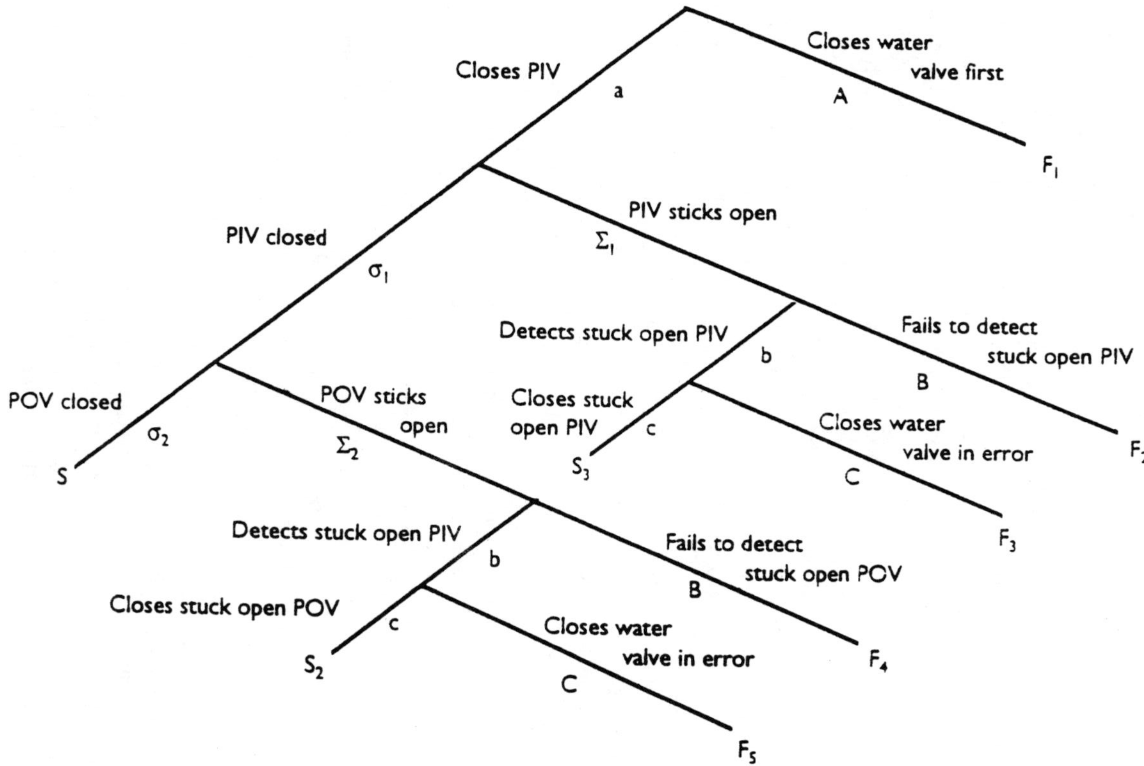

FIGURE 6-3. Example HFA Event Tree.

broader, more general policy statements at the corporate level, with the development of more specific interpretations of these goals at the operating unit (e.g., plant or site level). In either case, management support should be evident.

There are a wide variety of factors that should be considered when establishing detailed procedures intended to achieve the established goals of PHA. The more significant factors will be discussed in the following paragraphs. Some of these are prescriptively addressed in current and proposed government regulations.

When to Perform PHA—PHAs should be conducted on a proportional-to-need basis. In other words, some portions of a facility may require more frequent and rigorous review based upon either their demonstrated or perceived hazard potential. Similarly, some operations may be appreciably more subject to change in operating conditions or equipment design (see Chapter 10, *Management Of Change*). Consequently, a PHA schedule should be established for each facility to ensure that reviews are conducted at appropriate frequencies. The facility, and possibly individual processes, may be subdivided with each subdivision given a specific review frequency based upon its particular needs. For example, a powerhouse may require less frequent review than a polymerization reactor.

Another advantage to establishing such a schedule is that it permits planning and scheduling reviews in a manner that balances the PHA effort over an entire review cycle. An organization initially implementing a PHA program should establish a prioritized schedule of which operations to review in what order, along with a schedule for updating or revalidating previous reviews. Such schedule guidelines must be capable of addressing the changing needs of a process (e.g., in response to changes in the nature of the hazards present or marked changes in safety experience for that operation or similar operations). A statement on the derivation of the schedule may be included in the documentation record.

Technique Selection—As discussed briefly above, and covered in more detail in the CCPS book, *Guidelines for Hazard Evaluation Procedures, Second Edition with Worked Examples,* the various PHA techniques differ in approach and work product. Some are more rigorous and detailed, and some approach the PHA task from very different perspectives (e.g., contrast the difference in detail between WI and FTA techniques, or the inductive versus deductive approaches used by ETA and FTA analyses). Consequently, some organizations may want to establish guidelines for technique selection to ensure that a specific high-hazard operation receives a sufficiently rigorous review or that a too-rigorous technique is not wastefully applied to a low hazard process. Such guidelines should be just that—guidance rather than mandatory edicts. Often, the best judge of the most appropriate technique is the trained, experienced analyst who is responsible for the review. Such guidelines would form part of the documentation record.

Training and Qualifications—Some techniques like FTA and HRA require more detailed, specific training and practice in their application than is the case with a technique such as the CL review. An organization may wish to establish minimum levels of demonstrated competency for those responsible for the application of such techniques. An organization may consider documenting training requirements for the various personnel who participate in the reviews. Obviously, a team leader will need more training than someone who is merely participating in the review.

Team Make-up—For team-based techniques, it is important to get a proper mix of training, background, and expertise in the team make-up. Similarly, a proper mix of supervisory and nonsupervisory personnel helps ensure a balance of backgrounds and perspectives and contributes to more meaningful results. Organizations may wish to prepare guidelines for team composition to ensure that a proper mix of functional responsibilities (e.g., manufacturing, maintenance, technical, engineering, and R&D), background and training are provided on PHA teams.

PHA Results Documentation—Documentation of the results of the PHA will be discussed in Section 6.3.2. In the documentation of the PHA program, minimum standards for the form and content of PHA results documentation should be established. For example, what goes into the final, formal report? How is supporting information documented? Who is responsible for signing the final report? What degree of review is appropriate for the report before it is issued?

Recommendations and Their Resolutions—An important function of PHA is either to validate the safety of current equipment design and operating procedures or to propose necessary and sufficient corrective actions to ensure an appropriate level of process safety within the operation. Most PHA efforts result in a number of recommendations that may either define a specific course of action (do this . . .) or define the need for further evaluation (study this . . .). A documented system for tracking the status of these recommendations, and obtaining documentation of the implementation or resolution of each one, will help to ensure that the organization complies with its ethical responsibilities in this matter and achieves the full value of the effort that it has devoted to the PHA program.

Records Management—Retention and purge schedules for PHA reports and supporting documentation should be clearly established, documented, and followed. Legal and regulatory requirements should be considered. See Section 6.4 and Chapter 4, *Records Management*.

Communication of PHA Results—It was mentioned in the introduction to this chapter that PHA serves as the cornerstone to virtually all of the other PSM elements. The necessity of appropriately communicating the results of the PHA to other members of the organization must be carefully considered in establishing distribution lists for reports. Similarly, suitable storage locations for copies of the reports should be identified so that those not receiving a copy, yet having a need-to-know, can have access to this information. Suitable controls should be applied to ensure that these general access copies remain secure, accurate, and up to date.

PHA Program Responsibility—An organization should document the functional responsibility and accountability for implementation of the PHA program. Ultimate responsibility may reside with the site manager, but accountability for the actual implementation can be delegated. It is the responsibility of the program manager to interpret policy, establish frequency and scope of reviews, and examine reports for uniformity and quality of the reviews. In addition, members of the line organization will have their own responsibilities and roles in the implementation of the overall PHA program. These responsibilities, from the top of the organization to its lower levels, should be clearly established, documented, and communicated to all those involved.

6.3.2. Documentation Of PHA Results

A brief description of the type of results and documentation requirements specific to the more commonly used individual techniques was provided in the first part of this chapter. The following summary provides a more general perspective on broader issues related to documentation of the PHA effort, independent of which technique was used for the PHA.

The primary objective of PHA is to address the following basic needs:

- identify the hazards inherent in the process or activity;
- identify the credible human and/or equipment failures likely to lead to accident scenarios;
- evaluate the consequences and likelihood of the various accident/incident scenarios, that is, assess the risk;
- propose changes, where warranted, to equipment design, process conditions, operating procedures, etc. to mitigate risk to a tolerable level; and
- document the PHA study appropriately.

Unless a PHA is properly documented, there can be no assurance that the information gained from the PHA review will actually be applied to enhancing operational safety. Furthermore, future efforts may be less productive in that additional resources may be expended in relearning the same lessons. Finally, the results should be communicated in a way that allows for adequate scrutiny of the technical quality of the review.

The results of a PHA are typically documented and communicated to those affected via a formal summary report. The content of the report will vary somewhat based upon the scope of the review and the technique used, but the basic structure of the report typically can, and should, be standardized for the organization. Table 6-6 shows one suggested outline for such a summary report.

As shown in the table, the report typically contains the following information:

- a clear, concise identification or description of the facility or process under review including a brief description of the process chemistry and the hazardous properties of the materials involved;

TABLE 6-6
Items to Consider Including in PHA Study Reports

Item	*Description*
Study objectives	Primary purpose and ancillary needs
Physical and analytical scope	Specific process areas, systems, and equipment covered. Hazards analyzed along with situations beyond the study's scope
Analysis team composition	Participants in the review (categorized by organization and job/experience)
Meeting dates and duration	When, where, for how long the team met and attendance lists
PHA technique descriptiuon.	Rationale for selection of the PHA techniques. Description of the techniques employed
Process/activity description	Description of the system analyzed. Explanation of the procedure steps used in batch operations. Listing of important process conditions. Descriptions of safety systems. Listing of materials and hazards.
Drawing, specification, and procedures list	Documents (with dates and revision numbers) used as the basis for the analysis
Conclusions	Team consensus as to the safety of the facility or operation
Action items/suggestions	Recommendations made to management for risk reduction with clear delineation of responsibilities and required timing
Detailed documentation of any review meeting	Minutes of meetings, summaries of topics covered
Technique-specific lists, forms, or graphical models	Checklists, lists of What-If questions, worksheets, FMEA or HAZOP tables, fault trees, event trees, cause–consequence diagrams, HFA event treesa

- careful definition of the scope and limitations of the review (e.g., was only off-site risk considered, does the review address environmental damage as well as personnel injuries);
- clear statements defining, and justifying, the need for recommendations, namely, what specifically must be done and what is to be achieved, recognizing that some recommendations may merely define the need for additional study before a course of action can be proposed;
- clear statements defining responsibilities and required timing for resolution of recommendations;
- documentation of actions that were taken immediately to enhance operational safety, along with explanations of why they were necessary;
- identification in the report of those parties that participated in the review and their affiliations, attendance rosters for each meeting, where multiple meetings were required;

- documentation of the technical content, including the identification of the PHA technique used (which may be described in either the PHA report or a PHA methodology/protocol, or other referenced documents) with some explanation as to the rationale for its choice.
- key results and conclusions of the review should be clearly, unambiguously, and concisely summarized and documented and should be reviewed by all participants in the review, as an affirmation of their concurrence with the published conclusions.
- reference documents used during the review should be listed (e.g., drawing and/or procedure numbers along with the then current revision number or date, previous PHA reports or technical studies). Working papers, if retained, should be filed together in a secure location and indexed in the report.

The main body of the report will usually be comprised of output generated during the PHA. Typical outputs of the individual techniques were described earlier in Section 6.2. In general, however, there are two broad categories of information that would typically be documented. The first is primarily text, which may or may not be presented in a tabular (or columnar) format. This sort of output typically results from techniques such as WI, CL, HAZOP, or WICL.

The second common type of information will be diagrams, e.g., process flow diagrams, simplified piping and instrument diagrams, FTA and ETA charts, HFA charts, facility layout diagrams. These diagrams are usually accompanied by some text describing their content.

Some techniques, such as relative ranking, produce a variety of output including charts, graphs, tables, and standardized forms.

Readers wishing a broader background may find further descriptions of many of the concepts discussed in this section by referring to Chapters 2 and 7 of the CCPS book, *Guidelines for Hazard Evaluation Procedures, Second Edition with Worked Examples.*

6.3.3. Resolution of PHA Recommendations

The final decision with regard to adoption of PHA recommendations is typically a management responsibility and the status of PHA recommendations may not be determined before the PHA summary report is issued. Once a recommendation has been made, it must be resolved and documentation of the final resolution of all recommendations is the last task in documenting a PHA. Dates by which the recommendations should be implemented, or otherwise resolved, should be clearly communicated and documented in the PHA summary report. Responsibility for recommendation follow-up should be clearly assigned.

Resolution is not synonymous with adoption; not all recommendations will eventually be implemented as originally proposed. Circumstances change, some recommendations may ultimately be seen to be inappropriate, or a better means of achieving the same results may become known. In summary, the various ways that a recommendation could be resolved include:

- Accepted and implemented;
- Rejected for cause:
  —would have introduced a greater hazard; or
  —change in circumstances, no longer required.
- Accepted in principle:
  —equivalent protection obtained via another route.

In any event, the method of final resolution of all recommendations should be documented, either in the summary report, in an addendum to the report, or in a separate follow-up report. The rationale for not implementing the recommendation as originally proposed, as well as any alternative course of action intended to achieve the objective, should be clearly documented. The date of resolution, as well as the name of the person accountable should also be documented.

6.4. Records Management

6.4.1. Records Management Program

In addition to the description of the PHA program given in Section 6.3.1, the organization's PSM program documentation should define PHA records management practices. This should address documentation gathered, generated, and utilized during reviews as well as the reports, recommendations, and documents relating to follow-up activity stemming from the PHA reviews. The records management practices should provide access by the PHA team members to the information required for their evaluation and reviews, while at the same time ensuring that appropriate control and retention procedures exist to preserve historical or sensitive data, and to gather, store, and disseminate documentation of pertinent interim and final reports.

6.4.2. Media and Methods

Depending on the size of the organization or facility, the nature and scope of the review, and PHA technique chosen, the media used to document meetings and results can range from text information recorded by a meeting scribe on a notepad, to computer-based records. Since many of the techniques are team-based, a media visible to the team as a whole would be an appropriate choice. In its simplest form, this could be large chart pads displayed on meeting room walls, to permit the entire team to refer to information discussed or conclusions previously reached. Similarly, worksheet transparencies shown on overhead projectors, could permit team members to view, revise, and record data during meetings. Alternatively, the scribe might utilize a personal computer, commercial software and a wall screen projection system to show the data, allowing editing or reformatting as the analysis is being performed and results formulated.

Electronic media is attractive for records management since it is versatile in displaying, recording, and rapidly producing hard copies, if needed. It is easily updated and increasingly economical to implement.

Specialized software exists for generating some of the records that are presented in diagram form (especially for FTA, ETA, and HFA diagrams). Alternatively, CAD programs or simply desktop publishing, drafting, or presentation graphics can be used. Even the simplest drafting programs offer specialized *clip art* features for addition to drawings, as needed, thus allowing creation of specialized FTA templates.

6.4.3. Responsibility and Accountability

Someone in the organization, by job title, should be identified as having the responsibility and accountability for managing PHA related documentation within the organization's normal records management function. Someone on each PHA team should be designated to ensure that this PHA documentation is properly completed and entered into the records management system.

6.4.4. Distribution, Access, and Retention

Storage locations, access controls, and revision and purge schedules should be in accordance with the organization's established PSM records management procedures. It should be remembered that PHA serves as the foundation for many other PSM elements. Thus, it is important that ready access to PHA documentation be provided to personnel implementing other PSM functions. Documentation should be retained at least until the PHA is next updated. Note, however, that current regulatory requirements dictate that PHA reports for covered processes be retained for the life of the facility. See Chapter 4, *Records Management* for additional guidance.

6.5. Auditing

The PSM program should establish the requirements for auditing PHA. The first step in auditing PHA is to verify that a documented program exists for implementing PHA throughout the facility life cycle. This program would include elements outlined in Sections 6.3.1 and 6.4 that are pertinent to the needs of the organization or facility.

The second step in auditing PHA is to verify that the results of implementing PHA have been documented in the manner prescribed by the program. Particular emphasis should be placed on auditing the documentation of the resolution of recommendations.

Any deficiencies identified during the audit must be documented and promptly resolved. See Chapter 14 and the CCPS book *Guidelines for Auditing Process Safety Management Systems* for more guidance on auditing PSM.

6.6. Examples

6.6.1. Runaway Reaction in a Polymerization Reactor

A severe cooling problem in a polymerization reactor was believed to be due to fouling from polymer deposits in the cooling system. The operations management team decided to empty the reactor, refill it with refined reactant and circulate this in the reactor cooling system to dissolve the polymer deposits and thus restore efficient heat transfer. This operation had been in progress for more than an hour when an incident occurred:

- the emergency high pressure alarm went off;
- the relief valve began to discharge into the blowdown system;
- the pressure recorder in the control room went off scale;
- a loud noise was heard from outside;
- the unit gas detection alarm sounded; and
- a field report came in of hydrocarbon vapors escaping from a ruptured blowdown drum.

Emergency procedures were immediately implemented and response teams set up water curtains to disperse the hydrocarbon vapor cloud. Fortunately the damaged drum was safely isolated from the flare and no further consequences occurred.

Investigation revealed that residual catalyst in the reactor had apparently initiated a polymerization reaction. Since heat removal from the reactor was impeded, the bulk temperature of the reactants had exceeded the acceleration temperature, producing a runaway reaction and rapid pressure buildup. This pressure exceeded the relief valve design basis and capabilities, and rose too fast to allow a built-in reaction kill system to respond.

Fundamental process changes had been introduced by the revised cleaning procedure:

- changing the polymerization conditions; and
- changing the reactant concentration for the cleaning operation.

No formal, systematic, or critical process hazard analysis (PHA) had been done to enable operations personnel to manage and understand the potential hazards associated with these process changes. In fact, the plant's technology group was not contacted about the modifications.

Performance of a timely and well documented PHA, based on known and documented plant experience on cooling system fouling and known data on the probability of increased hazards of runaway reactions at the changed operating conditions, would have informed the operations team of the hazards and could have prevented this incident.

7

Chemical Process Quantitative Risk Analysis

7.1. Overview

7.1.1. Introduction

The concept of risk is not new, and has always been a feature of our daily life and activities. Risk is everywhere around us. Most people are probably aware that there is risk involved in being on the roads, whether as a motorist or pedestrian, and that every day people are killed in accidents. Most people who live in areas close to hurricane-prone coasts realize that they risk losing their lives as well as their homes. Nevertheless, most people accept these or similar potential risks. This may be due to lack of awareness of the real potential. They may believe, having only speculation and anecdotal history as guides, that nothing adverse will happen to them. Alternatively, this may reflect a willingness to accept a higher level of risk when there is a perceived benefit in doing so.

A number of major industries pose potential hazards to those who work in or near them. These vary from coal mining and quarrying, to modern travel and the nuclear industry. Some of these industries can have high potential for extensive loss of life in a single incident. In order to assist in risk evaluation and control, quantitative risk analysis has been developed to provide logically derived information that can be used directly to support decision making. The nuclear industry, which dates back to the mid 1940s, was one of the leaders in using this approach. Risk management started in the 1960s in the process industries. In the chemical process industry (CPI) this approach is known as Chemical Process Quantitative Risk Analysis (CPQRA).

This chapter reviews the type and extent of the documentation needed to support this type of analysis.

7.1.2. Goals and Benefits

Both the goals and benefits of the application of CPQRA within the CPI flow from the basic definition of the phenomenon known as risk.

One definition of risk is the potential of an activity or operation to cause harm (e.g., death, injury, damage, or pollution). Conceptually, risk (R) is often represented as the product of the consequence (C) of an incident and probability of occurrence (P):

$$R = C \times P$$

The number of injuries, fatal or nonfatal, is commonly used to characterize severity or consequence of the incident; although, the overall cost of damage can also be used. The frequency of the incident can be used to represent probability of occurrence. Risk can be:

- voluntary, where the participant willingly and knowingly undertakes the high-risk activity, such as:
  —recreational adventures (e.g., SCUBA diving or rock climbing) or
  —occupational risk, where people work in a job or location where they have a vested interest (e.g., remuneration via wages or salaries);
- involuntary, where the person involved has little or no choice and no benefits in undergoing the risks, such as:
  —natural disasters; or
  —where toxic releases, fires, and/or explosions, may endanger those in close proximity.

Risk can be quantified, but special expertise may be required. CCPS has published a book *Guidelines for Chemical Process Quantitative Risk Analysis* as a resource document on how to conduct a CPQRA. Some engineering and construction firms and some major organizations in the process industry employ reliability engineers and/or risk analysts to assist in preparing these assessments. Where such expertise is required but not available, organizations may seek assistance from qualified consultants.

The goals of applying CPQRA within the CPI are:

- to identify and quantify, insofar as possible, the key factors that contribute to the total risk related to using a particular chemical process, or a facility incorporating that process; and
- to systematically evaluate and analyze this data so that the risk can be reduced, managed, or negated by intelligent choices of design, construction, operation, and control aspects of the specific processes or facilities involved.

The benefits achieved from CPQRA are derived from the ability to objectively evaluate all types of risk (e.g., environmental, economic, and exposures of humans to hazards). For example, the ability to evaluate risk permits the selection of cost-effective risk reduction measures for existing facilities. Further, it permits assessing the feasibility of applying new chemical processes before significant amounts of financial and human resources are invested in new facilities.

7.2. Description of CPQRA

7.2.1. Objectives

There can be many reasons for performing a CPQRA. Some of these are addressed in the following paragraphs.

Layout or Siting. Traditionally, the layout of a plant site has been determined using experience and judgment to establish interplant and interunit separation distances, as well as distances to centers of employee concentration like control rooms, workshops, and offices. Some organizations have established historically based guideline tables of minimum acceptable separation distances. However, some organizations and some authorities consider the guideline approach as inadequate, and they require more extensive quantification of the actual risks involved. A CPQRA can address factors such as overpressure, thermal radiation, or toxic effects when estimating the risks associated with proposed facility layouts, permitting a quantitative approach to the selection of alternatives.

Community Impact. The close proximity of a community to a high hazard installation can pose problems that are best resolved before construction. Subsequent correction may be difficult or impossible without shutdown or elimination of the activity. CPQRA provides a mechanism for evaluating the risk posed by a facility to the surrounding communities. CPQRA may also be used in emergency planning to establish priorities or to determine credible impact zones.

Comparison of Risk Reduction Alternatives. One of the common uses of CPQRA is for the comparison of risk reduction alternatives, to identify the lowest risk alternative where there are several possible designs, layouts, or processes. In this comparative mode, CPQRA simplifies the decision making, since:

- no comparison needs to be made with absolute standards or criteria, so this complex issue does not enter into the arguments, at least at this stage;
- comparisons can be limited to the areas of divergence between the alternatives (of considerable benefit in shortening an involved procedure, and reducing the necessary documentation);
- some of the assumptions that are made may be common for all of the alternatives, thus it is not critical that these assumptions be extensively researched to provide statistically high levels of confidence; reasonable data can be used, knowing that it may no longer be a critical item; and
- organizations that perform CPQRA find that the more limited comparative CPQRAs facilitate decision making by identifying the lowest risk alternative, without clouding the issues with intricate details.

Regulatory Requirements. CPQRA can be required to achieve regulatory compliance. There are few instances where CPQRA regulatory requirements currently exist in the United States; however, they are more common in some western European countries.

7.2.2. Performance of CPQRA

CPQRA cannot be simply described with a listing of its various analysis techniques, as can be done for PHA (Chapter 6) or Human Factors Analysis (Chapter 9). Many different approaches exist for estimating the frequency of occurrence for an undesired event. Similarly, there are a variety of ways in which the consequences of an event can be manifested, such as fires, explosions, and toxic gas exposures. For each of these types of event, there exists a multitude of ways to model, or estimate, the magnitude of the consequences. Reference should be made to the CCPS books and other publications listed in Section 7.6 for detailed descriptions of the various elements and tools that comprise CPQRA.

7.3. CPQRA Documentation

7.3.1. CPQRA Program Documentation

Organizations wishing to implement CPQRA should include in their PSM program documentation guidance on the goals of CPQRA as well as procedures outlining how it is to be implemented. The PSM program documentation may include statements relevant to CPQRA including:

- who (by function) is to be responsible and accountable for any needed CPQRA;
- the practices to be followed in deciding if CPQRA is to be performed;
- the organization of the CPQRA itself, including selection of study scope, team, analysts, and methodology;
- criteria to be used for assessing CPQRA results;
- form of reporting the CPQRA results;
- resolution of recommendations resulting from the study;
- auditing the implementation and documentation of CPQRA; and
- auditing of the organization's CPQRA program, itself.

Further elaboration of some of these topics is provided below:

CPQRA Analysts. Considerable training and experience is required to properly perform some of the more sophisticated methodologies used in CPQRA. Some organizations may find it helpful to establish and document minimum qualification requirements for CPQRA analysts.

Scope. Performance of CPQRA can be time consuming and, therefore, expensive. The corporate PSM program documentation may contain guidance to assist in identifying those situations warranting CPQRA.

Methodology. As mentioned previously, there exists a multiplicity of CPQRA modeling methodologies, many requiring the use of sophisticated computer programs for their effective use. Some organizations may find it appropriate to standardize upon, and support, a limited subset of these methodologies. Appropriate guidance might be included in the CPQRA program documentation.

Criteria for Acceptance. The CPQRA program may document tolerable levels of risk for the various categories and types of risk (e.g., on-site, off-site, individual, or societal).

CPQRA Report. The CPQRA program documentation should identify key pieces of information that should be documented in the CPQRA report. One such list is given in Section 7.3.2, another example is given in APPENDIX C of the CCPS book *Guidelines for Chemical Process Quantitative Risk Analysis.*

Resolution of Recommendations. As in other elements of the PSM program, there should be a documented requirement for the prompt resolution of recommendations, including requirements for the assignment of specific responsibilities and accountabilities.

Auditing. The procedures for auditing CPQRA documentation should be documented.

7.3.2. Documentation of CPQRA Results

CPQRAs will vary considerably in scopes and objectives; consequently the type and style of documentation is likely to vary from one study to another. However, there are certain aspects of a CPQRA that should be carefully and clearly documented in order to provide adequate transparency; i.e., the ability of a reader to see and understand how the CPQRA was performed.

It is important in preparing the final report to remember who the potential readers may be. Some potential readers include:

- senior corporate officials;
- federal, state, or local government officials;
- consultants, including experts in CPQRA;
- other plant personnel; and
- the public.

Sometimes organizations perform CPQRA for internal purposes, and only their own management personnel are expected to see the report. Alternatively, CPQRAs may be called for by federal, state, or local government regulatory agencies, in which case the reports will be in the public arena for inspection. Since confidential internal documents sometimes become public (especially after incidents), the writer should remember who the potential readership may be. A less technical summary, in addition to the executive summary, may then be appropriate.

The various aspects of the documentation that should be considered for inclusion in the CPQRA report are, by section name and content:

- *Table of Contents,* to include also a list of Appendixes.
- *Executive Summary*, to summarize the conclusions of the CPQRA, which should be given in lay language since some readers may be unfamiliar with CPQRA and its jargon;
- *Objective*, to state why the CPQRA was performed, and the objectives or goals to be met;
- *Scope*, to state what parts of a facility were included and explicitly what parts are not. Alternatively scope could be limited to hazards of a particular type (e.g., those due to flammable or toxic materials);
- *Definitions* (especially, for any potentially unfamiliar terms or terms whose use differs from their usual definition)
- *System Description*, to describe the process, the facility, and the equipment studied;
- *Hazard Scenario Identification*, to include:
  —listing of all hazard scenarios identified (with their likelihood to be considered later in *Probability Assessment*); and
  —statement of methods used to identify and develop scenarios and the reasons for those selected (see Chapter 6, *Process Hazard Analysis*).
- *Incident Consolidation*, to give an indication of which events and probabilities have been consolidated to produce a cumulative total probability (e.g., often a series of simple incidents like pipe flange leakages can be grouped to reduce the complexity of the modeling);
- *Selection of CPQRA Models*, to document the rationale used for selecting particular methods to mathematically model different aspects of the CPQRA;
- *Consequence/Impact Assessment*, to document how the selected models produce results that quantify the consequences of an incident, and how people and/or property are likely to be impacted;
- *Probability/Frequency Assessment*, to document how the probability of the incident is determined, to include:
  —source of the data bases for equipment failure or human error and frequency;
  —any critical assumptions made; and
  —any adjustments or departures from the referenced data bases and their justification
- *Risk Assessment,* to document the calculation of risk and the presentation of the basic results (bringing together *Consequence/Impact* and *Probability/Frequency* assessments);
- *Risk Analysis*, to document the derivation/selection of sets of values for risk, their meaning, and how they are to be compared with criteria/targets, or with other facilities, and to ensure that:

—realistic comparisons are made;
—effects of uncertainties, or confidence limits, are addressed; and
—conclusions drawn are the best ones, in terms of accuracy or reliability, which the models will support.

- *Recommendations for Risk Mitigation*, to recommend ways of addressing risk, if the assessed risk value was uncertain or not tolerable; and
- *Appendixes*, to ensure that the report itself is concise and not excessively interrupted by technical detail.

Appendixes can be very helpful in organizing that data presented in the report. The subject matter for each appendix should be kept separate, and appendixes should be included in a logical order. An Appendix index and/or a table of contents would be advisable, in addition to cross referencing in the text. Some examples of documentation that can be placed in appendixes are as follows:

- population statistics (demography);
- meteorological statistics (stability, wind speed and direction);
- relevant historical accidents (plant and worldwide);
- tabulations of comparable risks;
- equipment failure rate data base;
- human errors data base;
- physical property data;
- toxicological data;
- graphical results (e.g., F–N curves and Iso-risk contours);
- references cited;
- diagrammatic results (e.g. FTA); and
- maps and drawings.

This may not be a complete list, and some items may not be relevant to a particular CPQRA. Alternatively, some organizations may choose to document the more voluminous parts of this information, not in appendixes to a report, but as a separate report with a limited distribution.

The CPQRA documentation needs to be thorough, clear, and concise. Proper documentation helps to prevent guesswork on the part of the reader as well as to provide the full basis for the assessment, which can be invaluable for any future updates. Refer to Appendixes 7A through 7C for additional guidance on documentation of CPQRA.

7.3.3. Resolution of CPQRA Recommendations

It is common, as a result of a CPQRA, to formulate recommendations intended to reduce the risk of operating a process or facility. It is important to promptly and fully resolve each recommendation, just as was the case for PHA recommendations in Chapter 6. Reference should be made to Section 6.3.3 for a discussion of documentation of recommendations and their resolution.

7.4. Records Management

The records management task for CPQRA is quite similar to that for PHA. Indeed, there is considerable overlap in that much of the activity involved in hazard identification, scenario development, and frequency estimation for CPQRA is actually PHA using the techniques described in Chapter 6. Since there are no records management issues unique to CPQRA, reference is made to Section 6.4 and Chapter 4 for general guidance.

7.5. Auditing

An audit of CPQRA should first verify that a documented program for implementing CPQRA exists. This program would include elements outlined in Section 7.3.1 and records management practices referenced in Section 7.4 that are pertinent to the needs of the organization or facility.

The audit should next verify that the results of implementing CPQRA have been documented in the manner prescribed by the program. Particular emphasis should be placed upon the documentation of the resolution of recommendations.

Any deficiencies identified during the audit must be documented and resolved in a timely manner. It should be remembered when auditing, that the intent is not to confirm the technical accuracy or adequacy of the contents of the documentation but, rather, to confirm that the documentation practices conform to the established program. Auditing may, however, reveal areas of potential improvement in the implementation or the documentation of the CPQRA program, itself.

7.6. References

In addition to the references in this section, see those listed in Appendixes 7A, 7B, and 7C.

AIChE–CCPS, *Guidelines for Chemical Process Quantitative Risk Analysis,* 1989

AIChE–CCPS, *Guidelines for Hazard Evaluation Procedures, Second Edition with Worked Examples,* 1992

AIChE–CCPS, *Guidelines on Process Equipment Reliability Data with Data Tables*, 1989

AIChE–CCPS, *Tools for Making Acute Risk Decisions: with Chemical Process Applications*, 1994

Environmental Protection Agency, *EPA Blue Book, Technical Guidelines for Hazard Analysis,* EPA, Washington, D.C.

Industrial Risk Insurers, *Plant Layout and Spacing for Oil and Chemical Plants,* IM.2.5.2. June 1993

International Organization for Standards, *Analysis techniques for system reliability—Procedure for Failure Modes and Effects Analysis (FMEA)*, ISO-CEI/IEC 812:1986

International Organization for Standards, Fault Tree Analysis, ISO-CEI/IEC 1025: 1990

International Organization for Standards, *List of basic terms, definitions, and related mathematics for reliability,* ISO-CEI/IEC 271:1974

Appendix 7A. Graphical Presentation of CPQRA Results

7A.1. Overview

A number of ways exist to present the results of a CPQRA in graphical form. These are briefly described here to illustrate that portion of the CPQRA documentation task.

7A.2. Iso-Risk Contours

One method of displaying individual risk is to plot on a map of the site and surrounding areas, contours that demarcate areas exposed to equal levels of risk. CPQRA computer programs will normally calculate the individual risk level for each point in a map grid. Grid points exposed to approximately equal levels of risk can then be linked graphically by drawing contours on the grid. These are known as Individual Risk Contours or Iso-Risk Contours. A more detailed discussion of Iso-Risk contours can be found in the CCPS book, *Guidelines for Chemical Process Quantitative Risk Analysis.* An example of an Iso-Risk Contour plot is given in Figure 7-1.

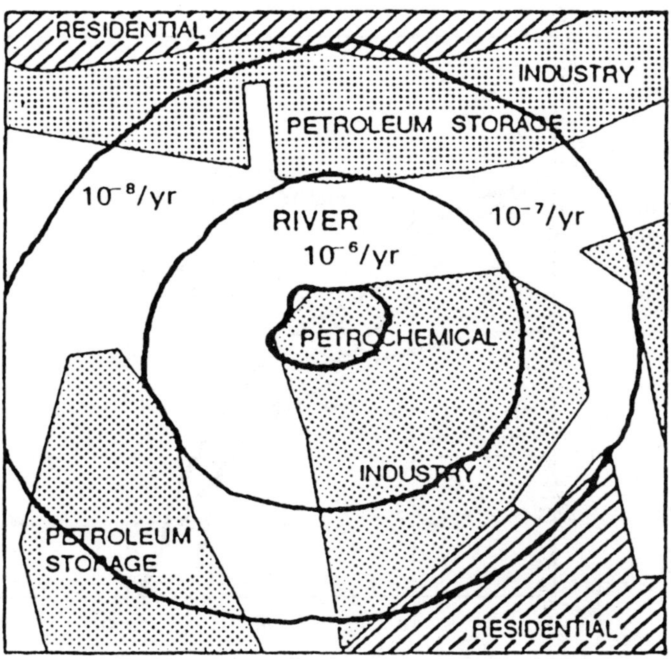

FIGURE 7-1. Example of an Individual Risk Contour Plot

7A.3. F–N Curves

Society has generally taken a view that multiple deaths in a single incident are far worse (i.e., less tolerable) than an equivalent number of accidents involving single deaths. In other words, the frequency level at which such multiple fatalities would be tolerated is lower than for the same number of single fatalities. In order to address this perspective on tolerable risk to large groups of people, societal risk is often presented graphically, on logarithmic scales, by plotting the *number of fatalities* (*N*) in an incident; versus the *frequency of occurrence* (*F*) causing *N* or more fatalities. This type of risk plot is called an F–N curve. Figure 7-2 illustrates F–N curves for a number of hazardous activities. The CCPS book *Guidelines for Chemical Process Quantitative Risk Analysis* provides more guidance on societal risk and the preparation of F–N curves.

7A.4. References

AIChE–CCPS, *Guidelines for Chemical Process Quantitative Risk Analysis,* 1989.

Rijnmond Public Authority, *Risk Analysis of Six Potentially Hazardous Industrial Objects in the Rijnmond Area—A Pilot Study*, Reidel, 1982.

Sanders, Roy E., *Management of Change in Chemical Plants*, Butterworth-Heinemann, 1993.

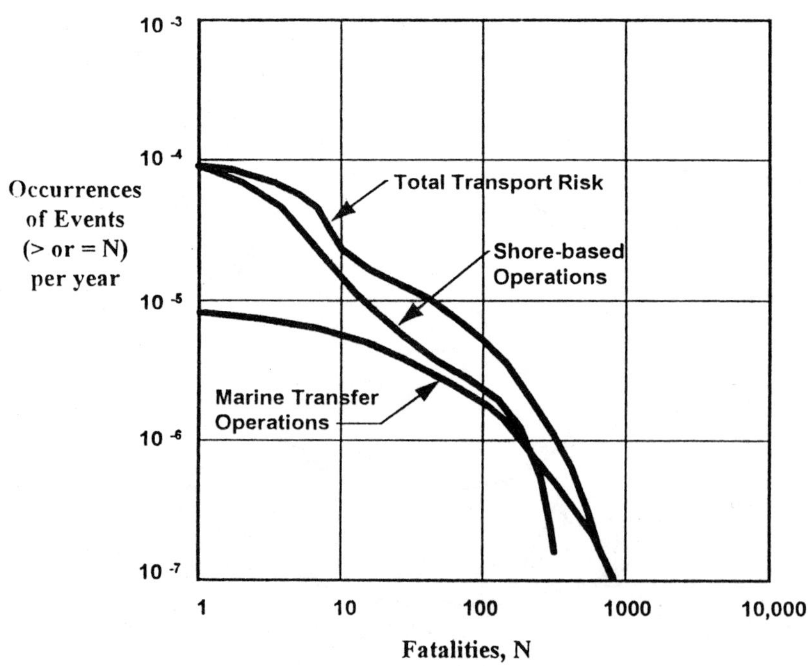

FIGURE 7-2. Example of a Societal Risk F–N Curve

Appendix 7B. Documentation of Supporting Data

7B.1. Overview

CPQRAs are often based upon a considerable amount of data. This data should be appended to, or its origin should be identified or referenced in the summary report to enhance the creditability of the study and to facilitate auditing. This section suggests some types of supporting data that might be addressed when documenting CPQRA.

7B.2. Basic Process Data

Basic process descriptive data may be found in the following:

- material safety data sheets (MSDSs);
- process flow diagrams (PFDs);
- piping and instrumentation diagrams (P&IDs), engineering flow diagrams, utility flow diagrams;
- operating procedures;
- maintenance procedures;
- instrument test schedules;
- layout and site drawings; and
- results of any Process Hazard Analysis (PHA).

7B.3. Specialized Data

In addition, some specialized data may be required. These could include:

- meteorological (weather) data;
- demographic (population) data;
- equipment failure rate data;
- human error rates; and
- historical accident data.

Each of the above will be described in more detail, below.

Meteorological Data. These data are prepared from routine observations taken at meteorological stations, typically airports. The raw data, or processed data, can be provided from national meteorological sources. When preparing for a consequence analysis, it is not uncommon to use a 10-year collection of data to assess mean probabilities for the different combinations of wind speed, direction, and stability class. Ten years of hourly readings produce nearly 88,000 sets of observations. For each set of observations, at least four pieces of data (time, wind speed, wind direction, and cloud cover) are required to calculate a joint frequency distribution of wind speed and stability class. The method used to reduce this mass of data to the required frequency distribution should be documented together with the source of the raw data, and a summary of the reduced data. Justification for the choice of the method of reduction should also be documented.

Demographic Data. The principal sources of demographic data are census statistics produced for local areas, together with detailed local maps. An analysis will be required to relate population to defined areas, for example to a grid of appropriate dimensions, typically, 200 m, 500 m, or even 1000 m. The source of raw data, the derived demographic data and the method used to prepare them should be documented.

Equipment Failure Data. For an incident to occur, there must be an initiating event. While a natural disaster (e.g., tornado or earthquake) might cause a process incident, the overwhelming majority of initiating events are either equipment failures (discussed in this section) or human errors (discussed in the next section).

Historically, equipment failures result from some identifiable cause or causes. For instance, overpressure can cause a pipe or vessel to burst and the cause of the overpressure can be identified through the application of PHA (e.g., HAZOP or FTA). Alternatively, equipment can fail due to some inherent defect or due to deterioration during service caused by stress, corrosion, or wear. Over a large number of installations and a large number of years, a history of such failures can be developed, and average failure rates can be determined. Also, if the data are extensive enough, their confidence limits can be calculated.

When documenting a CPQRA it is often important to document, for each set of data used:

- their origin (in the raw form) by quoting the reference;
- what adjustments have been made to them, for instance, removal of inappropriate failures; and
- confidence limits, where available, or comments on uncertainty where appropriate.

For additional guidance on the treatment of failure rate data, see the CCPS book *Guidelines on Process Equipment Reliability Data, with Data Tables*, and the references contained in that book.

Human Error Rate. This is a part of many CPQRAs because human beings may be less consistent and less reliable than the equipment they operate or maintain. The human being is subject to complex emotional issues, unrelated to the job, like domestic or financial problems, as well as job-related emotional issues such as those caused by stress or emergencies. Emotional issues and ergonomic issues constitute performance shaping factors (PSF) that should be addressed in the CPQRA; consequently, the reason for selection of each human error rate used should be documented in the CPQRA. The assessment is more credible if the logic underlying human error rate estimates is properly documented.

Refer to Chapter 9, *Human Error* and the AIChE–CCPS book, *Guidelines for Chemical Process Quantitative Risk Analysis,* for additional guidance on the treatment of human error.

Historical Accident Data. There are several objectives in documenting the accident history of the plant being considered and, where possible, of similar plants in the industry.

As discussed previously, CPQRA involves estimating the likelihood and potential consequences of serious incidents. Historical data can serve to help calibrate these estimates. Additionally, analysis of actual incidents may suggest event scenarios not otherwise identified in the CPQRA.

Pertinent accident data should be considered for inclusion, or reference, in the CPQRA documentation.

Appendix 7C. Other Aspects of CPQRA Documentation

7C.1. Selection of Cases to Consider

One of the continuing challenges in CPQRA is selection of the basic scenarios to model. In the real world there is usually an endless spectrum of scenarios to consider and such an infinite number of combinations and permutations of events cannot be modeled. Instead, a finite set of representative cases is considered, the integration of which is intended to represent the extent of the actual situation. Describing and justifying the selection is not difficult, but requires some experience in recognizing realistic parameters. Guidance is given in Chapter 1 of the CCPS book, *Guidelines on Chemical Process Quantitative Risk Analysis.*

Documentation should indicate which scenarios have been selected and the rationale for selection. The documentation might include:

- a brief description of the selected scenarios;
  —identity of material released;
  —size of hole or other source of loss of containment;
  —mass release rate, of each phase;
  —fallout/rainout/stayout;
  —nature of dust, aerosol or vapor cloud, and its dispersion; and
  —ignition scenarios.
- the assumptions that have been made;
- the possible uncertainty in these assumptions; and
- reference to the data bases used, and any adjustments made to raw data.

7C.2. Methodologies for Consequence Assessment

The calculation of risk for an installation typically addresses a wide variety of potential events and requires the estimation of the consequences to people or equipment that can result. For example, accidental releases can lead to fires, explosions, or toxic impact. The consequence assessment methodology used for each type of event differs, and there is a need for a reader of a CPQRA to know which methodology was used, why it was selected (if there are alternatives), how it was used, and what confidence levels were assigned to the results. The following discusses some of the more important aspects of consequence modeling and suggests documentation requirements for each.

7C.2.1. Dispersion Modeling

In assessing the consequences of the release of hazardous (e.g., flammable or toxic) materials, it is necessary to model the travel and dilution (i.e., the dispersion) of the cloud. Reference should be made to the CCPS book *Guidelines for Use of Vapor Cloud Dispersion Models* and to the CCPS *Workbook of Test Cases for Vapor Cloud Source Dispersion Models* which give extensive details on the methodologies and computer programs available.

Extensive documentation is often required for dispersion modeling and might include the following:

- description of scenarios being modeled;
- reasons for selection of model used;
- validation tests of model where required (references may be adequate);
- assembly of basic input data (e.g., release rates, material physical properties);
- assembly of support data (e.g., meteorology, terrain, atmospheric data);
- assumptions made;
- selected output data (e.g., examples of printouts, tables, and graphics);
- summary output tables;
- description of results;
- sensitivity of results;
- conclusions; and
- references.

7C.2.2. Ignition, Fire, and Deflagration

Releases of flammable materials can result in a variety of undesired events ranging from pool fires to devastating vapor cloud explosions. Reference should be made to the CCPS books *Guidelines for Evaluating the Characteristics of Vapor Cloud Explosions, Flash Fires, and BLEVEs* and *Guidelines for Chemical Process Quantitative Risk Analysis*.

In evaluating the potential consequences of an airborne release of flammable materials, it is often necessary to estimate the size of the resultant flammable cloud; dispersion calculations have been discussed in the previous section.

The variety of potential events make it impractical to specifically address the documentation requirements for each type of consequence modeling. In general, however, the documentation requirements parallel the categories of information presented in the previous section on dispersion modeling.

7C.2.3. Modeling the Consequences of Exposure to Toxic Gases

The physiological response to a toxic gas is often a complex function of both the concentration and the duration of the exposure. Modeling of the consequences of toxic gas exposure, therefore, requires estimation of both concentration and duration, calculation of an appropriate "dose" and relating this "dose" to the estimated physiological response (e.g., death or illness).

One common modeling approach is based upon the concept of the probit function. The probit function relates the exposure concentration and duration to the probability that a stated level of health effect will be experienced. Probit relationships are typically derived from animal exposure test results, that have been extrapolated to equivalent human health effects. An example is the following probit equation for fatalities due to exposure to chlorine:

$$Pr = -8.29 + 0.92 \,\mathrm{Log_e}\,(C^2 t)$$

where C is the concentration in parts per million and t the duration of exposure in minutes. The Pr value is 5.0 for a 50% probability of fatality. For full details of this type of calculation see the CCPS book on *Guidelines for Chemical Process Quantitative Risk Analysis*.

Other approaches exist for modeling the consequences of exposure to toxics. Whatever the methodology chosen, its basis for selection should be described in the documentation of the CPQRA. It is also important to note that there is a very large number of assumptions inherent in the modeling of toxic dose consequences, that can have a profound influence on the results. These include the protocol used for the original tests to produce the animal toxicity data, the translation of results to human beings, the absence of concentration fluctuation, and the variation in human susceptibility. Where pertinent, discussions of key assumptions, and their significance, should be included in the CPQRA documentation.

7C.3. References

AIChE–CCPS, *Guidelines for Chemical Process Quantitative Risk Analysis*, 1989

AIChE–CCPS, *Guidelines for Evaluating the Characteristics of Vapor Cloud Explosions, Flash Fires and BLEVEs,* 1994

AIChE–CCPS, *Guidelines for Use of Vapor Cloud Dispersion Models*, 1987

AIChE–CCPS, *Workbook of Test Cases for Vapor Cloud Source Dispersion*, 1989

8

Process Equipment Integrity

8.1. Overview

8.1.1. Introduction

The chemical process industry (CPI) utilizes a broad range of equipment types and systems to contain, convey, control, and process chemical substances that may be of a hazardous nature. Since such equipment constitutes the main body of the process plant, its design, fabrication, installation, operation, and maintenance are of primary importance in assuring plant safety. Through adequate attention to all of these areas and recognition of their interdependence, premature failures, which could result in accidental releases or other incidents, may be reduced in frequency and severity.

The integrity of equipment begins with an accurate, comprehensive, written definition of design, operating, and maintenance parameters. This information is utilized throughout the life cycle of the equipment by plant designers, equipment manufacturers and plant personnel. This documentation satisfies a number of technical and operational needs and helps fulfill applicable requirements of federal, state, and local regulatory agencies.

8.1.2. Goals and Benefits

The goals of the Process Equipment Integrity element are to ensure that process equipment will perform as expected and that adequate documentation exists to support its operation, maintenance, or modification in a reliable, safe manner.

Benefits to be derived from proper documentation include:

- preserving a record of design conditions and materials of construction for new and existing equipment;
- establishing records of maintenance activities to provide information for improving performance by planned preventive maintenance or similar programs;

- providing support for incident investigations;
- providing sources of information for training of operations and maintenance personnel;
- establishing proof that due diligence has been followed to support any legal or regulatory needs; and
- providing a means of improving equipment design based on historical performance.

8.2. Description of Process Equipment Integrity

8.2.1. Objectives

The main objectives in establishing a documentation system for process equipment integrity are to:

- provide procedures to ensure that key safety aspects of equipment design, fabrication, installation, and operation have been considered;
- provide data to allow safe, efficient equipment operation; and
- furnish adequate information to support equipment maintenance or modification throughout its useful life.

8.2.2. Process Equipment Integrity as Part of Process Safety Management

The Process Equipment Integrity element ensures that consideration of equipment reliability and suitability for safe performance of function is addressed throughout the following phases of the equipment life cycle:

- equipment definition and categorization;
- equipment design and/or selection;
- procurement;
- fabrication, initial inspection and testing;
- equipment installation;
- initial start-up;
- useful service life;
  —maintenance, repair, modification, and replacement;
  —testing and inspection during life cycle;
  —use in training; and
- other special tasks such as, decommissioning, demolition, and removal.

Maintaining up-to-date and accurate data for equipment is an important element of any facility process safety management (PSM) effort. Typically, the amount of PSM related equipment information grows as an equipment item progresses through its life cycle. The phases above are presented are in chronological order and each phase requires, to some degree, the information generated in earlier phases for its safe, successful completion. The generation, compilation, and protection of this information, so that it is available to the appropriate personnel, should be addressed for each phase of the equipment life cycle.

8.3. Process Equipment Integrity Documentation

8.3.1. Process Equipment Integrity Program Documentation

An organization's PSM program documentation should establish the activities required and assign related responsibilities for implementing the process equipment integrity element. The documentation challenges posed by this element are quite similar to those previously described for the *process knowledge* element, since both elements are broad in scope and involve a number of diverse activities. Additionally, the tasks requiring documentation are conducted by many people, over a staged life cycle, and the responsibilities for the resultant documentation must be transferred from one organization or individual to another at key milestones throughout that life cycle. For example, process equipment integrity documentation related to design and procurement may initially be administered by an engineering organization and turned over to the operations group at start-up. Refer to Chapter 5, *Process Knowledge*.

The PSM program documentation should also establish requirements for the routine auditing of the process equipment integrity documentation, and for promptly resolving any inadequacies that might be detected.

8.3.2. Records from Implementing Process Equipment Integrity Element

The balance of this section suggests typical records that might be generated during the documentation of the process equipment integrity element. This discussion follows the equipment life cycle stages defined in Section 8.2.2 and addresses some of the more common types of process equipment that are usually found in process facilities. The treatment described here can be extended to other equipment that may be identified as important to process safety.

Equipment Definition and Categorization. The initial step in documentation of process equipment integrity is to compile and categorize a list of equipment for inclusion in the program. This list would typically cover the following types of equipment and systems:

- mechanical equipment, such as:
  —pressure vessels and storage tanks;
  —heat exchangers;
  —process piping;
  —rotating equipment; and
  —processing equipment (filters, conveyors, columns, etc.).

- electrical equipment, including:
  —motors and motor controls;
  —transformers, switchgears, load banks, etc.;
  —power distribution and cabling systems; and
  —other electrical systems (e.g., heating, ventilation, and air conditioning (HVAC), lighting, communications).
- instrumentation and related systems encompassing:
  —instrument transmitters, meters, and indicators;
  —distributed control systems (DCS), computers, and peripherals;
  —emergency shutdown systems;
  —control valves of all kinds; and
  —other measurement and control equipment.
- safety systems, including:
  —relief and vent systems;
  —rupture disks, safety relief/control valves;
  —alarms and interlocks;
  —fire protection systems; and
- other process safety related items or systems.

Design Basis for Equipment Selection. Equipment integrity begins with specifications detailing equipment performance and physical characteristics, prepared by individuals and/or groups intimately familiar with individual process requirements. This information is typically based on the process knowledge and documentation outlined in Chapter 5, *Process Knowledge*.

Using this process documentation, mechanical details are usually defined by an equipment specialist with input and review by other disciplines (structural, electrical, instrumentation, etc.) and by qualified operating and maintenance personnel. This information becomes the primary document for determining the type and size of equipment selected and is commonly referred to as an *equipment data sheet*.

Typically *equipment data sheets* include the following information:

- process data pertinent to the particular item of equipment or system being described, defining any special constraints or limits;
- mechanical data with description of equipment and applicable specifications and fabrication requirements;
- electrical data providing interface information, power requirements, area classifications and any special features or limiting characteristics that apply;
- instrumentation and control systems data and specifications related to the subject equipment; and
- general requirements including codes and standards, tests, and inspections required, operating and maintenance requirements or conditions.

The *equipment data sheet* becomes the document that serves as the basis for equipment selection and design to ensure that it:

- meets or exceeds the specification of its design;
- is properly constructed;
- is suitable for its intended application;
- is designed in accordance with *good engineering practice*; and
- will meet the goals established for productivity, quality, safety, and protection of the environment.

This documentation should be available for the life of the equipment. Management of Change procedures must be followed to make any proposed changes. See Chapter 10, *Management of Change*.

Procurement. Procurement documentation will generally include the following:

- inquiry and bid documentation including reference specifications or data sheets, standards and terms and conditions;
- inspection requirements and acceptance protocols;
- quotation analyses outlining technical and economic acceptability of vendor proposals;
- approvals;
- purchase orders; and
- change orders.

Documentation of the correct method of packing and shipping from the manufacturer or supplier should be defined as well as applicable interim storage procedures.

Equipment Fabrication, Inspection, and Testing. A quality assurance procedure should be in place to assure that equipment as built will perform as designed. The following documentation should be considered for retention where applicable:

- welding procedures used by the fabricator;
- welder's qualifications;
- material certificates and heat numbers for alloyed materials indicating chemical analysis, physical properties and country of origin;
- documentation of a quality assurance program covering traceability of materials, worker identification and responsibility and test procedures;
- qualifications of inspectors;
- nondestructive testing such as radiographic, magnetic particle, dye penetrant, ultrasonic and high voltage spark test;
- hydrostatic, pneumatic, or vacuum leak test reports;
- material properties test reports;
- rotor balance, overspeed test reports;
- mechanical run-in test reports;
- other performance test reports;
- inspection reports, intermediate and final; and
- repair procedures.

Equipment Installation. Installation and test procedures should be documented. To adequately define areas of responsibility for the period between shipment and start-up, the following documentation should be addressed:

- field receiving reports, to document equipment shortages, overages, and/or damaged condition for insurance claim purposes and to define any field repairs or replacement that may be required;
- interim storage requirements;
- erection, rigging procedures; vendor documentation detailing critical lift points and special requirements (sub-contractor rigging diagrams may be required for work in existing plant areas.);
- installation procedures;
- alignment records (cold and/or hot), to confirm that vendor recommendations and/or plant maintenance tolerances have been met;
- instrument calibration records, to confirm instrumentation is operable and readings are accurate;
- as built/as installed dimensional records;
- protective system test records, to confirm that these systems are operational and meet expected design criteria;
- flushing, cleaning, blowdown records, to confirm that lines and equipment are free and clear of debris;
- dryout records, to confirm systems have been dried of moisture if its presence may prove deleterious to the system component or the material handled, accelerate corrosion or create an undesirable chemical reaction;
- lubrication schedule, to follow vendor furnished recommendations to maintain equipment integrity; and
- system pressure test records.

Start-up, Initial Operation. Initial start-up operating logs should be kept to confirm the design requirements and to document any early failures, abnormal operating conditions, and corrective actions. Appropriate documentation of any changes required, including authorizations, should be retained and incorporated into baseline descriptions using Management of Change procedures. See Chapter 10, *Management of Change*.

Maintenance, Repair, Replacement. Documented maintenance procedures should be prepared by qualified persons, intimately familiar with the types of equipment and systems to which they will be applied. Consideration should be given to standardizing such procedures throughout the process or plant, where possible, by class, type, and criticality of equipment function.

Maintenance procedures and records generated by following these procedures, such as logs of maintenance and repairs, are essential to promoting and maintaining equipment integrity and may be useful in improving maintenance programs. Documentation developed during these activities may be unique to a process and complementary to those which are in the equipment suppliers' standard operating and maintenance manuals. Maintenance procedures may also be the basis for special training requirements. See Chapter 12, *Training*.

Inspection and Testing. Most equipment, with or without moving parts, is subject to some form of deterioration during service, for example, due to frictional wear, thermal cycling, or corrosion. Procedures for detecting and measuring equipment problems that degrade performance or safety of process equipment must be developed. Inspection and testing procedures should be thoroughly evaluated and documented, as previously described for maintenance procedures, since they supplement, interact with, and lead to appropriate enforcement and application of the maintenance procedures. Consideration should be given to standardizing the methods, criteria, and applicability of inspection and testing procedures throughout the facility or organization. Records of inspections and test results should assist in predicting useful life, allowing replacement of equipment before it fails and leads to downtime or incidents. Such records may also provide information that assists in identifying ways of improving equipment performance.

Training Records. As part of a training procedure, documented information on specific process equipment is helpful in familiarizing individuals with a process in an expeditious manner. See Chapter 12, *Training*, for recommendations on training personnel from operations, maintenance, and other functions. This should contribute to safe, efficient process and plant utilization and help ensure equipment integrity.

Other Special Tasks. When other special tasks, not included in the above categories, are performed on process equipment in any way that changes its short or long term characteristics and could impact equipment integrity, the activity and results should be documented and the records made available to any functional groups affected. For example, decommissioning, demolition and removal of units or equipment may change physical relationships, location, interfaces, or other aspects of neighboring process equipment. Proper documentation allows those charged with the responsibility for operation and maintenance of the adjacent processes to be informed of the change (i.e., what has been done and why) and provides the information necessary for them to determine how to safely cope with any aspects of the change that might impact their job. Documentation and communication of results are also important to allow appropriate complementary modifications to operating procedures, maintenance procedures, and perhaps to the emergency response plan.

The appendixes to this chapter contain examples of procedures and typical documents for equipment common to most process plants. At the end of each appendix is a *sample equipment data* sheet that illustrates some of the data that can be maintained for each equipment type.

- Appendix 8A, **Example of Documentation Requirements for Mechanical Equipment;**

- Appendix 8B, **Example of Documentation Requirements for Electrical Equipment;**
- Appendix 8C, **Example of Documentation Requirements for Instrumentation Equipment; and**
- Appendix 8D, **Example of Documentation Requirements for Safety Systems Equipment**.

8.4. Records Management

In addition to the description of the Process Equipment Integrity program outlined in Section 8.3.1, the organization's PSM documentation should also define pertinent Process Equipment Integrity records management practices. It should be noted that the records management task for this element is similar in scope and complexity to that previously described for the Process Knowledge element. See Chapter 5, *Process Knowledge*, and Chapter 4, *Records Management*, for additional guidance on records management issues.

8.4.1. *Who and When*

When an effective equipment integrity program predates the equipment, the necessary documentation is generated by those parties responsible for implementing the various life cycle activities listed in section 8.2.1. Conversely, the complexity of the process equipment integrity documentation task can be greater for existing facilities. As was the case for process knowledge, if equipment was designed, procured, installed, and operated prior to the implementation of a formalized process equipment integrity program, the organization may have to retroactively generate the information necessary to document that the equipment is suitable for its service.

Once an effective program has been implemented, and operations begun, the responsibility for administering existing information and for generating additional information (e.g., operating and maintenance histories, change records, or incident reports) is often shared by the operations and maintenance organizations.

8.4.2. *Where Maintained*

Information created during the design, fabrication, installation, and start-up phases of a project (e.g., original equipment descriptions and specifications, purchasing records, initial performance test results) might be stored in a central repository or in a long-term archive, depending on the frequency of need to access the information. More frequently used equipment information (e.g., maintenance manuals, periodic test procedures, detailed equipment drawings, or expendable copies of such information) might be maintained in or near offices of maintenance and/or operations personnel having day-to-day responsibility for the facility.

8.4.3. Updating Documentation to Reflect Equipment Changes

Equipment changes typically affect a variety of process equipment integrity records. For example, replacement of a pump with one of a larger size might generate a new equipment specification, require a P&ID revision and a new maintenance procedure, and necessitate the reevaluation and documentation of the sizing of a downstream relief valve. *Management of Change*, including revision of related documentation, is discussed in detail in Chapter 10.

8.5. Auditing

The PSM program documentation should establish requirements for auditing of the process equipment integrity documentation. Audits should seek to determine whether a documented program, such as described in Sections 8.3.1 and 8.4, exists and whether it is being administered as described. The records described in Section 8.3.2 should be audited, not for technical accuracy but, rather, to determine if they are being compiled and managed as required by the program. Such audits may reveal potential improvements to be made in the program itself.

Any deficiencies identified during the audit must be documented and promptly resolved. See Chapter 14 and the CCPS book *Guidelines for Auditing Process Safety Management Systems* for more guidance on auditing.

8.6. Examples

8.6.1. Inadequate Documentation of Equipment Integrity Requirements and Management of Change Procedures

The major explosion at a chemical process plant at Flixborough in the U.K. in 1974, which resulted in 28 fatalities and $48 million in damage, was determined by the investigating team to have been due to lack of adequate equipment integrity documentation coupled with lack of proper technical oversight of equipment modifications. A poorly designed temporary pipe and bellows system, installed to route process chemicals around a reactor vessel that was being repaired, flexed, resonated, and ruptured, releasing flammable, toxic chemicals which ignited and exploded.

Effective analysis of the design of the temporary system versus equipment integrity requirements could have prevented the incident. Effective *Management of Change* procedures could also have prevented the system change and thus avoided the incident.

8.6.2. Inadequate Documentation of System Status

A processing unit, which had been removed from service a number of years previously, was scheduled for demolition. A review of the records documenting the preparation of the facility for shutdown did not reveal how, or if, the equipment had been decontaminated. Since an ether had been used as a solvent in one portion of the process, there was concern that residual ether, had it been present, might have formed potentially explosive peroxides.

Considerable cost was incurred in analyzing the situation and in planning and implementing a means of safely flushing the suspect equipment. These additional costs might have been avoided had proper documentation existed confirming that the equipment had been adequately decontaminated prior to shutdown of the facility.

8.6.3. Similar or Identical? Inadequate Equipment Integrity Documentation

A company used an inspection system that was based on the premise that equipment that is safe under certain conditions will be safe under similar conditions. Every two years a vessel from a group of vessels in similar service was selected for corrosion evaluation. If the inspected vessel showed an acceptable level of corrosion, the other vessels in the group were not inspected until their turn came around, which could be as long as twelve years.

An absorption tower on a nitric acid plant leaked, several years before its inspection was due. Closer evaluation revealed that the leaking tower operated at 100–125°C while other towers in the inspection group operated at 90°C. The higher temperature had increased the rate of corrosion.

The documentation of the inspection program was inadequate in that it did not explicitly define criteria for evaluating similarity of service conditions. A potentially dangerous incident occurred, which could have been avoided by more thoroughly documenting the test requirements.

8.7. References

The following information sources may be useful in the analysis, design, fabrication, operation, and maintenance of equipment and systems to contribute to process equipment integrity throughout the life cycle.

AIChE–CCPS, *Guidelines for Auditing Process Safety Management Systems,* 1993

AIChE–CCPS, *Guidelines for Hazard Evaluation Procedures, Second Edition with Worked Examples,* 1992

AIChE-DIERS, *Emergency Relief System Design Using DIERS Technology,* 1992

AIChE–CCPS, *Guidelines for Process Equipment Reliability Data, with Data Tables*, 1989

AIChE–CCPS, *Guidelines for Safe Storage and Handling of High Toxic Hazard Materials*

AIChE–CCPS, *Guidelines for Technical Management of Chemical Process Safety,* 1989

American National Standards Institute, *Safety in Welding and Cutting, ANSI-Z49.1,* New York, N.Y.

American National Standards Institute/National Fire Protection Association, *National Electrical Code,* ANSI/NFPA 70, New York, N.Y.

American Petroleum Institute, *Safe Welding and Cutting Practices in Refineries, Gasoline Plants and Petrochemical Plants,* Publication 2009, Washington, D.C.

American Petroleum Institute, *Management of Process Hazards,* Recommended Practice (RP) 750, Washington, D.C.

American Society of Mechanical Engineers, *Fabrication and Inspection Codes,* ASME, New York, N.Y.

Canadian Chemical Producers Association, *Essential Components of Safety Assessment Systems,* 1985

Instrument Society of America, *Shutdown Systems*, ISA-SP84, Triangle Park, N.C.

International Organization for Standardization, *Analysis Techniques for System Reliability—Procedure for Failure Modes and Effects Analysis*, 150-CEI/IEC 271:1974

Kletz, T.A., *Critical Aspects of Safety and Loss Prevention,* Butterworths, Guildford, UK, 1990

Pew, R.W., Miller, R.C. and Feeher, C.E., *Evaluation of Proposed Control Room Improvements Through Analysis of Critical Operator Decisions,* EPRI/NP-1982, Bolt, Beranek and Newman, Inc.

Appendix 8A. Example of Documentation Requirements for Mechanical Equipment

8A.1. Introduction

The purpose of this appendix is to provide examples of the types, levels, and areas of coverage of the documentation needed to support mechanical equipment through typical life cycle phases in chemical process plant applications. Of the many varieties of equipment categorized as mechanical (see Section 8.3.2), the example utilized here is a pressure vessel. Pressure vessels of many kinds, classified by material, type of service, construction techniques and other applications criteria are in use in the CPI. The documentation requirements illustrated herein are adaptable to most such vessels. An example vessel form is included as Figure 8-1.

8A.2. Design Phase

During the design phase, an *equipment data sheet,* sometimes referred to as a *process data sheet,* should be generated to describe and specify the pressure vessel intended for this particular application. It should contain the basic information pertaining to the vessel. This information could include, but not be limited to, the following:

- equipment identification number and description;
- contents and specific gravity;
- design and operating pressure and temperature;

- overall dimensions;
- materials of construction;
- design codes and standards used (API, ASME, ANSI, ISO, ISA, NEMA, NACE, etc.) with revision number;
- type of support;
- nozzle schedule;
- corrosion allowance;
- any special considerations such as:
  —fabrication specification;
  —jacketing and tracing;
  —post weld heat treatment;
  —lethal service;
  —cyclic conditions; and
  —wind and earthquake loads as well as any other loads contributed by ancillary equipment such as agitators.
- testing procedures; and
- painting and insulation requirements.

Outline drawings may also be produced to supplement the above information.

8A.3. Fabrication and Inspection

The vendor should be required to submit drawings for approval. Any alterations required by the organization to the drawings submitted by the vendor should be marked and returned to the vendor. The vendor should not be allowed to begin fabrication until these marked up drawings are received and agreed upon, unless otherwise authorized. The schedule for inspections, shop visits and hold points must be coordinated and the agreed upon schedule documented. Welding and inspection procedures and welder qualifications must be documented and approved. A report is issued after each inspection showing the results of the inspection and giving approval or rejection. Copies of these inspection reports should be retained. In the case of rejection, the cause may be waived if not detrimental to the vessel integrity. In any such cases, the waiver and its justification should be approved and documented.

The results of destructive and nondestructive tests that are performed should be recorded and reviewed, together with all material certificates and mill test reports. The complete set of test and inspection reports should be retained.

Examples of documents normally applicable for this fabrication and inspection stage are:

- vendor certified drawings;
- welding procedures;
- welder qualifications;
- mill certificates of coded, i.e. defined specification, materials;

- inspection reports;
- hydrostatic pressure test reports;
- nondestructive examination reports;
- Non-Destructive-Testing (NDT) inspector qualifications;
- code calculations;
- manufacturer's data report;
- manufacturer's certificate of compliance;
- heat treatment reports; and
- shipping, packing, and storage documents.

8A.4. Installation

Examples of documents for the installation stage are:

- shipping damage report;
- erection and installation procedures; and
- pressure testing records and similar documents.

8A.5. Start-up and Initial Operation

Run-in and final performance testing of the equipment occurs in this phase. The potential for identification of needed modifications is highest at this time. Such modifications must be properly analyzed, approved, and documented. Examples of documents for this stage are:

- start-up operation log;
- performance evaluation report;
- trouble shooting and modification records;
- maintenance, inspection, and operations procedures; and
- spare parts order records and inventory stocking records.

8A.6. Continued Operation and Maintenance

This phase constitutes the bulk of the useful life of the equipment. Selected data and documented information generated and used in the previous life cycle phases should be gathered and retained to support operation, maintenance, repair, and replacement during the long-term operational phase of the equipment. Documentation should be evaluated as to usefulness and applicability and periodically purged to keep the file documents current and to expunge material that is no longer useful, or required for regulatory or legal reasons. Examples of the types of documents that would be appropriate at this stage include:

- design and purchase specifications documents;
- documentation of as-built and installed status (e.g., materials certifications, fabrication, and code certifications, and as-built drawings);

- documentation of inspections and testing (during fabrication, installation, and initial operations);
- normal operation records or histories;
- detailed reports on periods of abnormal operation (e.g., overpressures or over temperature) together with analysis of subsequent vessel integrity; and
- maintenance schedules and records.

8A.7. Example

The example shown on the facing page, Figure 8-1, contains selected parts of a typical requirements specification. This specification is provided for illustration purposes only and should not be used as an actual source of information.

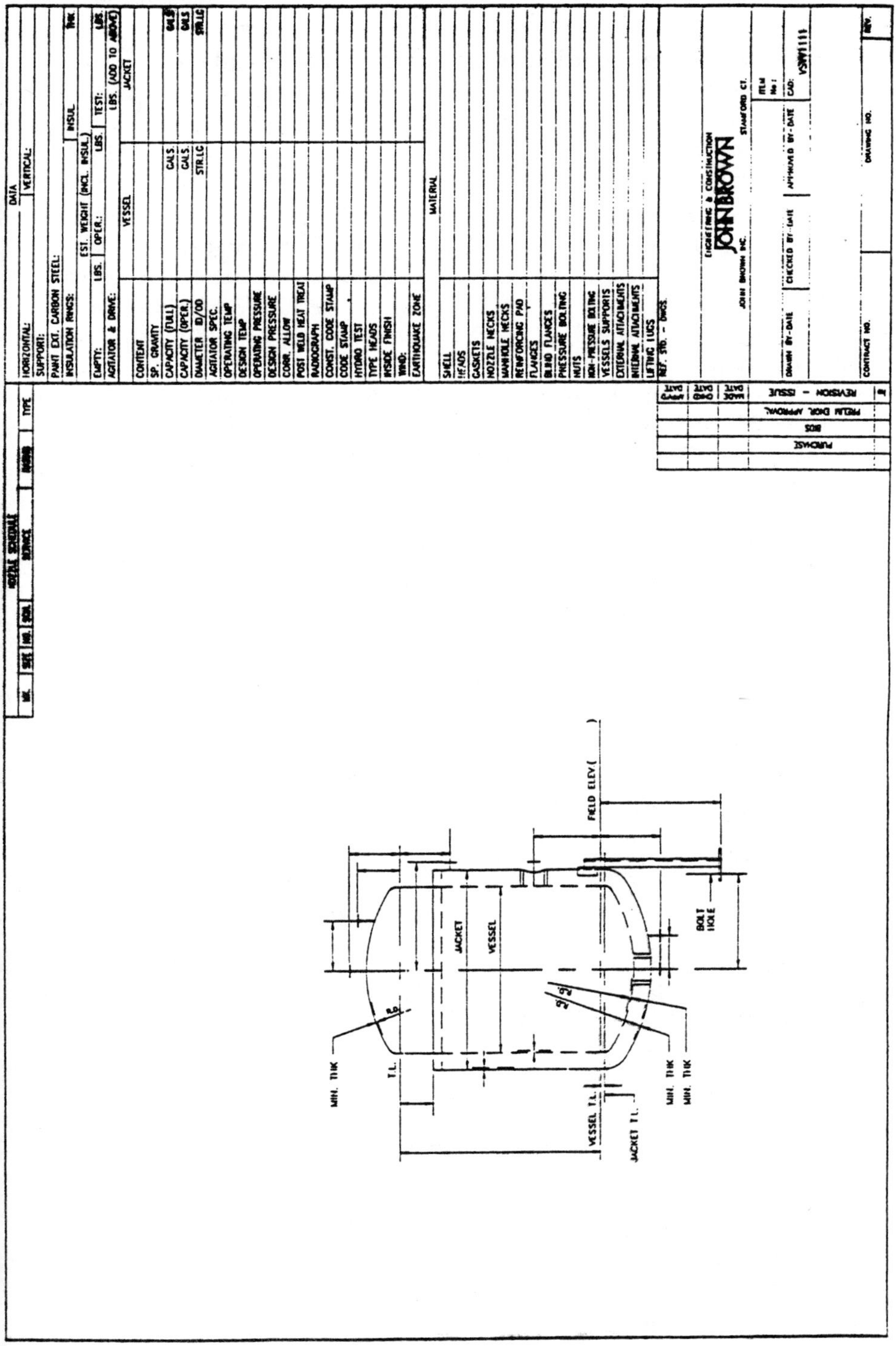

FIGURE 8-1. Example of Process Vessel Specification

Appendix 8B. Example of Documentation Requirements for Electrical Equipment

8B.1. Introduction

The purpose of the appendix is to provide examples of the types, levels, and areas of coverage of the documentation needed to support integrity of electrical equipment through typical life cycle phases in chemical processing plant applications. Of the numerous kinds of equipment categorized as electrical (see Section 8.3.2) the example utilized here is a low voltage (LV), i.e., 480 volts, motor control center to be located in an electrical equipment room in nonhazardous (unclassified) area. Many other types and different applications of electrical equipment occur within the CPI. However, the documentation requirements illustrated herein are adaptable to most such applications.

8B.2. Design Phase

During the design phase, an *equipment data sheet,* sometimes referred to as a *product specification*, should be generated to design and specify the LV motor control center selected for this particular application. It should contain the basic information pertaining to the motor control equipment.

Before the detailed specification can be generated, either the organization or its contractor will need to generate and document the following information:

- power system study (fault calculations, coordination data);
- master one-line diagram;
- motor list/load list; and
- applicable codes and standards.

The detailed product specification could include, but not be limited to, the following.

- general description and functional loop/block diagrams;
- ratings (voltage, current, fault bracing);
- table of components;
- general layout, controls, and interface information;
- MCC one-line diagram;
- reference to applicable codes and standards (NEMA, IEEE, NFPA, NEC, UL, etc.); and
- particular design details, such as:
  —enclosure type;
  —class of wiring; and
  —component types.
- test requirements; and
- painting requirements.

8B.3. Fabrication and Inspection

The vendor should be required to submit drawings for approval. Any alterations required by the organization to the drawings submitted by the vendor should be marked and returned to the vendor. The vendor should not be allowed to begin fabrication until these marked-up drawings are returned and agreed upon, unless otherwise authorized.

The schedule for inspection of the finished product should be coordinated and documented. The manufacturer's test results should be on hand at the time of the inspection. A report should be issued after the inspection showing the results of the inspection, indicating approval or rejection and detailing any required re-test or re-inspection. In any case of rejection, the cause may be waived if not detrimental to integrity and safety of the relief system. If a waiver is allowed, it should be justified and documented. The complete set of test and inspection reports should be retained. Examples of documents for this stage are:

- manufacturer's certified drawings;
- manufacturer's test reports;
- inspection report; and
- shipping/packing/storage documents.

8B.4. Installation

This phase includes the following:

- setting the equipment in place and assuring it is level;
- connecting and testing incoming line and load circuits; and
- setting all protective devices in accordance with the design documents.

Examples of documents for this stage are:

- installation specifications;
- installation drawings;
- manufacturer's certified drawings and instructions; and
- inspection and test reports.

8B.5. Start-up and Initial Operation

See Appendix 8A, Section 5.

8B.6. Continued Operation and Maintenance

See Appendix 8A, Section 6.

8B.7. Example

Figure 8-2, pages 140–144, contains selected parts of a typical requirements specification. This specification is provided for illustration purposes only and should not be used as an actual source of information. Only selected pages from the entire specification are presented.

JOHN BROWN

STANDARD SPECIFICATION

SS 501.20/0

LOW VOLTAGE MOTOR CONTROL CENTERS

TABLE OF CONTENTS

(APPENDICES B THRU E FOR INTERNAL USE ONLY)

ISSUED BY: ______________________
Electrical Engineering
Department Manager

APPROVED BY: ______________________
Vice President - Engineering

AUTHORIZED BY: ______________________
Vice President - Operations

FIGURE 8-2. Example of Electrical Equipment Specification (Page 1 of 5)

JOHN BROWN

STANDARD SPECIFICATION

SS 501.20/0 C
1 of 9

LOW VOLTAGE MOTOR CONTROL CENTERS

1.0 GENERAL

1.1 Scope

This specification together with the requisition covers the requirements for Low Voltage Motor Control Centers.

1.2 Standards and Codes

All wiring and equipment shall conform to the requirements of the editions of the following codes and standards as in force on January 14, 1993:

- American National Standards Institute (ANSI)
- Institute of Electrical and Electronic Engineers (IEEE)
- National Electrical Code (NEC)
- National Electrical Manufacturers Association (NEMA)
- Underwriters Laboratories (UL), where applicable

1.3 Manufacturer's Standard Equipment

It is intended that equipment covered by this specification be manufacturer's standard including normally available options. Ratings and features specified are to be interpreted minimum acceptable ratings. Higher ratings and additional features are acceptable to meet manufacturer's standards.

2.0 CONSTRUCTION DETAILS

2.1 Structure

2.1.1 The type enclosure shall be as specified on the requisition sheets.

2.1.2 The control center shall consist of vertical sections approximately ninety (90) inches high by twenty (20) inches wide by twenty (20) inches deep.

2.1.3 Vertical sections shall contain no more than the equivalent of six NEMA size 1 combination motor starters for front-only construction or six such units on each side for back-to-back construction.

2.1.4 Continuous channel iron floor sills and lifting angle shall be provided.

2.1.5 Sections shall have doors with concealed hinges for each unit in front and covers or doors in the rear.

2.1.6 Each section shall be complete with horizontal and vertical wiring troughs. Vertical wiring troughs shall have separate hinged doors.

2.1.7 Where front only configuration is specified, the control center is intended for installation against a wall, and no rear access shall be required after the equipment is set in place.

FIGURE 8-2. Example of Electrical Equipment Specification (Page 2 of 5)

JOHN BROWN

STANDARD SPECIFICATION

SS 501.20/0 C
9 of 9

LOW VOLTAGE MOTOR CONTROL CENTERS

7.0 VENDOR DOCUMENTATION

7.1 Bidders' proposals shall include descriptive brochures and drawings or sketches showing proposed motor control center layout and overall dimensions.

7.2 After purchase order the following documents are required for approval and for record in the number of copies specified in the requisition.

	FA	FR
Descriptive brochures and catalog cuts	(X)	(X)
Outline dimensions and shipping weights	(X)	(X)
Foundation dimensions and anchor bolt locations	(X)	(X)
Plan and elevation drawings with all dimensions	(X)	(X)
Schematic and connection drawings	(X)	(X)
Complete list of components	(X)	(X)
Arrangement drawings	(X)	(X)
Reproducible curves of all protective devices	( )	(X)
Installation, start-up, and operating instructions	( )	(X)
Recommended spare parts list (priced)	( )	(X)

FA = For Approval
FR = For Record

FIGURE 8-2. Example of Electrical Equipment Specification (Page 3 of 5)

JOHN BROWN

STANDARD SPECIFICATION

SS 501.20/0 C
APPENDIX A
1 of 7

LOW VOLTAGE MOTOR CONTROL CENTERS

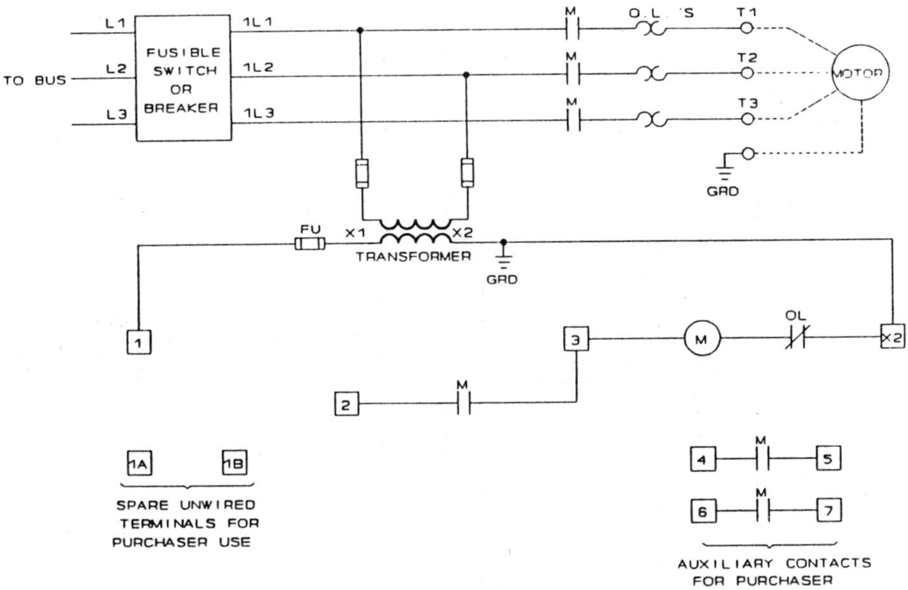

SCHEMATIC DIAGRAM
STARTER UNIT SYMBOL 1 (FVNR)

FIGURE 8-2. Example of Electrical Equipment Specification (Page 4 of 5)

JOHN BROWN

STANDARD SPECIFICATION

SS 501.20/0 C
APPENDIX A
7 of 7

LOW VOLTAGE MOTOR CONTROL CENTERS

TERMINAL COMPARTMENT

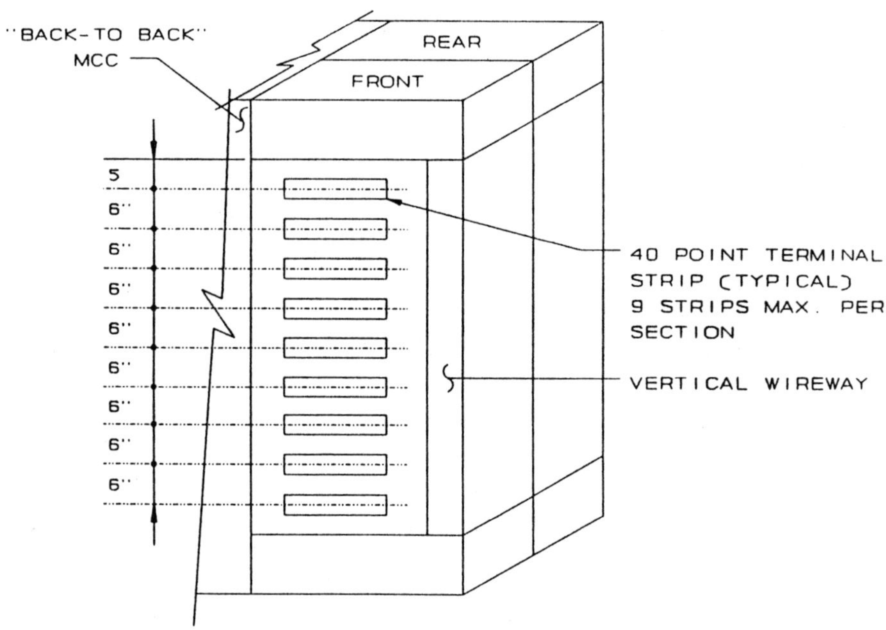

NOTES:

1. TERMINAL BLOCKS SHALL OCCUPY FRONT RIGHT VERTICAL SECTION.
2. MCC SHALL HAVE THE CAPABILITY FOR FUTURE EXTENSION TO THE RIGHT OF THE TERMINAL BLOCK SECTION.
3. MASTER TERMINAL BLOCKS SHALL BE PROVIDED ONLY FOR THOSE MCC's AS SPECIFIED IN THE REQUISITION.

MODIFICATION SYMBOL G

FIGURE 8-2. Example of Electrical Equipment Specification (Page 5 of 5)

Appendix 8C. Example of Documentation Requirements for Instrumentation Equipment.

8C.1. Introduction

The purpose of this appendix is to provide examples of the types, levels, and areas of coverage of the documentation needed to support instrumentation equipment through typical life cycle phases in chemical process plant applications. Of the numerous kinds of equipment categorized as instrumentation (see Section 8.3.2), an explosion proof modulating globe control valve, to be used outdoors in a hazardous environment is the example utilized here. The documentation requirements shown are adaptable, to the wide variety of instrumentation equipments found in the CPI.

8C.2. Design

During the design phase, an *equipment data sheet,* sometimes referred to as a *product specification*, should be generated to design and specify the modulating globe valve selected for this particular application. See Figure 8-3. It should contain all of the basic information pertaining to the valve assembly. This information could include, but not be limited to, the following:

- general description, characteristics, dimensions, and operating functions, (may be text and drawings or both);
- physical properties of the fluid;
- design and operating temperature, flow, pressure, and other pertinent conditions;
- required materials of construction;
- hazardous area classification indoor/outdoor area, i.e., NEMA 4;
- signal types (4–20 mA, 3–15 psig);
- trim characteristics (equal percentage, linear);
- failure position;
- shutoff pressure;
- pressure drop;
- allowable seat leakage rating;
- maximum allowable valve noise;
- required valve size (C_V);
- connection and interface information; and
- references to codes and standards.

8C.3. Fabrication and Inspection

The vendor should be required to submit drawings for approval. Any alterations required by the organization to the drawings submitted by the vendor should be marked and returned to the vendor. The vendor should not be allowed to begin fabrication until these marked-up drawings are returned and agreed upon, unless otherwise authorized.

A procedure for inspection should be established and documented. This procedure may include a shop inspection. However, all instruments should also have a complete inspection once they have arrived at the facility, including a bench calibration.

Examples of the documents for this stage are:

- instrument specification;
- purchase order;
- vendor's certified drawings;
- vendor's connection diagrams;
- vendor's calibration report;
- vendor's hydrostatic test report, if applicable; and
- shipping, packing, and storage documents.

8C.4. Installation

An installation procedure should be generated, consistent with the vendor's installation specifications, to define where and how installation in the facility is to be performed and specifying the methods to be used to verify that the installation is correct (including the correct location and orientation within the process and with the appropriate physical orientation). This procedure should also provide instructions to ensure that the installation allows accessibility for operations and maintenance and that the connections are made in accordance with requirements.

The documentation effort is facilitated by giving each instrument a unique tag number and establishing an instrument list. The tag number allows tracking of the instrument through the life cycle of the facility. The numbering system and other symbology used as the standard in the industry for describing an instrumentation system is defined by the Instrument Society of America (ISA) in their Standards and Practices.

The instrument index lists all of the devices by their unique instrument tag number. In addition, the following information about the instrument may also be listed: purchase order number, P&ID number, instrument specification number, installation detail number, loop diagram number, as well as a brief service description of the instrument. This list records and tracks the instruments at the facility.

Documentation for this stage may include:

- installation procedure and details;
- instrument specifications that include functional characteristics (e.g., design and operating pressures, flows, temperatures, set points, size, and materials);

- instrument vendor documentation (e.g. installation, operations, and maintenance manuals, electrical and connection drawings); and
- calibration and testing records; and
- loop diagrams.

8C.5. Start-up and Initial Operation

See Appendix 8A, Section 5.

8C.6. Continued Operation and Maintenance

See Appendix 8A, Section 6.

8C.7. Example

The example that follows, Figure 8-3, contains selected parts of a typical requirements specification. This specification is provided for illustration purposes only and should not be used as an actual source of information.

JOHNBROWN MATERIAL REQUISITION	REQUISITION NUMBER			PAGE 8 OF 59		
				DATE:		BY:
	P.O. NUMBER:			VENDOR:		
	REV. NO.	DATE	BY	REV. NO.	DATE	BY
CUSTOMER:						
LOCATION:						
MATERIAL:						

REV				
	1. TAG NO.	PV-123-05	FV-123-06	FV-123-07
	2. SERVICE	NITROGEN	COLD SYLTHERM	HOT SYLTHERM
	3. LOCATION OR PIPE SPEC	123 / A1	123 / A1	123 / A1
	4. LINE SIZE ¦ SCHED. NO.	1 ¦	6 ¦ STD	6 ¦ STD
	5. TYPE OF BODY	GLOBE-SINGLE PORT	GLOBE-SINGLE PORT	GLOBE-SINGLE PORT
	6. BODY SIZE ¦ PORT SIZE	1 IN ¦ 3/8	4 IN ¦ 2.50	4 IN ¦ 2.5
	7. GUIDING ¦ NO. OF PORTS	TOP ¦ ONE	TOP ¦ ONE	TOP ¦ ONE
	8. END CONN. & RATING	150# RF	300# RF	300# RF
	9. BODY MATERIAL	CARBON STEEL	CARBON STEEL	CARBON STEEL
	10. PACKING MATERIAL	TEFLON	TEFLON	TEFLON
	11. LUBRICATOR ¦ISOLATING VLV.	N/A ¦ N/A	N/A ¦ N/A	N/A ¦ N/A
	12. BONNET TYPE	PLAIN	PLAIN	PLAIN
	13. TRIM FORM	EQUAL PERC.	EQUAL PERC.	EQUAL PERC.
	14. TRIM MTL. SEAT ¦ PLUG	316 S.S. ¦ 316 S.S.	316 S.S. ¦ 316 S.S.	316 S.S. ¦ 316 S.S.
	15. TRIM MTL. SHAFT	316 S.S.	316 S.S.	316 S.S.
	16. REQUIRED SEAT TIGHTNESS	CLASS IV	CLASS IV	CLASS IV
	17. MAX.ALLOW. SOUND LEVEL dBA	90 dBA	90 dBA	90 dBA
	18. ACTUATOR MODEL NO. & SIZ			
	19. TYPE OF ACTUATOR	PISTON	PISTON	PISTON
	20. CLOSE AT: ¦ OPEN AT:	4 mA ¦ 20 mA	4 mA ¦ 20 mA	4 mA ¦ 20 mA
	21. FLOW DIRECTION TO:	OPEN	CLOSE	CLOSE
	22. FAILURE POSITION	CLOSED	CLOSED	CLOSED
	23. HANDWHEEL & LOCATION	N/A ¦ N/A	N/A ¦ N/A	N/A ¦ N/A
	24. POSITIONER MFR & MODEL NO.			
	25. FILTER ¦ GAGES ¦ BYPASS	NO ¦ YES ¦ NO	NO ¦ YES ¦ NO	NO ¦ YES ¦ NO
	26. INPUT SIGNAL	4-20 mA	4-20 mA	4-20 mA
	27. OUTPUT SIGNAL			
	28. AIR SUPPLY PRESSURE	60 PSIG (MIN) - 100 PSIG	60 PSIG (MIN) - 100 PSIG	60 PSIG (MIN) - 100 PSIG
	29. TRANSDUCER MFR & MODEL NO.	N/A	N/A	N/A
	30. INPUT SIGNAL			
	31. OUTPUT SIGNAL			
	32.			
	33.			
	34.			
	35. FLOW UNITS	SCFM	USGPM	USGPM
	36. FLUID	NITROGEN	COLD SYLTHERM	HOT SYLTHERM
	37. QUANT. MAX ¦ Cv MAX	80.0000 ¦ 1.9732	725.0000 ¦ 98.3823	675.0000 ¦ 92.5111
	38. QUANT. OPR. ¦ Cv NORM	80.0000 ¦ 1.9732	363.0000 ¦ 49.2590	337.0000 ¦ 46.1870
	39. VALVE Cv ¦ VALVE Fl	3.5000 ¦ 0.9000	118.0000 ¦ 0.8000	118.0000 ¦ 0.8000
	40. NORM INLET PRESS ¦ DELTA P	57.7000 ¦ 47.7000	82.0000 ¦ 50.9000	66.3000 ¦ 40.6000
	41. MAX. INLET PRESSURE	100.0000	90.0000	90.0000
	42. MAX SHUT-OFF DELTA P	100.0000	90.0000	90.0000
	43. TEMP. MAX ¦ TEMP. OPR.	150.0000 ¦ 80.0000	320.0000 ¦ -5.0000	320.0000 ¦ 300.0000
	44. OPR. SP. GR. OR MOL. WT.	0.9660 ¦ 0.0000	0.8800 ¦ 0.0000	0.7160 ¦ 0.0000
	45. OPR. VISC. ¦ % FLASH	0.0000 ¦ 0.0000	3.1200 ¦ 0.0000	0.3710 ¦ 0.0000
	46. % SUPERHEAT ¦ % SOLIDS			
	47. VAPOR PRESS.¦ CRIT. PRESS.		8.0000 ¦ 192.0000	7.6000 ¦ 191.0000
	48. PREDICTED SOUND LEVEL dBA	66	80	78
	49. VALVE MANUFACTURER			
	50. MODEL NO.			
	51. REFERENCE NOTES		1,2,3	1,2,3

FIGURE 8-3. Example Specification for Modulating Globe Control Valve

Appendix 8D. Example of Documentation Requirements for Safety Systems Equipment

8D.1. Introduction

The purpose of this appendix is to provide examples of the types, levels, and areas of coverage of the documentation needed to support safety systems equipment through typical life cycle phases in chemical process plant applications. Of the many types of equipment categorized as safety systems (see Section 8.3.2), the example utilized here is a conventional relief valve. The documentation requirements shown are adaptable to most varieties of other safety systems found in the CPI.

8D.2. Design

During the design phase, an *equipment data sheet,* sometimes referred to as a *product specification*, should be generated to design and specify the relief valve selected for this specification application. It should contain the basic information pertaining to the relief valve. See Chapter 5, *Process Knowledge* and Table 5-4 for information on establishing relief system design basis. This data should be developed and used as the criteria upon which the specification is based. The information normally included in the specification could include, but not be limited to, the following:

- general physical and functional descriptive documents (text and drawings);
- available pressure range, with maximum and minimum settings identified;
- temperature operating range (maximum and minimum);
- physical properties and phases of relieving material;
- size;
- design codes and standards applicable (ANSI, API, etc.);
- connections;
- materials of construction, including:
  —body material;
  —trim material;
  —soft seat pallet insert (or metal/metal data);
  —pallet weights; and
  —compression spring.
- flow capability at design basis (equivalent to relieving rate); and
- testing requirements.

8D.3. Fabrication and Inspection

The vendor should be required to submit drawings for approval. Any alterations required by the organization to the vendor drawings should be marked and returned to the vendor. The vendor should not be allowed to begin fabrication until these marked-up drawings are received and agreed upon, unless otherwise authorized.

The schedule for inspection of the finished product should be coordinated and the agreed upon schedule documented. The manufacturer's test results should be on hand at the time of the inspection. A report is issued after the inspection stating the results of the inspection, indicating approval or rejection. In any case of rejection, the cause may be waived if not detrimental to integrity and safety of the relief system. If a waiver is allowed, it should be justified and documented. The complete set of test and inspection reports should be retained.

Examples of documents for this stage are:

- manufacturer's dimensional drawings;
- code and standard certifications;
- manufacturer's test reports;
- inspection report; and
- shipping, packing, and storage documents.

8D.4. Installation

This phase includes the following:

- installation of item, including any required bracing or supports;
- checking to ensure that the item is properly tagged;
- inspecting to make sure item is appropriately sealed and wired; and
- checking discharge and supply piping and other connections to ensure that the relief valve was installed in accordance with drawings.

Examples of documents for this stage are:

- installation specifications;
- manufacturer's certified drawings and instructions;
- results from any on-site test; and
- piping and connection drawings.

8D.5. Start-up and Initial Operation

See Appendix 8A, Section 5.

8D.6. Continued Operation and Maintenance

See Appendix 8A, Section 6

8D.7. Example

The example that follows, Figure 8-4, contains selected parts of a typical requirements specification. This specification is provided for illustration purposes only and should not be used as an actual source of information.

JOHN BROWN MATERIAL REQUISITION	REQUISITION NUMBER			PAGE 5 OF 7		
				DATE:		BY:
	P.O. NUMBER:			VENDOR:		
	REV. NO.	DATE	BY	REV. NO.	DATE	BY
CUSTOMER:						
LOCATION:						
MATERIAL: RELIEF VALVES						

REV				
	1. TAG NUMBER			
	2. SERVICE			
	3. LINE NO. OR VESSEL NO.			
	4. FULL NOZZLE/SEMI NOZZLE	SEMI NOZZLE	SEMI NOZZLE	FULL
	5. SAFETY OR RELIEF	RELIEF	RELIEF	SAFETY
	6. CONV.,BELLOWS,PILOT OPR.	CONVENTIONAL	CONVENTIONAL	CONVENTIONAL
	7. BONNET TYPE	CLOSED	CLOSED	CLOSED
	8. SIZE: INLET ¦ OUTLET	0.75" X 1.0"	0.75" X 1.0"	1.5 in ¦ 3 in
	9. FLANGE RATING OR THREADED	THREADED - NPT	THREADED - NPT	150# INLET ¦ 150# OUTLET
	10. TYPE OF FLANGE FACING	N/A	N/A	RAISED FACE
	11. BODY & BONNET MATERIAL	316 S.S ¦ CARBON STEEL	316 S.S ¦ CARBON STEEL	CARBON STEL ¦ CARBON STEEL
	12. SEAT & DISC MATERIAL	316 S.S ¦ 316 S.S.	316 S.S ¦ 316 S.S.	316 S.S ¦ 316 S.S.
	13. RESILIENT SEAT SEAL MAT'L	N/A	N/A	N/A
	14. GUIDE & RINGS MATERIAL	316 S.S ¦ 316 S.S.	316 S.S ¦ 316 S.S.	316 S.S ¦ 316 S.S.
	15. SPRING MATERIAL	HIGH TEMP. ALLOY RUST PROOF	HIGH TEMP. ALLOY RUST PROOF	CARBON STEEL RUST PROOF
	16. BELLOWS MATERIAL	N/A	N/A	N/A
	17.			
	18. CAP : SCREWED OR BOLTED	SCREWED	SCREWED	SCREWED
	19. LEVER : PLAIN OR PACKED	NOT REQUIRED	NOT REQUIRED	NOT REQUIRED
	20. TEST GAG	NO	NO	NO
	21. OTHER OPTION	------	------	------
	24. CODE	API RP 520	API RP 520	API RP-520
	26. SIZING BASIS	THERMAL RELIEF	THERMAL RELIEF	VALVE FAILURE
	28. FLUID OR STATE	LIQUID	LIQUID	NITROGEN
	29. REQUIRED CAPACITY	------	------	200 SCFM
	30. MOL. WT. ¦ OPR. SP. GR.	¦ 1.0	¦ 1.0	28 ¦ 0.966
	31. OPR. PRESS. ¦ SET PRESS.	80 ¦ 100 PSIG	80 ¦ 100 PSIG	5 ¦ 15 PSIG
	32. OPR. TEMP. ¦ RELIEF TEMP.	40 ¦ 40 DEG.F	40 ¦ 40 DEG.F	70 ¦ 70 DEG.F
	33. BACK PRESSURE: CONSTANT	------	------	------
	34. VARIABLE	------	------	------
	35. TOTAL	------	------	------
	36. % ALLOWABLE OVERPRESSURE	10%	10%	10%
	37. OVERPRESSURE FACTOR	------	------	
	38. COMPRESSIBILITY FACTOR	------	------	1
	39. LATENT HEAT OF VAPORIZATION	------	------	------
	40. RATIO OF SPECIFIC HEATS	------	------	1.4
	41. OPERATING VISCOSITY	1.0 cP	1.0 cP	
	42. BAROMETRIC PRESSURE	14.7 PSIA	14.7 PSIA	14.7 PSIA
	43.			
	44.			
	45. CALCULATED AREA SQ. IN.	------	------	0.3471 in2
	46. SELECTED AREA SQ. IN.	0.06 in2	0.06 in2	0.785 in2
	47. ORIFICE DESIGNATION	------	------	H
	48. MANUFACTURER			
	49. MODEL NUMBER			
	50. REFERENCE NOTES			

FIGURE 8-4. Example Specification for Conventional Relief Valve

9

Human Factors

9.1. Overview

9.1.1. Introduction

Human factors (HF) is one of several terms used to describe the discipline that considers the involvement of the human being in design and operation of equipment and systems. Other terms can be used for some aspects of human factors; for instance, human reliability is primarily associated with the study of human errors, and ergonomics is associated with the design of systems involving human beings, often also referred to as the man/machine interface (MMI) or user interface.

Human error can occur in all situations where the human being is involved. The effects are found throughout the life cycle of a process from early design through construction, start-up, operation, maintenance, and even in the decommissioning of the facility. At every stage in the life cycle, consideration should be given to the prevention of human error, and to the effects and control of situations caused by human error. Documentation of the way HF has been considered and applied during the life cycle of a process is important to ensure that the potential for incidents from this cause has been minimized.

9.1.2. Goals and Benefits

The principal goals to be achieved from the study of human factors as applied to process safety management (PSM) are:

- an improved understanding of HF in all stages and all aspects of the process;
- the early identification and correction of potential human error situations in the design phase of an installation;
- the identification and correction of possible human errors in man/machine interfaces;

- the identification and correction of potential human error in operating and maintenance procedures and associated training;
- the identification of those situations where a more detailed human factors analysis (HFA) may be required; and
- an improved workplace, leading to safer and more efficient performance.

The benefits to be obtained from a human factors program include:

- improved plant safety performance;
- reduction of the human error rate;
- more thorough incident investigations;
- improved productivity and product quality;
- the documentation of the way HF has been applied in PHA and PSM for future reference; and
- the ability to achieve and demonstrate regulatory compliance in PSM.

9.2. Description of Human Factors

9.2.1. Objectives

As was discussed in Chapter 6, *Process Hazard Analysis,* humans are known to be fallible. Their conduct is affected by a wide variety of performance influencing factors (PIFs); for example, stress, training level, experience, and environmental factors such as noise. Error-likely situations (ELS) are work situations where the PIFs are not compatible with the capabilities, limitations, and needs of the worker. These situations lead to an enhanced likelihood of errors. The basic objectives of a HF program include the identification and correction of such error likely situations.

Error-likely situations (ELS) can involve both hardware (equipment) and software (such as procedures or training). Some types of ELS should be obvious and have typically received due consideration (e.g., missing, incorrect or ambiguous numbering or lettering of equipment or missing notices or signs on dangerous equipment or locations). Less obvious ELS, such as procedures that are prone to misinterpretation or that prompt unsafe shortcuts, are also causes for concern and should be more thoroughly studied. Effort should be made to identify ELS at an early stage in the project life cycle.

9.2.2. Human Factors in the Life Cycle of a Process Facility

There are numerous phases or activities in the life cycle of a process facility where consideration of HF is required as part of a quality PSM program. Situations where consideration of HF is appropriate are briefly discussed as follows.

Design Phase. During the design, or modification of a process, individuals apply their skills, training, and experience; however, mistakes can be made, and may go undetected. A review of HF considerations may suggest controls to prevent or detect such errors. For example, the advisability of an independent review of key equipment specifications by an engineer other than the design engineer might be identified by an HF study. Or, an HF review of a calculation sheet (e.g., for relief valve sizing) might reveal the potential for mistakes, either due to ambiguity of instructions, poor arrangement, or lack of cross checks.

The design of the man/machine interface should not allow, nor contribute to, the opportunity for human error in operation or maintenance activities. To achieve this, it is necessary to be attentive to HF requirements when addressing the large number of interface issues that frequently include potential ELS. Some examples are:

- the arrangement, location, labeling, and numbering of valves, pumps, vessels, other equipment, and their controls;
- access to all manually operated valves, sampling systems and related manually operated equipment; and
- control room layouts, including panels, control instruments, visual display units (VDUs), printers, keyboards, communication systems, etc.

Inspection (during manufacture, construction, and reassembly after maintenance). Inspections are performed to ensure that what is fabricated, constructed and/or installed is in accordance with drawings and/or specifications and to detect errors that may have been made. Inspections can be physical inspections of equipment, or inspection of the documentation. In either case, an HF review may identify ELS upon which the inspection program might be focused.

Operations. During the performance of PHA, human error is commonly identified as one of the causes of potential incidents, and steps may be taken to eliminate these causes or mitigate their consequences. Some PHA methods that typically lend themselves to consideration of HF are:

- What If (e.g., what if the operator were to do something different?);
- Checklist (may require a HF checklist);
- HAZOP;
- FMEA; and
- FTA or ETA.

In practice, a large majority of ELS are related to either the lack of correct operating procedures, procedures that are insufficiently detailed or too vague, the use of outdated or incorrect procedures, or improper training. Thus, HF considerations should also be addressed in procedure preparation and training.

Maintenance. Many processes have been designed without adequate consideration of maintenance activities. This could arise, for example, due to a process change that leads to an unexpected degree or type of maintenance (e.g., frequent blockage of equipment or excessive corrosion). Unless the maintenance procedures and tasks are examined in detail for HF considerations, especially during design, it is possible that they may be error prone.

Incident Investigation. Many major incidents have been attributed to human error, when in fact the root cause was an undiscovered ELS. For example, an incident may have resulted when personnel followed the operating procedure, but the procedure was in error. Since many incidents involve the actions of operating or maintenance personnel, HF should be considered and the results documented in the investigation report.

9.2.3. Performance of Human Factors Analysis (HFA)

A variety of methods is available for analysis or review of HF. The reader is referred to the CCPS book *Guidelines for Preventing Human Error in Chemical Process Safety* for an extensive description of these. Experienced human factors specialists are best qualified to select the appropriate method to use and to know how to apply it effectively. However, other individuals may have obtained some experience in human factors analysis, either directly through participation in HFAs led by an experienced HF analyst, or perhaps through formalized training. These individuals may be able to apply some of the less detailed techniques.

The scope of the HFA should be defined in terms of the study objective, scope, technical approach (HFA methodology), measurement criteria, applicable constraints, and resource requirements. These should be agreed upon before the HFA starts and included in the documentation.

Although the potential for human error exists in all activities involving human participation, it is not always necessary to perform an in-depth HFA. For example, the results of a PHA, may indicate whether the understanding of the causes and effects of human error is important enough to the particular process to develop this understanding in more detail. In addition, other PSM activities such as incident investigation or procedure preparation may indicate the need for a HFA to be developed in more detail.

Normally the HF specialist will not complete the HFA alone. The specialist usually works in conjunction with plant management, operations personnel, and if appropriate, the designer. These are the people who can provide the background information that is needed by the specialist to answer questions raised in the analyses.

Appendix 9A to this chapter contains details of documentation aspects for a selection of these HFA methodologies.

9.3. Human Factors Documentation

9.3.1. Human Factors Program Documentation

Organizations implementing HF should include intheir PSM program documentation a discussion of the goals of HF as wellas any general procedures that outline how HF will be implemented.

There are a wide variety of factors that should be considered when establishing detailed procedures intended to achieve the established goals of HF. The more significant factors will be discussed in the following paragraphs. These factors are, in many ways, similar to the same considerations discussed in Chapter 6, *Process Hazard Analysis*. The reader may wish to refer to this earlier chapter for a more detailed discussion.

When to Perform Human HFA. HFA may be conducted on a periodic basis in support of continuing programs such as PHA or procedure preparation, or it may be performed on an as-needed basis; e.g., to support incident investigations. The PSM program documentation may describe the role that HF plays in supporting the implementation of other PSM elements. Additionally, it may provide more specific guidance for determining when an HFA should be performed.

Technique Selection. As discussed in Appendix 9A, the various HF techniques differ in approach and work product. Some organizations may wish to include, in their program documentation, guidance on selection of techniques.

Training and Qualification. Many HF techniques require detailed, specific training and practice in their application. An organization may wish to establish minimum levels of demonstrated competency for those responsible for application of such techniques. Even a highly trained and skilled HF analyst requires the support of facility personnel who provide detailed input as to operating practices, working conditions, equipment details, etc. An organization may wish to define minimum training requirements for such HFA participants.

HFA Results Documentation. Documentation of the results of HFA will be discussed in Section 9.3.2. In the documentation of the HF program, minimum standards for the form and content of HF results documentation should be established.

Recommendations and Their Resolution. As with PHA, the performance of an HFA is likely to result in one or more recommendations for remedial action. Documented procedures for tracking, and ultimately resolving, each recommendation should be established.

Records Management. Pertinent records management requirements should be included in the program documentation. See Section 9.4 and Chapter 4, *Records Management*.

HF Program Responsibility. Functional responsibility and accountability for HF should be clearly defined in the program documentation.

9.3.2. Documentation of HFA Results

A brief description of the types of results and documentation requirements specific to the more commonly used HF techniques is provided in Appendix 9A. The following summary provides a more general perspective on broader issues related to documentation of the HF effort, independent of which technique was used for HFA.

The results of an HFA are typically documented and communicated to those affected via a formal summary report. The content of the report will vary somewhat based upon the scope of the review and the technique used. A typical format for such a report could be as follows:

- an introduction, which explains briefly the objective of the report;
- a summary of the report, its conclusions and major recommendations, preferably written in terms to enable those without technical knowledge to understand the conclusions;
- the scope of the study;
- identification of the HFA specialist, the analysis team and their qualifications;
- a short description of the selected method of analysis;
- a description of the process, operation, or task being studied;
- the details of the analysis, including reference to the location where supporting documentation is filed;
- the assumptions made in performing the analysis;
- the data used (if extensive they could be placed in an appendix);
- appendixes for calculations, diagrammatic results and/or tabular data/information;
- detailed conclusions;
- detailed recommendations;
- the assignment of responsibilities and specification of target completion dates for resolution of recommendations; and
- the follow up of the recommendations made by the HFA, including recording of the completion of the actions (or their resolution). See Section 9.3.3.

Many of the basic elements of an HFA report will be similar to the corresponding elements of a PHA report and the reader may wish to refer to Chapter 6, *Process Hazard Analysis*, for additional perspective.

An HFA will often be conducted as part of a PHA, or in support of another PSM element, such as *Procedures*. The documentation of these other elements should show how HF was involved, including the documentation of pertinent HF results.

9.3.3. Resolution of HFA Recommendations

It is common, as a result of HFA, to formulate recommendations intended to minimize the potential for human error. It is important to promptly and fully resolve each recommendation, just as was the case for PHA recommendations. Reference should be made to Section 6.3.3 for a discussion of documentation of recommendations and their resolution.

9.4. Records Management

In addition to the description of the HF program given in Section 9.3.1, the organization's PSM program documentation should define pertinent HF records management practices. It should be noted that the HFA records content is very similar to that of PHA records. The output from both include a mix of text, tabular, and graphic results. The supporting documentation for both types of analysis are similar, and sometimes overlapping. In fact, as discussed previously, HFA is often performed as an integral part of PHA. It should not be surprising, therefore, that the records management task for HFA is quite similar to that for PHA. Since there are no records management issues unique to HFA, reference is made to Chapter 4, *Records Management*, for general guidance.

9.5. Auditing

An audit of HF documentation should first verify that a documented program for implementing HF exists. This program would include elements outlined in Section 9.3.1 and referenced in Section 9.4 that are pertinent to the needs of the organization or facility.

The audit should next verify that the results of implementing HF have been documented in the manner prescribed by the program. Particular emphasis should be placed upon the documentation of the resolution of recommendations.

Any deficiencies identified during the audit must be documented and resolved in a timely manner. See Chapter 14, *Auditing*, and the CCPS book *Guidelines for Auditing Process Safety Management Systems,* for more information on auditing PSM programs.

9.6. Examples

9.6.1. Types of Human Error

Human error is frequently quoted as the sole cause, or one of the principal causes, of incidents and accidents. The following examples are representative of some of the classes of human error. They are listed in summary form to illustrate the ways in which human error can play an unwanted role.

Man/Machine Interface. The examples listed here illustrate failure to transfer the correct information from the machine or process to the operator. The operator is frequently faulted for these situations, but often the logistics of the equipment results in an ELS. Such ELSs include:

- valves not numbered or labeled, causing confusion;
- logical numbering system not followed, e.g., pumps installed in a row: 1 2 3 5 4;

- lack of indication instruments local to the valve being turned or no radio contact system;
- excessive information from computer alarm printouts which overloads the operator or hides the key alarm indicating the cause of the upset; and
- inconsistent or misleading instrument scales (e.g., combinations of pressure gauges in barg, atmospheres, or psi, and other scales in 0–100% instead of engineering units).

False Assumptions. The examples below illustrate the failure to transfer information correctly, for instance, assuming incorrect information to be true, or correct information to be false. They include:

- assuming that an unusual reading must be a faulty instrument;
- assuming that a panel light for valve position actually means the valve is fully open or shut, when it may actually mean that the system is energized to open or shut, without feedback to confirm that this was achieved; and
- assuming that job preparation has been effectively completed when signing off a permit to work.

Poor Operating Procedures. Procedures frequently cause problems when poorly written. Such problems include:

- do special instructions for special tasks override all normal procedures, or just some of them? This should be made clear; and
- two steps are included together in the same sentence when in fact they cannot be performed at the same time.

Chronic Violations. These are so-called because the written procedure is consistently disregarded or violated, for any of a variety of reasons, including:

- the procedure is extremely difficult to do in the way it is written. Corrective action should be to either change the instruction or improve the process to regain conformity;
- An attitude of *production comes first,* as opposed to the appropriate attitude of *safety comes first*. If safety is a significant obstacle to production, there is a need to perform some type of review to establish the facts and improve the situation;
- steps in the procedure omitted. Often an operator will omit what are considered to be unnecessary steps, for example, he will use a single valve instead of a full block and bleed facility. Lack of awareness of the reason for safety-oriented procedures or inadequate training are usually contributory causes; and
- not wearing all specified personal protective equipment. In some circumstances an operator considers the personal protective equipment too cumbersome or unnecessary.

Ineffective Organization of Work. Poor work organization contributes to human error, for example:

- permit-to-work systems that are not fully documented as to their status at the end of the day or the end of the shift, can lead to errors overnight or on the next shift. (refer Chapter 18, *Permit-to-Work Systems*); and
- checklists not used when preparing or performing tasks. It is difficult to remember *everything*, so these lists can be invaluable.

Failure to Explicitly Allocate Responsibility. Misunderstandings can lead to human error; for example, where there is dual or multiple responsibility, there must be clear documentation on who is responsible and to what extent, and interfaces should be fully defined so that key steps are not omitted.

Inadequacies in the Organization's PSM Program. If there is no leadership, human error is not likely to be adequately treated; for example, where a corporate safety practice has not been established. Procedures for the safe use of nitrogen, electric power tools, double block and bleed, block valves under relief devices, etc., are typical examples. Follow-up and audit are keys to ensuring that the PSM program is sufficiently comprehensive and is actually followed.

9.6.2. Examples of Incidents

Missing Instructions. A plant experienced two similar incidents four years apart. Each involved the neutralization of an acidic effluent with an alkaline slurry in a tank with a covered manhole. On each occasion, the agitator stopped. When an operator found the agitator stopped, he switched it on again, causing a violent reaction, which ejected material through the manhole. No injuries resulted.

After the first incident, an instruction was written and issued detailing the action to be taken when the agitator stopped. The operator in the second incident had not seen this instruction, or had forgotten it, and no copy could be found. This is an example of the need for written procedures and for their continual use in operations and training.

Unclear Instructions. A day foreman left instruction for the night shift to clean a reactor. He wrote: *Agitate with 150 liters nitric acid solution for 4 hours at 80°C.* However, the foreman did not provide explicit instructions to fill the reactor with water first, as was usually required. A violent reaction took place, followed by the rupture of the reactor. Written instructions must be complete and unambiguous if incidents are to be avoided.

9.7. References

This partial list of references applicable to HF applications is supplemented by those shown in Appendix 9A which are keyed to specific types of HFA methods.

AIChE–CCPS, *Guidelines for Auditing Process Safety Management Systems,* 1993.

AIChE–CCPS, *Guidelines for Chemical Process Quantitative Risk Analysis*, 1989.

AIChE–CCPS, *Guidelines for Hazard Evaluation Procedures, Second Edition with Worked Examples,* 1992.

AIChE–CCPS, *Guidelines for Preventing Human Error in Chemical Process Safety,* 1994.

EFCE, *A Report of the Human Factors Study Group to the Loss Prevention Working Party of the European Federation of Chemical Engineers*, 1990.

Institution of Chemical Engineers, *Loss Prevention Bulletin*, Rugby, U.K.

Kletz, T.A., *An Engineer's View of Human Error,* 2nd Edition, Institution of Chemical Engineers, Rugby, U.K., 1991

Kletz, T.A., *Learning from Accidents in Industry*, Butterworth's & Co. Ltd., 1988.

Kletz, T.A., *Lessons from Disaster: How Organizations Have No Memory and Accidents Recur,* Institution of Chemical Engineers, Rugby, U.K., 1993.

Kletz, T.A., *Plant Design for Safety,* Hemisphere Publishing Co., 1991

Kletz, T.A., *What Went Wrong? Case Histories of Process Plant Disasters,* 2nd Edition, Gulf Publishing Co., 1988.

Lees, F.P., *Loss Prevention in the Process Industries,* Vol. 2, Appendix 3—Case Histories, Butterworth's & Co., Ltd., 1980.

Lorenzo, D.K., *Improving Human Performance in the Chemical Industry: A Managers Guide to Reducing Human Errors*, Chemical Manufacturers Association, Inc., Washington, D.C., 1990.

Mill, R.C., *Human Factors in Process Operations*, I Chem E, Rugby, UK, 1992.

UKAEA-SRD, *Human Error in Risk Assessment*, SRD R510, February, 1990.

UKAEA-SRD, *Human Reliability Assessors Guide*, October, 1988.

U.S. Nuclear Regulatory Commission, *Accident Sequence Elevation Program*, NUREG/CR-4772, February, 1987.

U.S. Nuclear Regulatory Commission, *Procedure for Conducting a Human Reliability Analysis*, NUREG/CR-2254, Sandia Labs, May, 1993.

U.S. Nuclear Regulatory Commission, *Handbook of Human Reliability Analysis*, NUREG/CR-1278, Sandia Labs, August, 1983.

Appendix 9A. Typical Technical Documentation Of HFA

9A.1. Introduction

This appendix provides a brief description of several HFA methodologies. Also provided is a description of the typical output yielded by each technique, as well as the typical documentation requirements for the technique. This comprises the core of an HFA study report. The main part of Chapter 9 provides more background on how the report is to be written and what it should additionally contain.

Details of the various methods are to be found in a variety of books and papers. These are referenced at the end of each method. An extensive review, with details of the methods and examples of the documentation is to be found in the CCPS

Book *Guidelines for Preventing Human Error in Chemical Process Safety*. The reader should consult these references for further details.

The scope of the documentation of HFA was discussed in Section 9.3.2. The descriptions that follow refer only to the specialized documentation appropriate for some specific HF methods; they are largely graphical or statistical presentations.

9A.2. Action Oriented Analysis

9A.2.1. Hierarchical Task Analysis (HTA)

HTA was originally developed for analyzing potential errors in process control, but now has wider application. There are two basic parts to the record.

First, the overall goal of a task is analyzed using a top-down chart, similar to an event tree. This should clearly show the subtasks necessary before a task can be successfully completed. The boxes should give brief but concise descriptions of the subtasks.

Second, this diagram can then be analyzed in multicolumn format, with the columns typically recording:

- task steps;
- inputs (cues);
- outputs (actions);
- feedback;
- communications;
- time characteristics;
- task dependencies;
- secondary duties (or distractions); and
- comments.

Examples of an HTA diagram and its tabular analysis are illustrated in Figure 9-1 and Table 9-1, respectively.

In addition to manual generation, both stages can be constructed using software that includes typical flow sheet and spreadsheet programs. Specialist software includes THETA by Human Reliability Associates, Wigan U.K.).

References

AIChE–CCPS, *Guidelines for Preventing Human Error in Chemical Process Safety*, 1994.

Duncan, K.D., "Analytical Techniques in Training Design, in *The Human Operator in Process Control.* Taylor and Francis, 1974.

Sheppard, A., "Heirarchial Task Analysis and Training Decisions." *Programmed Learning and Educational Technology*, 22:162–176, 1985.

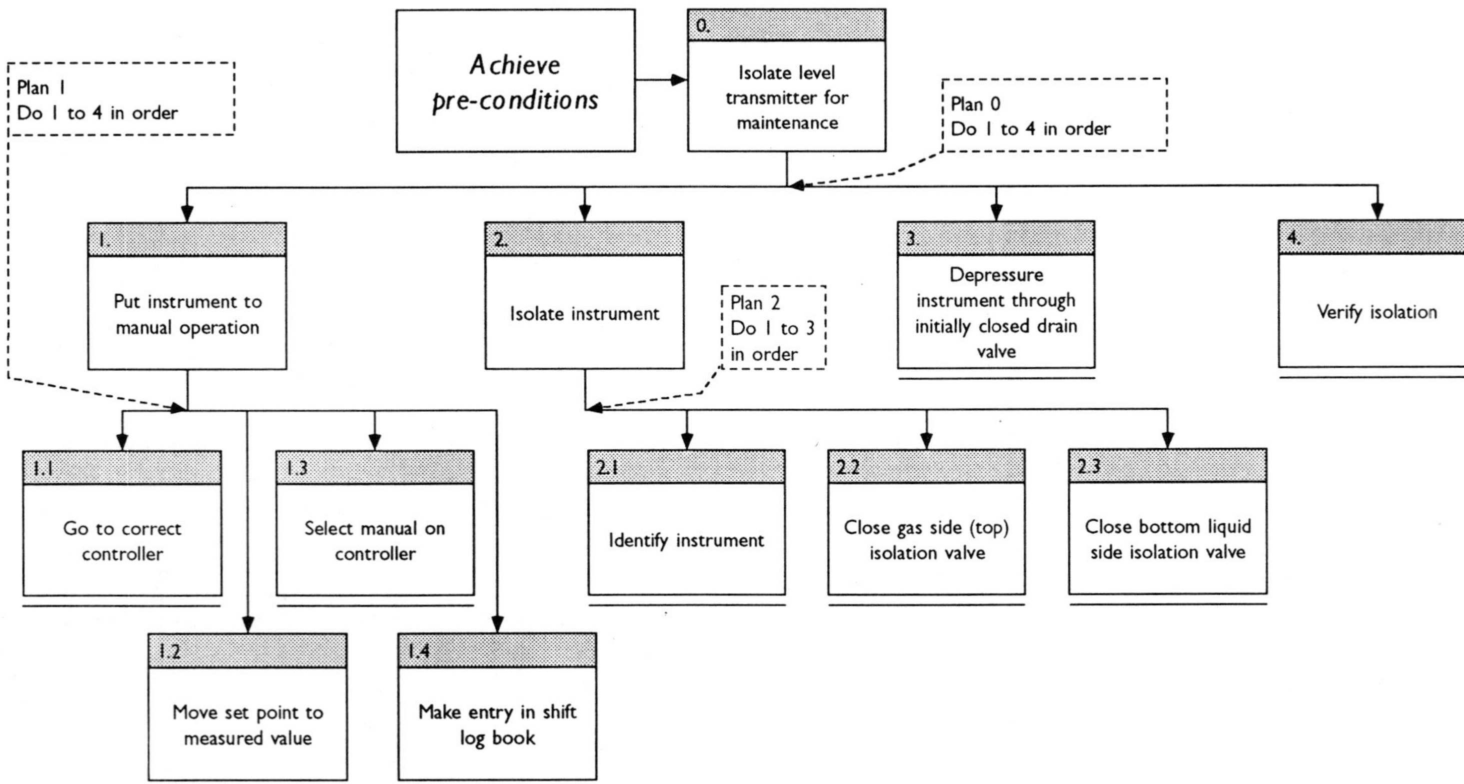

FIGURE 9-1. Example of HTA Diagram

TABLE 9-1
Example HTA Tabular Analysis

TASK STEPS	INPUTS (CUES)	OUTPUTS (ACTIONS)	FEEDBACK	COMMUNICATIONS	TIME CHARACTERISTICS	TASK DEPENDENCIES	SECONDARY DUTIES, DISTRACTIONS	COMMENTS
4 Optimize pressure value (PR5) in column within 1–1.5 atm.	Pressure recorder indicates PR5 > 1.5 atm (No alarm indications)	If PR5 > 1.8 atm increase cooling rate in condenser (4.2) and decrease heating rate in reboiler (4.3) If PR5 < 1.8 atm do 4.2 only (CRO monitors recorders and OSO adjusts valves on site)	Pressure recorder Temperature recorder Condensate level in reflex drum	Radio communications between control room operator (CRO) and outside operator (OSO)	Optimization should start **not later than** 2 min from the initiation of the abnormal event	Fluctuations of temperature will degrade quality of products	CRO busy with other tasks in the control room	CRO can fail to detect PR5 increase OSO can omit to adjust rates of cooling to the required levels Hazards: Danger of explosion due to accumulation of vapors inside column OSO should wear protective clothing

9A.2.2. Operator Action Event Trees (OAET)

OAETs resemble event tree diagrams and are constructed in a similar fashion. Each node in the tree represents a particular task or subtask, with the outcome typically being *success* or *failure.* If alternatives or omissions are to be considered, separate trees would need to be constructed to include these variants, so there can be a very large number of trees for any single facility. These should be carefully titled and indexed to facilitate ease of identification of a particular tree. A variety of specialist FTA and ETA software exists that is also suitable for OAET. See Chapter 6.

References

AIChE–CCPS, *Guidelines for Preventing Human Error in Chemical Process Safety*, 1994

Hall, R.E., Fragola, J., Wreathall, J., *Post Event Decision Errors: Operator Action Tree/Time Reliability Correlation,* NUREG/CR-3010, U.S. Nuclear Regulatory Commission, Washington, D.C., 1982.

Kirwan, B. & Ainsworth, L.K., *A Guide to Task Analysis,* Taylor & Francis, 1992.

9A.2.3. Decision/Action Flow Diagrams

These diagrams are very similar to the flow charts used for computer program development and are based on binary decision logic. They have only a single level of task description and become unwieldy if the task is complex. They can be easy to understand, with the decision boxes that record the question posed usually leading to *yes* or *no* answers, with recorded consequences, or perhaps to another question. An example of a typical decision/action flow diagram is shown in Figure 9-2. Typical software for flow diagrams can be used.

References

AIChE–CCPS, *Guidelines for Preventing Human Error in Chemical Process Safety*, 1994.

Duncan, K.D. & Gray, M.J., "Scoring Methods for Verification and Diagnostic Performance in Industrial Fault-Finding Problems." *J. Occupational Psychology,* 48:93–106, 1975.

9A.2.4. Operational Sequence Diagrams (OSD)

OSDs are essentially flow charts, drawn from top to bottom, with limited branching and looping. They are similar in approach to Decision/Action flow diagrams. Various symbols can be used, appropriate to their purpose, and these need to be identified in the diagrams. Extensive text is needed to support these diagrams. Where complex tasks are involved, the diagrams can become extensive and, perhaps, incomprehensible. In these situations, the diagrams should be subdivided and accompanied by explanatory text to enable the reader to follow them. Typically three types of OSD can be used:

- temporal—based on time relationships;
- partitioned—based more on sequential operations with any associated interface; and

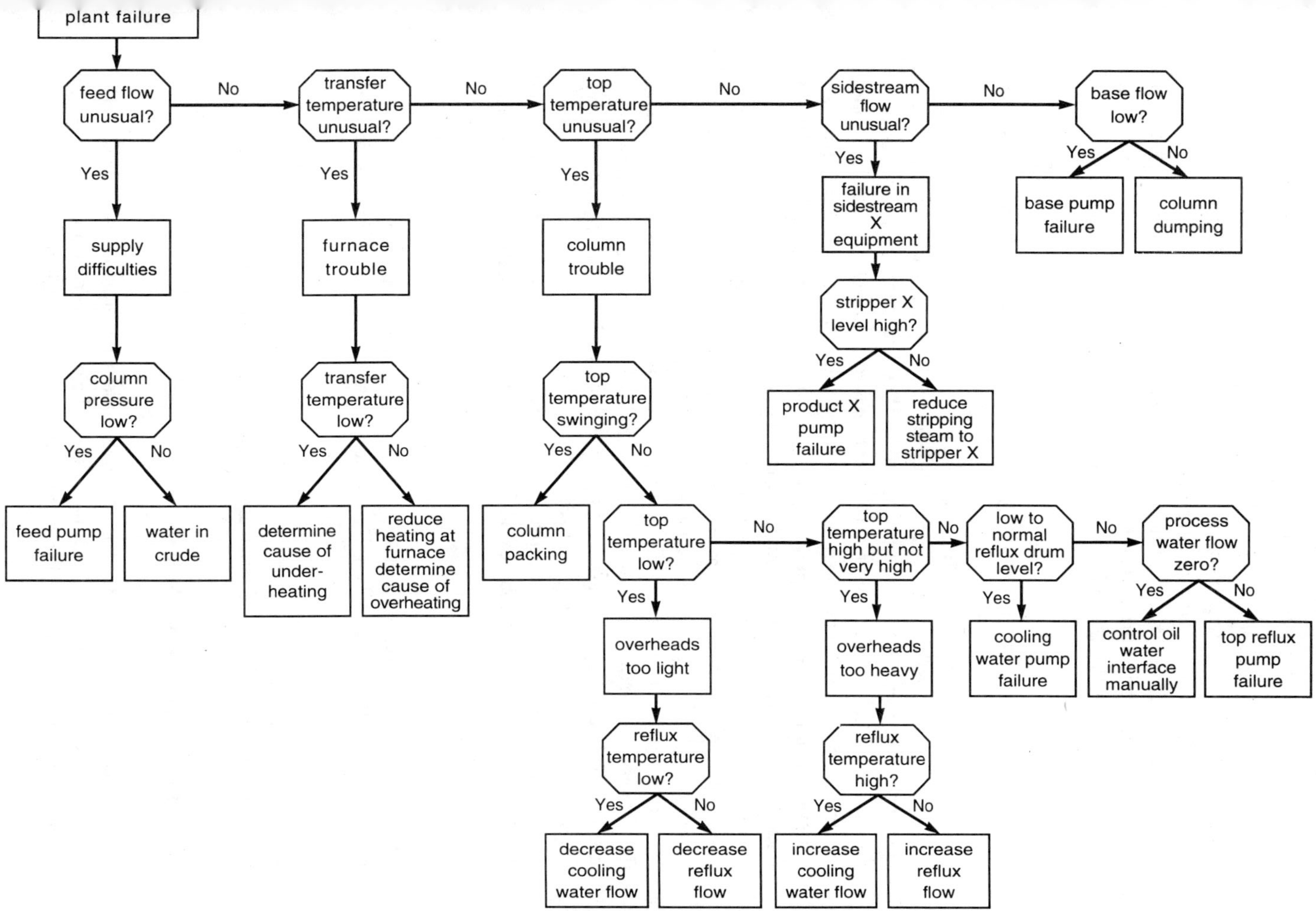

FIGURE 9-2. Example of Decision/Action Flow Diagram

- spatial—this is overlaid on a map of the equipment, and can be useful for studying teams engaged in a task.

The method study approach is followed, where columns can record the interactions, or movements of information between, for example: Operator; Helper; Machine; or Operator; Interface; Computer.

The resultant tables resemble those generated when studying batch operations, where materials are transferred between vessels and treated; however, in the case of OSD, it is the thought process or activities that are studied.

Reference

AIChE–CCPS, *Guidelines for Preventing Human Error in Chemical Process Safety,* 1994

9A.2.5. Signal-Flow Graph Analysis (SFG)

SFGs are similar to flow diagrams and are used to illustrate the link between operators and the process, the information they are seeking, and what they can do with it. SFG links the material flow patterns to the stages at which an operator has an involvement. They can also identify the information flow and control measures available to the operator, thus helping to understand how the system actually works, rather then how it should work.

Two types of flow diagram can be used. The block diagram can illustrate the basic sequence of events in a process. To support this, a more complex flow diagram can be prepared which illustrates the links between, for example, the operating instructions and the various tasks or subtasks involved in each stage of the process. These diagrams are useful when designing the man/machine interface, and for evaluating the understanding of a system. Illustrations of both type of diagram are provided in Figure 9-3.

Several SFGs may be required for a basic operation in order to cover the various possibilities that may exist. This requires text to explain the background to this analysis.

References

AIChE–CCPS, *Guidelines for Preventing Human Error in Chemical Process Safety*, 1994

Beishon, R.J., "Problems of Task Description in Process Control." *Ergonomics,* 10:177–186, 1967

Available Software

Use can be made of typical software for preparation of flow diagrams.

9A.3. Cognitive Task Analysis

9A.3.1. Critical Action and Decision Evaluation Techniques (CADET)

The documentation for this technique should clearly record the various stages of the study together with the role of the study in the overall analysis.

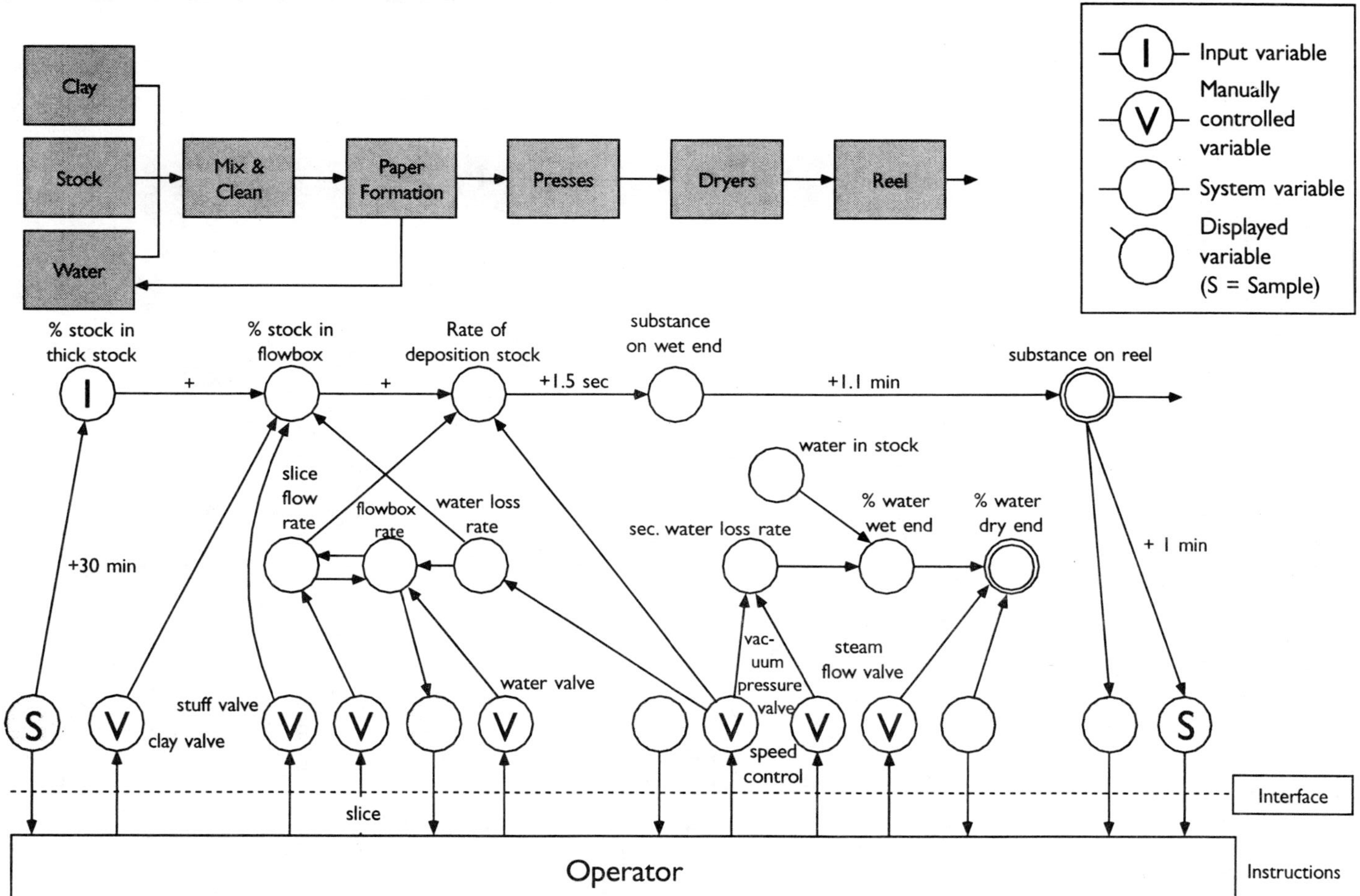

FIGURE 9-3. Example of Block Flow Diagram and Associated Signal-Flow Graph

CADET is based on a stepladder model, where there are two legs. One, the ascending leg, is associated with the operator's approach to the identification of the features of the operation that lead to the goal; i.e., the selection of the problem to be tackled. The second leg descends from this goal and concerns the way the operator tackles the operating problem; i.e., planning, executing, and analyzing the feedback. In CADET the emphasis is on the critical actions or decisions that require operator attention. The analysis of these decisions can be recorded in multicolumn format, listing in sequence the steps of the ladder, up then down. The columns list:

- Decision/Action Element;
- Objectives; and
- Typical Error Patterns.

For a fault-diagnostic task, the columns would list:

- Time (in sequential order of real time);
- Signal Detection;
- Data Collection;
- Identification;
- Interpretation; and
- Goal Selection.

An example of a CADET analysis is provided in Table 9-2.

References

AIChE–CCPS, *Guidelines for Preventing Human Error in Chemical Process Safety,* 1994.

Lucas, D. & Embrey, D.E., *Critical Action and Decision Evaluation Technique;* Human Reliability Associates, Wigan U.K., 1974.

Pew, P.W., Miller, D.C, Feeher, C.E., *Evaluation of Proposed Control Room Improvements Through Analysis of Critical Operator Decisions*, EPRI/NP-1982, Bolt Beranek & Newman Inc.

9A.3.2. Influence Modeling and Assessment Systems (IMAS)

The IMAS technique deals with subjective cause–consequence models (SCCM) of process abnormalities as obtained from personnel. This technique is important since well-trained and experienced operators have a better appreciation of the potential consequences of a plant abnormality and so are better prepared to deal with it. The SCCM is similar to a flow diagram, and represents the thought and knowledge process of personnel. Causes and indications are linked leading to the decision making point. SCCM prepared using experienced personnel can be used to identify training requirements for particular situations, and to monitor personnel training accomplis.

The SCCM can also be used to improve processes, when considering possible diagnostic error by operators in CPQRA exercises, and when examining emergency procedures.

TABLE 9-2
Example of CADET Analysis Form

TIME	SIGNAL DETECTION	DATA COLLECTION	IDENTIFICATION	INTERPRETATION	GOAL SELECTION	CADET ANALYSIS
t_1	Column temperature alarm		Not a complete indication at this stage. It may be a spurious alarm	Cross-examine related indicators		**Data collection:** Can operator acquire irrelevant or insufficient data? Can operator fail to cross-check for spurious indications? **Identification/Interpretation:** Can operator fail to consider all possible system states and causes of problem? Can operator fail to perform a correct evaluation? Can operator fixate on the wrong cause? **Goal Selection:** Can operator fail to consider possible side-effects? Can operator fail to consider alternative goals? Can operator fixate on the wrong goal?
t_2		TR14 = High (new) TR15 = Very High (check)	Inadequate cooling of column or thermal conditions of input are disturbed	Distinguish between the two. Examine flow rate and temperature of input		
t_3		FI1 = Normal (new) FR15 = Normal (check) TRC8 = Normal (new)	Conditions are as specified. It must be inadequate cooling of column	Possible causes: • Cooling water pump failure • Top reflux pump failure		
t_4		LIC3 = High (new) Drum sight glass = High (check)	Conditions are as specified. It must be inadequate cooling of column	Level in drumis high, thus condensation is OK. It must be failure of the top reflux pump		
t_5		FIC8 = No Flow (new)	Conditions are as specified. It must be inadequate cooling of column	Top reflux pump failure (confirmed)	**Alternative goals:** • Reduce heating in reboiler • Reduce flow rate of input • Increase cooling in condenser	

TR14, TR15 = Column Temperature; LIC3 = Level in Reflux Drum; FIC8 = Reflux Flow; FI1, FR15 = Crude Flow at Entry Point; TRC8 = Crude Temperature at Entry Point.

The record, in addition to a description of the background of the exercise and its objective, would consist mostly of a series of SCCM flow charts.

References

AIChE–CCPS, *Guidelines for Preventing Human Error in Chemical Process Safety*, 1994.

Embrey, D.E., and Humphreys, P., "Support for Decision-Making and Problem Solving in Abnormal Conditions in Nuclear Power Plants." In Methlieh, L.B. & Sprague, R.H., *Knowledge Representation for Decision Support Systems.* North Holland, Amsterdam, 1985.

Embrey, D.E., *Human Reliability,* in Serra, A., Barlow, R.E., Theory of Reliability, North Holland, Amsterdam, 1986.

Available Software

IMAS (LINKCC)(EXPLORE)(Human Reliability Associates, Wigan U.K.).

9A.4 Human Error Analysis

9A.4.1. Predictive Human Error Analysis (PHEA)

The starting point for this analysis is usually the output of a HTA. It can either be a plan analysis, considering the errors identified in the HTA plan, or a task element analysis, where the procedure systemically identifies a range of error modes.

Following the HTA, the analysis can take the form of a multicolumn format, with columns representing:

- task step (in sequence);
- task type (see checklist in CCPS Guidelines Book referenced below);
- error type (see checklist in CCPS Guidelines Book referenced below);
- description (of the error);
- consequences (of the error);
- recovery (from the error); and
- error reduction strategy.

The next step is to check which of the above error modes are possible, so as to develop means to deal with them.

The record can either be made manually, or use can be made of a computer program like THETA.

References

AIChE–CCPS, *Guidelines for Preventing Human Error in Chemical Process Safety*, 1994.

Embrey, D.E., *Top-Down Human Error and Task Analysis (THETA),* Software Manual, Human Reliability Associates, 1990

Lucas, D. and Embrey, D.E., *Critical Action and Decision Evaluation Technique,* Human Reliability Associates, Wigan, U.K., 1987.

Available Software

Use can be made of typical word processors providing in multicolumn editing capability and specialized software written for more detailed applications. An example is: THETA (Human Reliability Associates, Wigan, U.K.).

9A.5. Quantitative Techniques

Several techniques have been developed to assist in cases where quantification of human error rates is necessary, whether for Chemical Process Quantitative Risk Analysis (CPQRA) or for evaluation of human error improvement schemes. The descriptions that follow refer only to the form of technical documentation and drawings that result. More extensive descriptions of these techniques are to be found in the references.

9A.5.1. Technique for Human Error Rate Prediction (THERP)

The recording of this analysis centers around an event tree that follows certain specific conventions, e.g., branches to the left represent success; branches to the right, failure. These conventions should be clearly defined in the supporting text. These event trees will be very complex where tasks are complex. In these cases, it is preferable to prepare multiple trees in order to ensure that the analysis is adequately clear. Where cognitive errors are involved, use is usually made of a time-base, where the probability of error is related to the time available for diagnosis or for the task. Human failure rate data are available for problems of this type. There are also other types of error that can be included, for example, noncommission, wrong selection, wrong sequence, wrong time, wrong amount, wrong material, etc. Recovery potential is included usually as a dotted line in the tree, and also, where important, Performance Influencing Factors (PIFs) which are known in THERP as Performance Shaping Factors, are identified.

Often, it will be necessary to use symbols or letters to represent the probability of a task failure. In these cases a tabulation is prepared of what the symbol represents, in terms of description, estimated probability, and data source. This table can include both hardware and operator failure rates.

As in most of these analytical methods, it is important to ensure that, in addition to the diagrams and tables, there is sufficient description of the scope and objective of the analysis and of the results and conclusions. The experienced reader must be able to follow it throughout. Use can be made of ETA graphical techniques, including software, subject to the conventions required for THERP. Special HRA trees, which are usually quantified, can also be generated. The degree of breakdown in the tasks often follows the level of quantitative data on those tasks which is available to the analyst. An example of an HRA event trees was provided as Figure 6-3.

References

AIChE–CCPS, *Guidelines for Chemical Process Quantitative Risk Analysis*, 1989.

AIChE–CCPS, *Guidelines for Preventing Human Error in Chemical Process Safety*, 1994.

Bell, B.J., Swain, A.D., *A Procedure for Conducting a Human Reliability Analysis for Nuclear Power Plants*, NUREG/CR 2254 U.S. Nuclear Regulatory Commission, Washington, D.C., 1983.

Lorenzo, D.K., *Improving Human Performance in the Chemical Industry: A Manager's Guide to Reducing Human Errors,* Chemical Manufacturers Association, Inc., Washington, D.C., 1990.

Swain, A.D., Guttmann, H.E., *Handbook on Human Reliability Analysis with Emphasis on Nuclear Power Plant Applications* NUREG/CR-1278 U.S. Nuclear Regulator Commission, Washington, D.C., 1983.

Swain, A.D., *Accident Sequence Evaluation Program Human Reliability Analysis Procedure,* NUREG/CR-4772, U.S. Nuclear Regulatory Commission, Washington, D.C., 1987.

9A.5.2. Success Likelihood Index Method (SLIM)

This is a flexible method, suitable for quantification, which can be applied at any level of task or subtask. The probability of error is considered to be associated with the PIFs for that situation. These are mostly the direct effect PIFs: training, quality of procedures, distractions, feedback, motivation etc. However, even indirect factors such as management policy could be considered. Good quality data from the plant or from experts is required.

The analysis needs to be fully described and argued to be successful. The first stage involves problem definition, where it is possible to use event trees to condense the analysis to the levels for which data are available, usually where common PIFs are applicable. Each event or operation is then rated for its PIF levels, and the results tabulated:

- Operation;
- Time Stress Rating;
- Experience Rating;
- Distractions Rating; or
- Procedures Rating.

These ratings are next weighted according to a recognized table and then, using a formula, the table is recast as rescaled ratings and Success Likelihood Indices (SLIs) in the following form:

- Operations:
- PIFs;
  —time stress;
  —experience;
  —distraction; and
  —procedure;
- SLIs.

Using a further formula, SLIs can be converted to error probabilities that are the data required for quantification of human error. The SLIM method is often used when improvements are to be made, allowing the improvement to be quantified.

References

AIChE–CCPS, *Guidelines for Preventing Human Error in Chemical Process Safety*, 1994.

Phillips, L.D., Embrey, D.E., Humphreys, P., and Selby, D.L., "A Socio-Technical Approach to Assessing Human Reliability." In *Influence Diagrams, Belief Nets and Decision Making: Their Influence on Safety and Reliability*. Wiley, New York, 1990.

9A.5.3. *Influence Diagram Approach (IDA)*

This method is based on the development of an influence diagram that is similar to other organizational and flow charts and can be produced using similar methods. These diagrams are backed up by sets of tabulations where the various estimates of performance are prepared and analyzed, mostly in a multicolumn format. Examples of the many tabulations produced by IDA are given in Appendix 5A of the CCPS book *Guidelines for Preventing Human Error in Chemical Process Safety*.

References

AIChE–CCPS, *Guidelines for Preventing Human Error in Chemical Process Safety*, 1994.

Embrey, D.E., Rea, K., Humphreys, P., Rosa, E.A., *SLIM-MAUD. An Approach to Assessing Human Error Probabilities Using Structured Expert Judgement,* Vols. I & II., NUREG/CR-3518, U.S. Nuclear Regulatory Commission, Washington, D.C., 1984

Kirwan, B., *Human Reliability Assessment,* in *Evaluation of Human Work, a Practical Ergonomics Technology,* Taylor & Francis, 1990.

10

Management of Change

10.1. Overview

10.1.1. Introduction

Change occurs continually in the chemical industry, and, if not carefully controlled, it can sometimes lead to incidents that could otherwise have been avoided. Change begins in the earliest phases of a project as, for example, in the evaluation of design alternatives. After start-up, changes are often made to improve the process. Additionally, unintentional changes can occur (e.g., a supplier might alter a raw material composition without notice). Such changes continue throughout the life of the facility and, as experience has shown, even a small or apparently minor change can have a significant impact on process safety.

When making changes, some people may only consider the direct objective of what they want to achieve and may overlook whether the change potentially has some undesired effect on other aspects of the design or operation. Safety reviews, such as Process Hazard Analysis (PHA), and audits can detect these effects. However, since these reviews are usually conducted intermittently, an incident might occur before the hazard could be identified in such a periodic review. The correct time to access the process safety significance of a change is prior to implementation of the change. This is the premise upon which the concept of Management of Change (MOC) is based.

10.1.2. Goals and Benefits

The primary goal of MOC is to evaluate the significance of all changes before they are made to ensure the changes do not compromise the safety and integrity of the process.

The benefits of a sound Management of Change system include:

- the reduction of incidents, injuries, and losses; and
- less process downtime and, as a result, more efficient operations.

10.2. Description of Management of Change

10.2.1. Objectives

The objective of MOC is to provide a system of administrative controls that ensures:

- the proper consideration of direct and indirect hazard potential *before* the change is implemented;
- that procedures are modified and retraining, where relevant, occurs *before* the modification is put into service;
- enhanced emphasis on the preparation and retention of records of each change; and
- the ability to audit the changes.

10.2.2. Management of Change as Part of Process Safety Management

For some, the concept of management of change is most readily associated with the mechanical details of a process or facility (e.g., in evaluating a change to a larger pump impeller). However, it is important to manage change in many process safety management (PSM) elements.

For example, there may be changes in process knowledge. These can occur all at once, because of something specific (e.g., a catalyst is changed) or progressively over a long period (e.g. an operating variable is gradually changed to optimize the process). Such changes also require thorough evaluation and documentation. Further guidance can be found in Chapter 5, *Process Knowledge*.

Similarly, operating procedures should detail the correct way to operate the process, and actual operating practices should comply with the content of the procedures. If this is not the case, the operating procedures should be revised, or the actual practices changed to conform with the written procedures. Such changes in the process often result from operators suggesting changes to improve performance. The PSM program should address how such changes are to be authorized and monitored, who is responsible for this ongoing activity, and how the records are to be maintained. Further discussion can be found in Chapter 11, *Operating Procedures.*

Thus, MOC is a tool that can be effectively used to ensure the continued integrity of the other PSM elements. However, MOC also requires that certain other elements (e.g., process knowledge and mechanical integrity) be in place and functioning; otherwise MOC will not be effective. For example, process hazard analysis (PHA) can be used to assess the safety of the current design and operating procedures; this establishes a baseline against which the significance of future changes can be assessed. At plants with a strong PSM program, MOC can be expected to play a key role in maximizing the value of the PSM program.

Older plants may present a greater challenge in implementing MOC since documentation of parts of their PSM program information may be out of date, or missing. As a result, it may be difficult to readily determine whether a change is safe to implement. In such situations, for a MOC program to be effective, priority attention should be given to updating documentation that supports the plant PSM program. Effective, up-to-date documentation is important to enable the facility to be operated safely and profitably.

Examples of occasions in which uncontrolled change has proved hazardous are given in Section 10.6.

10.3. Management of Change Documentation

10.3.1. MOC Program Documentation

The PSM program documentation should make clear to employees the conditions under which the Management of Change procedures are to be applied, and how they are to be implemented. This will assist employees who are required to institute or authorize change. It is important that the necessary culture be in place, not only in the document, but in the way the company manages its day-to-day business. The PSM program documentation may include the following:

- a short description of the organizational objective expected to be achieved by MOC;
- definition of what constitutes a *change*, including:
  —permanent changes;
  —temporary changes;
  —changes made during an emergency; and
  —aborted or reversed changes.
- definition of what constitutes a *replacement in kind* that does not require MOC:
- statement of scope, defining what facilities, processes, operations, and documentation are covered and any that are not covered;
- definition of the overall and individual responsibilities and accountabilities for Management of Change on a site, including the authority and any limitations on such authority, for making changes in any PSM element;
- requirements for written procedures detailing how changes in each PSM element are to be managed, including:
  —requiring that proposed changes be formally described;
  —requiring evaluation of proposed changes by appropriate personnel to assess any process safety impacts, before authorization;
  —providing sample documentation for requesting and approving a change;
  —requiring appropriate reviews of the change after it is implemented; and
  —requiring establishment of procedures for dealing with other effects of a change on all related elements of PSM (e.g., updating drawings, specifications, operating procedures and retraining).
- definition of methods of maintenance and control of records of change;

- communication of the change and its effects to affected personnel and documentation thereof; and
- requirements for the way the MOC program is to be audited (e.g., which type audit and the frequency).

These aspects are discussed in further detail as follows and in other chapters.

What Constitutes a Change? A definition is needed of what actually constitutes a *change*. The definition should include permanent change, temporary change, change made during emergency, and aborted or reversed change, all of which should come under the MOC program, but which have some unique characteristics.

Some examples of permanent change that should lead to a comprehensive MOC review are:

- substitution of a pump impeller with a larger one in the same case;
- use of an alternate material of construction for a valve;
- control software change; and
- operating procedures change.

Other types of changes include:

- temporary changes, planned for limited duration, which may become *permanent* if a procedure for their disposition is not in place and followed. The PSM program documentation should be clear on the approval of temporary changes and should require a closure or removal date. There should be a management system for dealing with the conversion of a temporary change to a permanent change, where applicable, as soon as reasonably possible;
- changes made during emergencies, which can also become *permanent* if not properly evaluated immediately following the emergency. Such changes, necessarily made in haste, may have been made without due consideration of the relevant safety issues. The procedure should cover their evaluation and disposition; and
- aborted changes, which do not always have zero impact on an installation, and may pose significant hazards. For example, consider a proposed equipment change that has been properly reviewed and authorized. The equipment change may have required collateral procedure changes that are already in place. Aborting the equipment change would result in a mismatch between the hardware and the procedures. Thus, aborted changes require proper review at the time the decision is made to abort. Similar considerations also apply to changes that are reversed. This is, in effect, changing back to the original situation, possibly because the change did not work out as intended. Reversed changes should be treated as a second change, and should comply with the standard procedures and documentation requirement.

Who Has Responsibility and Authority for Change? Change is initiated at every level, but it is necessary to control change and to limit the authority for approval of change. The PSM program may encourage requests for change, and there should be a clear procedure for dealing with these requests. The system should not be so cumbersome to use that it discourages safe and viable improvements, but must ensure evaluation of the safety aspects of any change.

Changes will vary in significance and organizations may find it appropriate to require higher levels of authorization for changes that are potentially more significant. Thus, it would be appropriate in describing the different levels of change to link these to the authority required for each. A tabular presentation of this authorization list may be helpful.

It should be remembered that the significance of a change should be evaluated on the magnitude of its potential impact, not on the size or dollar cost of the change.

Organizations may find benefit in a system requiring the approval and signature of a second, independent reviewer to ensure thorough review of the potential hazards of a change. For example, some sites may find appropriate a check and balance procedure requiring the approval of a manager in both the manufacturing and the plant technical organizations. Use of job titles rather than individual names is preferred when identifying those responsible for authorizing change.

When Are Changes Approved? Generally, planned or considered changes go through the facility's Management of Change system during normal daytime work hours. However, it should be recognized that some changes may be made when supervisory personnel are not normally present, e.g., at night, weekends or during absences. The MOC system should provide a means for obtaining approval at all times so that MOC is not circumvented when the usual approval authority is not present on site.

This also applies to changes made under emergency conditions. However, in some of these cases, it may not be possible to perform the desired level of review, nor to obtain approval from the appropriate authority, prior to making the change. In such situations, review and approval should be obtained and documented as soon as practical when conditions stabilize.

How Are Affected Personnel Notified? The program documentation should require notification of all personnel potentially affected by the change, as well documentation of the fact that this has been accomplished.

How Are Changes Authorized and Documented? Effective control over change is determined as much by the example of those in authority as by the formality of the system that is adopted. However, thorough, comprehensive documentation is required to make MOC effective. Many organizations utilize a structured change approval form, containing many listed checks to ensure that changes are properly reviewed and authorized by responsible parties and that related aspects of change are fully addressed and appropriate actions taken.

These change documents are vital to the auditing of the various PSM elements as they provide a record of what has changed from the original documentation, and why. For example, operating procedures could be audited against the accumulated change authorization documentation to verify that all required procedural changes were properly implemented. Auditing would be a tedious, and perhaps uncertain, task without these records.

What Related Changes Are Required? When a change is approved, related documents (e.g., drawings, process descriptions, MSDSs) must be updated. Some related documents (e.g., procedures) may require updating before the change is implemented. Others may be updated after the change, but on a timely basis. Organizations may wish to consider a documented program for the follow-up and resolution of all such change requirements. Also, any required retraining must be defined and performed prior to activation of the change. These requirements should be clearly defined in the MOC program documentation.

Why Was Change Made?. It is important to record the justification for the change. This record should address not only the objective of the change, but the technical basis and evaluation of why it is believed to be safe and effective. This record could also serve to inform those in charge of the plant in the future years why certain changes were made, or why changes were not made for safety or other pertinent reasons. This may avoid the need for future reevaluation of the same change.

10.3.2. Records from Implementing the MOC Element

Since changes can occur in so many different ways in a process, it is very important that a record of them be created and maintained. Even simple changes such as the substitution of a ball valve for a plug valve may lead to updating of at least two sets of records (P&ID plus operating instructions). Potentially affected documents should be identified and evaluated for needed revisions and the reader may wish to refer to other chapters in this book for further guidance.

This section describes in further detail several concepts introduced above; i.e., the change authorization form and the use of tracking systems for follow-up of related changes and personnel notification. Additionally, the important connection between MOC and PHA is discussed.

Authorization Form. A proven and reliable way of managing and documenting change is to utilize a *change authorization* or *change request* form. These have a variety of names and styles in industry but most share a common objective. Information that should be considered for inclusion on a change authorization form includes:

- a description and location of the proposed change;
- an explanation of the reason for the change (i.e., what was it intended to accomplish);

- technical basis for the change (i.e., how will it accomplish its intended goals);
- time limits of a temporary change;
- checklists to enable those preparing and authorizing the change to identify potential impacts and hence properly evaluate each of these *before* authorization;
- authorization signatures, drawn from those listed, by title, in the MOC program documentation concerning authorization procedures;
- a list of PSM elements requiring cross-checking as a result of this change; and
- signed statements by the persons responsible for implementing or incorporating the change that it has been completed (including changes in other relevant documentation).

Figure 10-1, Change Authorization Form—Examples 1, 2, and 3, depicts MOC forms that require much of the information discussed above. Examples 1 and 2 are more general and applicable to most of the changes that could occur in the workplace. These are generally equipment oriented. Example 3 would be appropriate for a change in technology.

Tracking System for Follow-Up Changes to Other Documents. Many changes have an immediate impact on other documents, thus requiring timely changes to them as well. The organization's MOC program may require the use of a tracking system for such related changes to ensure that they are properly implemented. Some organizations may find it helpful to incorporate this follow-up system as a checklist on the change authorization form. The documents requiring change can be identified, with the completion of their revision indicated by the signature of the responsible party and the date that the revision was completed.

Tracking System for Notification to Affected Personnel. Similarly, each organization may wish to consider a tracking system to ensure that affected personnel are notified of the change and its impact. Most affected personnel will be notified of the change before it is put into effect. However, the tracking system should have the flexibility to address the situation of personnel that are absent (e.g., on vacation) at the time the change is implemented. Such affected personnel must be notified of the change upon return and before they resume their duties.

Revision of Process Hazard Analysis. When a PHA is performed, it should be carried out using information that accurately describes the process or facility at the time of the review. The PHA report then documents the *baseline* against which the significance of future process and equipment changes can be evaluated. Subsequent changes have a potential to invalidate that PHA and must be carefully examined to establish whether or not they do have such an impact.

If a change invalidates the original PHA, the PHA (or the affected portion) should be reworked before the change is implemented. Additionally, the accumulated effect of smaller changes over time may invalidate the PHA. This is one

Initiation: (to be completed by originator)

Originator:________________________ Date:______________

Description of Project: (attach sketch, P&ID, etc.)

____

____

____

Process fluid:______________ Operating press:______________ Temp:______________

Approval: (to be submitted by field supervisor)

Ticket No.:______________

1. Operations assistant ______________________ Date ______________
2. Maintenance assistant ______________________ Date ______________
3. Technical team leader ______________________ Date ______________

cc: TSTL for files   TSO assigned______________

Process Review: (to be completed by TSO) TSO______________ Date______________

( ) Construction package completed
( ) Process impacts
( ) Scope defined (attach marked-up or revised P&IDs)
( ) Product quality impacts
( ) P&ID review (interdisciplinary)

Mechanical Review: (to be completed by PST/MEG/I/E as needed)

PST______________ MEG______________ I/E______________ Date______________

( ) MEG requirements
( ) Painting/insulation ______________
( ) Materials of construction
( ) Relief protection review
( ) I/E requirements
( ) Special inspection required
( ) Pipe spec: ______________ ( ) MAWP
( ) Gasket material: ______________
( ) Testing requirements:
  ( ) Hydrostatic test pressure: ______________ (hold at least 10 min)
  ( ) In-service test
  ( ) Visual
  ( ) Radiography

Construction: (to be completed by field supervisor and inspector)

( ) Construction drawings prepared
( ) Equipment folder updated
( ) P&IDs updated
( ) On-lines updated
( ) Car seal list updated
( ) Loop diagrams/folders updated
( ) RV list updated
( ) Fugitive emissions list updated
( ) Spare parts stocked
( ) Painting/insulation
( ) Steam tracing installed/list updated
( ) Appropriate bleeds/vents/drains
( ) Blind list updated
( ) SOME review
( ) Operating procedures
( ) HAZCOM updated
( ) Construction/testing complete and per design

Field supervisor ______________________ Date ______________

Inspector ______________________ Date ______________

Forward completed form to technical support team leader.

FIGURE 10-1. Change Authorization Form—Example 1

Notification of Process Change Checklist

Information about the Change:

Originator ____________ Date of Origination ____________

Proposed Date of Change ____________ Area ____________

□ Permanent □ Temporary From ____________ To ____________

Description and Location of Change (Scope) ____________

Technical Basis for Change ____________

Nature of the Change:

Change affects: □ Safety □ Loss Prevention □ Environment □ Health

Type of Change:
□ Alarm □ Shutdown Point □ Addition or Removal of Equipment
□ Piping Modification □ Chemical □ Process Computer Control
□ Job Procedure □ Instrument □ Equipment/Material Modification
□ Other ____________

Premodification Checklist:

Applicable	NA	Initials	
□	□	____	Consult piping and equipment specifications.
□	□	____	Perform reactive chemicals testing. □ In process?
□	□	____	Add involved materials to Toxic Substance Control Act (TSCA) inventory.
□	□	____	Calculate impact on F&EI and CEI.
□	□	____	Comply with Engineering Practices.
□	□	____	Comply with Technology Center guidelines.
□	□	____	Comply with Dow Environmental Protection Guideline for Operations.
□	□	____	Comply with Safety and Loss Prevention requirements.
□	□	____	Consult maintenance (name)____________.
□	□	____	Consult instrument and electrical technician (name)____________.
□	□	____	Consult parts technician (name)____________.
□	□	____	Evaluate and modify relief system (name)____________.
□	□	____	Consult Industrial Hygiene (name)____________.
□	□	____	Consult Process Engineering (name)____________.
□	□	____	Complete required reviews (name reviews)____________.
□	□	____	Other____________.

Postmodification Checklist (Before Startup):

Applicable	N/A	Initials	
□	□	____	Performed prestartup audit.
□	□	____	Completed or updated training program.
□	□	____	Wrote and obtained approval for job procedures.
□	□	____	Updated P&IDs process flow sheets and plot plans.
□	□	____	Trained personnel on the change.
□	□	____	Updated critical instrument checklist.
□	□	____	Changed computer code and documentation.

Approvals:	Name	Date
Originator	____________	____________
First Reviewer	____________	____________
Department Head/Superintendent	____________	____________

FIGURE 10-1. Change Authorization Form—Example 2

Test Authorization No. TA–C– ______________________ Date: __________
Product Name: ______________________ Product Code: __________
Area: ______________________ Pounds Proposed: __________
Title: ______________________ Building No.: __________
Prepared by: ______________________ Financial Risk: __________
Responsible for test and test conclusion: ______________________

Purpose of test:

Duration of test:

Standard involved:

CAS No.: ____________

TSCA inventory:
( ) No change ( ) New chemical ( ) Included on TSCA inventory ( ) Exemption or explanation attached*
Analytical control or specifications: ( ) No change ( ) Change described under test details
Process hazards review: ( ) Required before test—see attached ( ) Not required
Material safety data sheet: ( ) Revision required ( ) No change ( ) New raws (see attached)
Packaging and shipping classifications: ( ) Revision required ( ) No change
Waste minimization discussed: ( ) Attached ( ) Not applicable
Air permit or revision needed: ( ) Yes ( ) No
TCPA EHS: ( ) Yes ( ) No (refer to PHM guidelines, Section IX)
Technical basis for change:

Test details (include Safety and Occupational Health and Waste Disposal):

Have operating instructions been prepared and approved?
( ) Yes—see attached ( ) Not required

Distribution:

Approved by: Business Division

Area Env. Ctrl. Coord.	Date	Research Supervisor	Date	Business Manager**	Date
R&D TSCA Coord.*	Date	CWCL Lab Supv./Anal.Res.Supv.**	Date	Product Manager**	Date
Tech. Service Mgr.**	Date	Mechanical Supv.**	Date		
		PHM Coordinator***	Date		

Authorized by:

Area Manager	Date	Business Unit Manager**	Date	Plant Manager**	Date

* Required if new chemicals are used, made, or isolated which are not included on the TSCA inventory.
** Mark not required where applicable.
*** Required if EHS is involved, per PHM guidelines, Section IX.

FIGURE 10-1. Change Authorization Form—Example 3

reason why many organizations have adopted the practice of periodically revalidating their PHAs. This allows the incorporation of these accumulated changes into the PHA and establishes a new *baseline*.

It is important that this *baseline* be adequately documented. Refer to Chapter 6, *Process Hazard Analysis*.

10.4. Records Management

10.4.1. Records Management Program

In addition to the description of the MOC program given in Section 10.3.1, the organization's PSM program documentation should define MOC records management practices. This should cover not only the MOC authorization forms, but ancillary records, as well (e.g., records of change notification to affected personnel). The records management practices should provide access to all personnel having a job need for MOC information (e.g., an audit team, or a PHA review team), while at the same time ensuring that appropriate control and retention procedures exist to preserve this historical data. This section describes how such MOC records could be handled in order to ensure that the objectives of the MOC program are met.

10.4.2. Media and Methods

Change authorization forms have traditionally been paper copy. While a paper-based system will probably be satisfactory for most users (especially smaller organizations or facilities), some organizations have instituted computer-based systems for generating and maintaining MOC records. In such systems the originator can call up and fill in a copy of the MOC form on the computer. The completed form can then be printed out and the appropriate approval signatures obtained. Some organizations may find it more expeditious to electronically circulate the MOC form for review before printing it for signature. While signature approval of a paper copy may be required, electronic media may be used for subsequent record keeping (e.g., records of changes made and follow-up of consequential changes to other documents). Use of electronic media can make information more readily available to others who will be affected, including employees who may be directly affected and other managers who may need to institute collateral changes in other documents.

10.4.3. Responsibility and Accountability

Virtually all elements of a PSM program will be affected by change at some time. Normally a number of individuals, particularly at a large facility, will be responsible either for authorizing changes or monitoring their follow-up. Other individuals may be responsible for implementing the other potentially affected PSM elements. Thus, there is no single, correct system for MOC records management. Each system must be keyed to the unique needs of the facility or organization.

Some organizations may chose to hold completed MOC records in a single, centralized location. Alternatively, where this is more appropriate, records may be stored in several locations to meet the functional needs of the organization. In any event, responsibility and accountability for the MOC records management function must be clearly assigned. The designated storage location for each type of MOC record must be clearly identified for those individuals responsible for implementing MOC-related functions or other PSM elements.

10.4.4. Records Retention and Purge Schedules

The organization's PSM program documentation should establish the requirements for retention and purging of MOC records. These may include:

- establishment of a minimum retention period, at least up through the next PHA review, to enable all changes to be properly reviewed for aggregate effect;
- requirement that no purging to be carried out before the next routine audit;
- definition and identification of any documents that should be retained for regulatory purposes and should not be purged; and
- establishment of a schedule for purging consistent with other company policy constraints and objectives.

The reader may wish to refer to Chapter 4, *Records Management*, for more general guidance.

10.5. Auditing

The PSM program documentation should establish the audit requirements for the MOC element. The first step in auditing MOC documentation is to verify that a documented program exists for implementing MOC. This program would include elements outlined in Sections 10.3.1 and 10.4 that are pertinent to the needs of the organization or facility.

The second step is to verify that the results of implementing MOC have been documented in the manner prescribed by the program. Particular emphasis should be placed on documentation that consequential changes have been identified and resolved and that affected personnel have been notified.

Any deficiencies identified during the audit must be documented and resolved in a timely manner. Further details on auditing Management of Change are given in Chapter 6 of the CCPS book, *Guidelines for Auditing Process Safety Management Systems.*

10.6. Examples

10.6.1. Mislabeled Electrical Equipment/Inadequate Management of Change

In early 1993, workers performing inspections in a facility noticed several energized electrical breakers, labeled *SPARE*, feeding power to unknown equipment. The investigators determined that several pieces of safety-related equipment, including hydrogen analyzers and air monitors were powered by the *SPARE* breakers.

The incident investigation cited personnel error as the direct cause of the problem. Individuals making electrical changes failed to follow through on labeling and updating electrical drawings. The apparent root cause was an inadequate Management of Change control system and inadequate documentation of changes and their consequential effects.

This event illustrates how lack of or inadequate enforcement of a documentation system for Management of Change can pose a significant potential for serious, life-threatening incidents.

10.6.2. Change to Operating Procedure

In 1976, an animal feed additive manufacturer was batch drying an organic compound known to be suspectable to dust explosions. The dryer was fitted with dust explosion protection, including nitrogen inerting, explosion suppression devices, and explosion relief.

The material was left in the dryer overnight, as no instructions for discharging the material were available for the operators. More than 24 hours later, the dryer exploded.

Investigation discovered that self-heating, leading to decomposition, could occur if the material was held under adiabatic conditions. Apparently, the change in operating practice, leaving the material in the dryer rather than discharging it promptly, caused the explosion.

Adhering to the requirements of an MOC program (including, perhaps, a PHA) could have prevented this incident by identifying safe operating practices that could have been embodied in a properly executed revision to the operating instructions.

10.7. References

AIChE–CCPS, *Guidelines for Auditing Process Safety Management Systems*, 1993.

AIChE–CCPS, *Management of Change on CD-ROM,* 1994.

Canadian Chemical Producers Association, *Essential Components of Safety Assessment Systems,* 1985.

Kletz, T.A., *What Went Wrong? Case Histories of Process Plant Disasters,* 2nd Edition, Gulf Publishing, 1988.

Sanders, Roy E., *Management of Change in Chemical Plants,* Butterworth-Heinemann, 1993.

11

Operating Procedures

11.1. Overview

11.1.1. Introduction

Up-to-date written operating procedures that reflect current plant practice are necessary for the various modes of process operation to assure continuous, efficient, and safe operation of a facility. These documents should be readily accessible to employees who are responsible for operating a process and are a vital resource supporting the facility training program.

Incidents may occur when procedures are inadequate, incorrect, or could be misinterpreted. This chapter describes the documentation of policies and programs that support the preparation of complete and accurate operating procedures for each operating mode of a facility. It also covers ways to ensure that these procedures will be properly distributed, used, controlled, and updated at appropriate times.

11.1.2. Goals and Benefits

The goal of an operating procedure is to provide a clear set of instructions that include the correct steps, listed in the proper sequence, to safely perform a task. Some of the benefits of having effective and comprehensive operating procedures include reduced risks, higher product quality due to repeatability of operations, and smoother transition between operating shifts, all of which can contribute to reduced operating costs. Written procedures also serve as aids for training new employees and maintaining the competency of existing plant personnel.

11.2. Description of Operating Procedures

There are various ways in which operating procedures are written and styled. These can vary from precise step-by-step instructions that require initialing and/or data insertion by the operator, to procedures used only for reference when performing routine operations such as pump starting and stopping. In some cases, a separate (i.e., tear off) sheet is used for each operation. In others, a bound document or a computerized file of procedures may be used.

11.2.1. Objectives of Operating Procedures and Why They Are Required

Routine operating procedures should define safe operating limits and include the consequences of deviations from these limits. These documents should clearly detail the step-by-step execution of a task and should not be ambiguous. Documentation that meets these requirements minimizes variations in operations which, if left uncorrected, could have impacts ranging from poor product quality to an incident. To be effective, procedures should also be concise, accessible, and in an easily usable form.

Documented operating procedures for situations that occur infrequently, especially when an upset or emergency occurs, are important since the operators will have less experience in dealing with such situations. Thus, the correct response to an unusual situation can be ensured by anticipating the situation, developing the correct solution before it occurs, and training operators to use the solution when the situation arises.

11.2.2. Key Elements and Preparation of Operating Procedures

The key elements of written operating procedures are:

- operating steps for each operating mode:
  —initial start-up;
  —normal operations;
  —emergency operations;
  —normal shutdown;
  —emergency shutdown;
  —temporary operations;
  —start-up following a turnaround or after emergency shutdown; and
  —preparation for and return from maintenance.
- definitions of operating limits for each mode:
  —minimum and maximum pressure, temperature, flow, composition or other key parameters;
  —consequences of deviation from operating limits; and
  —steps to avoid or correct deviations.
- process health and safety concerns:
  —hazards and properties of chemicals involved;
  —steps required to prevent exposure;
  —treatment measures in the event of employee exposure;

—safety when opening process equipment;
—verification and quality control for raw materials; and
—other hazards, such as very high temperatures or pressures.

- safety systems and their functions:
  —alarms and interlocks;
  —emergency systems; and
  —employee response.

Operating procedures are prepared based on a careful review of each operating mode, recognition of the hazards involved, and the appropriate sequencing of safe operating steps. The process of developing and documenting procedures typically includes the following steps:

- develop an inventory of tasks to be performed for each mode of operation listed above. Consider that:
  —even if the same task is repeated during different operational modes, there may be subtle differences in implementation details; e.g., shut down procedures may differ between routine and emergency modes;and
  —logbook details recorded during shutdowns and turnarounds may prove helpful in determining the tasks to be performed.
- divide the job tasks into numbered, sequential steps. This will:
  —help in documenting the sequence by recording the activity as it is being performed;
  —promote consideration of safety precautions and other safety issues at each step; and
  —simplify checklist preparation (these are often used in batch operations, sometimes to record observations and steps as they are completed).
- write draft procedures. These may be written by operators, technical staff, technical writers, or a combination of these. Consider that:
  —integration of safety concerns is of prime importance;
  —procedures are more effective if the steps for each task are concisely stated, sequenced correctly and contain only one action per procedural step;
  —explanations should be kept separate from instructions;
  —definitions should be clear, to avoid ambiguities; and
  —general training or troubleshooting manuals are usually kept separate, but need to be consistent with the operating procedures.
- conduct a review of the draft procedure with input from management, technical staff and operators to ensure that it is readily understandable, includes all necessary elements, and is well documented so that:
  —responsibilities are defined;
  —the procedure is repeatable; and
  —it is consistent with the correct manner of operating the process.

11.3. Operating Procedures Documentation

11.3.1. Operating Procedures Program Documentation

Organizations implementing an operating procedures program should include in their PSM program documentation a discussion of the goals of the program as well as any general instructions that outline how the operating procedure program will be implemented. A written program is helpful in providing a uniform and adequate approach, defining responsibilities, and ensuring that proper documentation exists. Such a program would cover the tracking of documents, storage locations, approvals, revisions, and distribution. The program should also include a system for audits to verify compliance with this PSM element.

It can also be useful to include special procedures, such as lockout/tagout, confined space entry, opening of process equipment, hot work, and other safe work practices within the same administrative system (these procedures are discussed in Chapter 18, *Permit-to-Work Systems*).

Subsequent paragraphs in this section discuss in detail components of such an operating procedures program.

Responsibility and Accountability for Operating Procedures. Operating procedures should be written by an individual or a group familiar with the process. The writing may be done by technical staff, production supervisors, operators, or maintenance staff assigned to the facility. The need for input from operators and maintenance staff is crucial and it is important to have a cross-section of personnel involved in the writing and review process. One way to ensure that this occurs is to apportion writing and review responsibilities among the shifts and maintenance personnel, assigning one person or group to fully integrate the separate written portions into one coherent procedure document.

Allowing such participation in the development of procedures is intended to ensure that written procedures reflect actual practice and that those persons using the procedures will need less assistance in their interpretation. If procedures do not fit actual plant circumstances, operations and maintenance personnel will be faced with more than one set of ambiguous and conflicting instructions, enhancing the potential for incidents.

Technical staff input is necessary to highlight technical or quality concerns. Management input should ensure that the organization's policies are properly reflected in the procedures. Safety department input is recommended to ensure adequate consideration and integration of health and safety concerns.

The number of individuals participating in maintaining the operating procedures depends on the size of the facility and the degree of employee involvement in developing and reviewing the operating procedures. Responsibilities can be defined in a variety of ways that will result in an effective program. For a small facility, one individual may be responsible for writing, modifying, and maintaining the records for all procedures. For large facilities with multiple processes, responsibilities are often shared. In either case, the individual responsibilities and accountabilities for implementing the procedures program and for writing, modifying, maintaining, and authorizing operating procedures for each process unit should be clearly identified.

Preparation and Revision of Operating Procedures. The program documentation should identify both those activities requiring written procedures and the circumstances requiring modification of procedures. For example, operating procedures should be written before the initial start-up of a process and revised whenever process or equipment modification occurs. The procedures should be in place and checked for adequacy prior to start-up of the process. This may be confirmed as part of a pre-start-up safety review.

Operating procedures are sometimes modified to reflect minor process improvements with or without equipment modifications being necessary. This type of modification should be subjected to the Management of Change review process before implementation. See Chapter 10, *Management of Change*. In addition, Process Hazard Analysis (PHA) reviews, incident investigations, feedback from training activities, and audits may result in action items involving changes in operating procedures. The controls used to ensure that all operating procedures are properly updated should be documented. The revision process should consider the following:

- procedures should be reviewed at a predetermined frequency to ensure either that they are current or that necessary revisions are made;
- revisions should also be made when necessary as a result of changes in the process or equipment;
- procedures should be updated to reflect actual plant practice;
- the procedures documentation program should permit easy and clear identification of updates;
- the individual responsible for the maintenance of operating procedures should be identified;
- the individuals responsible for approval and/or authorization should be identified; and
- an index identifying the current revisions of procedures should be maintained so that, at any time, it can be determined that procedures in use are up to date.

Revisions should be reviewed by the operators who will use them, to ensure that all changes have been correctly implemented. Also, necessary training on the revised procedures should be completed prior to their implementation. Some facilities mark their procedures with an *effective date* rather than the *revision date* to clarify that they are to be used only after related preparations, such as operator training, are completed.

Operating Procedure Content. The key elements that should be considered in preparing operating procedures were detailed in Section 11.2.2. While there is no mandatory format, the organization and headings chosen for facility procedures should clearly indicate that operating procedures exist for each operating mode and that they include applicable elements presented in that section.

Each facility may have its own format for operating procedures and still meet the intent of this PSM element. The format can be customized to address the experience of the personnel assigned to the area, the hazards involved in the process, and the resources available to the facility (e.g., computer accessibility), as long as the necessary content is included. However, organizations may find it helpful for training and auditing if the format is consistent throughout the operating unit. The organization way wish to include guidance on procedure format and content in the program documentation.

Distribution and Use of Operating Procedures. Up-to-date operating procedures must be readily accessible to people who operate or maintain a process. At least three different systems exist for distributing procedures: documented hard copy, undocumented hard copy, and computer-based.

Some organizations prefer a rigidly structured system where paper copies of procedures are limited to a predefined distribution list. Copies may be sequentially numbered as an aid in inventory control. Typically, at least one copy is maintained at a normally staffed operating station or control room. Copies may also be kept at a location remote from the unit in case of an emergency when access to the control room copy may not be readily available. Undocumented copies, if permitted, are intended for reference only and are so marked to ensure that they are not used for actual operations.

Other organizations may choose to use a less structured hard copy system where undocumented copies are provided, as needed, to operating and maintenance personnel. Such a system imposes a higher degree of responsibility on the procedure user to confirm, perhaps against a master index, that the procedure about to be used is the most recent revision.

Finally, some organizations may prefer a computer-based procedure distribution system. In such a system, users gain access to the procedure via a computer or terminal. However, as will be discussed later, such systems are typically not paper-free, as some form of paper back-up copies of procedures must be available in the event of computer system failure.

Regardless of the media, it may be helpful to solicit input about the best locations for operating procedures from technical staff, management, and operations and maintenance teams. The program documentation may contain guidance on selection of media and related records management practices. See Section 11.4.

11.3.2. Records from Implementing the Operating Procedures Element

The most important records in this program are the master (or approval) copy and the distributed copies of the complete and up-to-date operating procedures. However, the procedures program should also consider tracking of other records related to management of operating procedures including:

- the operating procedures program described in Section 11.3.1;
- a listing of procedures with date of issue or effective date, revision number, and the names of the approvers for reference;
- distribution lists for controlled distribution procedures;
- a historical record of procedure revisions, for reference; and
- a record of any recertification process, for example, for a procedure issued with an expiration date.

Records of training given, both routinely and following change of procedure, are described in more detail in Chapter 12, *Training*.

11.4. Records Management

11.4.1. Records Management Program

The PSM program documentation should establish the practices to be followed for documenting operating and maintenance procedures and managing related records. In particular it should identify individuals (by organization and title) who will be responsible for certain defined units of the process and specify what documentation should be maintained. This section addresses pertinent records management issues. Refer to Chapter 4, *Records Management*, for more general guidance.

11.4.2. Where Are Records Maintained?

There are two types of records to be considered. The first type are the actual procedures intended for use by operating and maintenance personnel. These will be located as operational needs dictate. The individual assigned responsibility for maintaining the operating procedures for each process would typically also oversee the distribution, control, and updating of authorized copies, whether in electronic or hard copy media. This is necessary to ensure that copies are available to essential personnel; that copies are updated whenever procedures are revised; and that only current procedures are utilized. This provides the clear accountability necessary to audit the availability of procedures and proper distribution of updates.

The second type of records are the ancillary records that are generated during the implementation of the procedures element (e.g., master or approval copies of procedures or records of past procedure revisions). The location of these records depends on the needs of the facility or organization. The locations and individuals responsible for record keeping for each operating unit should be clearly defined. It should be considered that program execution may be cumbersome if the individuals responsible for approval, documentation, and record keeping are remote from the process. Personnel involved with operating the process may be discouraged from proposing modifications or suggesting improvements to the existing procedures and a less efficient system may result.

11.4.3. What Media Should Be Used?

The media used for the procedures themselves should promote access and use. Electronic media may allow certain advantages over hard copy, such as:

- more efficient distribution of draft procedures to reviewers and approvers;
- prompter implementation of new or revised procedures;
- less reliance on hard copy versions, simplifying the task of retrieving obsolete versions of a procedure when a revision is issued; and
- enhanced abilities for indexing, cross-referencing, and searching so that operating an maintenance personnel have greater assurance of locating pertinent procedures.

However, safeguards should be considered for the system to prevent unauthorized updates or revisions of procedures from occurring. Protection of such files was discussed in Chapter 4.

Access to procedures in a computerized system would normally be via a computer terminal. When primary access is provided at a terminal, it is advisable to provide hard copy backup for emergency situations such as when a major incident or a computer failure occurs. Hard copy may also be necessary for field activities.

Similarly, the media used to track the numerous ancillary records generated when implementing a procedures program can range from simple hard copy to computerized document management systems.

11.4.4. Retention/Purge Schedule for Records

Records are retained to meet operating unit or organization needs and regulatory requirements. Keeping all revisions of a procedure provides the most complete record, however, these records may eventually become voluminous. Also, dated information is subject to misinterpretation if taken out of the context of its original use. While there may be no general requirement to maintain historical revisions, selected outdated versions of procedures may be retained for technical reasons; for major incident documentation; for their relevance to a prior PHA; for documentation of an operating mode or product that may be resumed at a future date; or for the purpose of demonstrating responses to audit or PHA action items.

These outdated revisions should be clearly marked (e.g., by stamping them as being *Obsolete* or marking them with the last effective date for use) and filed in a manner that reduces the likelihood that they will be used in current operations.

A retention and purge schedule for PSM program records should be developed and reviewed by senior management and, perhaps, legal counsel.

11.4.5. Access Control

Control of access and distribution of procedures may be necessary to protect trade secrets and to minimize the potential for outdated copies of the procedures resurfacing and being used for operations. Access can be controlled by techniques such as limited and assigned distribution, computer passwords, combination lock-outs, and similar techniques.

11.4.6. Revision Control

To maintain accurate and up-to-date procedures, it is necessary to make sure that updates are promptly disseminated and that unauthorized and uncontrolled changes are prevented. One approach would be to:

- maintain a master or approval copy (clearly labeled as such);
- print the most recent effective date (or revision number) on the pages of procedures;
- check effective dates (or revision numbers) of field copies against the master copy during internal audits; and
- ensure that procedures accessed via electronic media cannot be revised except through the established revision protocol.

Management may review the operating procedures periodically to ensure that they are accurate and up to date. During the review process, it is advisable to label review copies of the procedure *Draft* until the new version is formally adopted. To provide structure for this review process, it may be helpful to provide a log for each procedure. The log could indicate when it was last reviewed and updated, when it is next scheduled for review, and, if it is currently being reviewed, the present stage of its review.

Personnel should not make revisions to the procedures outside the structured review and Management of Change protocols. For example, operators should not modify the control room copy and use it for operating the plant. However, operators should be encouraged to use the established revision procedures to propose changes once it becomes apparent that a revision is needed. One or more process-knowledgeable individuals should review and resolve proposed revisions on a timely basis.

Stringent controls are required to ensure that only authorized personnel are able to access and modify either hard copy master procedures or computer-based procedure files. Allowing uncontrolled revisions to the procedures in use bypasses the review process and could result in serious, unforeseen consequences. Refer to Chapter 10, *Management of Change,* for further discussion.

11.5. Auditing

The first step in auditing procedures documentation is to verify that a documented program exists for implementing the procedures element. This program would include features outlined in Sections 11.3.1 and 11.4 that are pertinent to the needs of the organization or facility. The second step in auditing is to verify that the results of implementing the procedures element have been documented in the manner prescribed by the program.

Any deficiencies identified during the audit must be documented and promptly resolved. Additional guidance on auditing can be found in Chapter 14, *Auditing*, and in the CCPS book, *Guidelines for Auditing Process Safety Management Systems*.

11.6. Examples

This section gives several examples of incidents or near misses that illustrate the importance of documenting clear, unambiguous, and up-to-date procedures. Some are taken from the writings of Trevor Kletz (see Section 11.7).

11.6.1. Temporary Operating Procedures

Any temporary operating procedures issued must be clear and unambiguous. Experience shows that those operations where temporary procedures are in effect are more likely to experience incidents than those with standard procedures, especially if the temporary procedures are inadequately documented.

As an example, a change was made in the procedures for taking a reactor off line for catalyst change in order to prevent recurrence of a fire. The change was recorded in the log book and in a handwritten note at the plant. The latter was eventually lost, and an operator resumed using the original procedure contained in the standard operating procedures manual. As a result, another fire occurred.

11.6.2. Abnormal Operations

Procedures should inform operators of the actions required when operating conditions deviate beyond defined limits. It is necessary to clearly indicate what operators are required to do, how and when it is to be done, and also to what degree operators are permitted to use their initiative.

Many accidents have occurred when operators have recorded abnormal readings on a log sheet and either have taken no further action or have taken incorrect action. An example of this is a case where failure of a water-cooled heat exchanger for chlorine gas occurred. A low-pH alarm sounded, the operator added soda ash to the cooling water system and the alarm went out. The alarm sounded again shortly thereafter. The operator had failed to understand the purpose of the alarm, which was to warn of tube failure. Operating procedures were rewritten to prevent a repeat of this incident.

11.6.3. Divided Responsibilities

When different plants or teams can use all or part of the same process equipment (e.g., blowdown or vent systems) the division of responsibilities should be clearly defined in the procedure. There should be no ambiguities as to the responsibility for executing each part of the procedure.

A foreman, concerned about the flammability of alcohol fumes from a atmospheric tank vent, connected a hose from the vent outlet to a drum of water before welders worked near the tank's roof. However, after the hose was connected, operations started to pump out the tank. The sealed vent allowed a vacuum to be created within the tank, causing the roof of the tank to partially collapse.

Thus, in this case, the apparent absence of adequate written procedures detailing the responsibilities of both the operations and maintenance personnel, as well as their lines of communications and authority, caused a costly and potentially dangerous incident.

11.6.4. Precise Requirements

It is important to ensure that procedures contains all necessary details. One of the reasons for the explosion in Pasadena, Texas in October 1989 was that the plant procedure had not incorporated the requirements for the use of double block and bleed procedures that had previously been identified and made part of the corporate policy. This breakdown in documentation contributed to the incident.

11.7. References

AIChE–CCPS, *Guidelines for Auditing Process Safety Management Systems,* 1993.

Kletz, Trevor A., *An Engineer's View of Human Error Second Edition,* Institution of Chemical Engineers, U.K., 1991.

Kletz, Trevor A., *Lessons from Disaster: How Organizations Have No Memory and Accidents Recur,* Institution of Chemical Engineers, U.K., 1993.

Kletz, Trevor A., *What Went Wrong? Case Histories of Process Plant Disasters,* Second Edition, Gulf Publishing Co., 1991.

Sutton, I.S., *Writing Operating Procedures,* Second Edition, Science Applications International Corp., 1992.

12
Training

12.1. Overview

12.1.1. Introduction

Training is basically the communication of knowledge and skills. In the chemical and hydrocarbon processing industries, the training function includes passing on process knowledge, skills in performing operating and emergency procedures, and many other aspects of process operation and maintenance to those who actually perform the tasks. Process safety management (PSM) programs should include safety-oriented training addressing operating, maintenance, and emergency response procedures. Industry codes and government regulations recommend, and, in some cases, require this training. Consequently, PSM programs need to identify and document training requirements covering all levels of management and employees, including contractor personnel, who manage, operate, maintain, and support the process units. The reader may also wish to refer to Chapter 17, *Contractor Issues*.

12.1.2. Goals and Benefits

The primary goal of training is to ensure that operating, technical and maintenance personnel, including contractors, are aware of important PSM-related concepts, such as:

- hazards present in the process;
- the significance of their actions relative to process safety;
- how to operate and maintain the process within safe operating limits as specified in established process safety information; and
- how to handle potential emergencies in the process area.

Effective training programs can reduce the number and severity of incidents arising from process operations, and may be instrumental in preventing small problems from developing into significant incidents. Training programs accomplish this by helping employees understand the nature and causes of potential and historical problems arising from process operations. They also increase employee awareness of process hazards.

Documentation of the training element is required to ensure that the training is appropriate to the need, that it is carried out, and that it is properly understood. The system should be flexible to allow for the ongoing, changing nature of training requirements, especially those due to refresher training, changes in the process, and changes in personnel.

12.2. Training Program Design

Employees, including maintenance and contractor employees, involved with hazardous chemicals should understand the safety and health hazards of the chemicals and processes they work with for the protection of themselves, their fellow employees and residents of nearby communities. Hazard communication training will assist employees in understaning the information contained in Material Safety Data Sheets (MSDSs). However, additional training in subjects such as operating procedures, safe work procedures and practices, emergency evacuation and response, routine and nonroutine work activities, and other subjects pertinent to process safety and health should also be covered by an employer's training program. Refer to Appendix 12A for a typical listing of training topics.

In documenting their training program, employers should clearly identify the employees to be trained, what subjects require coverage, and the goals and objectives they seek to achieve with the training. Careful consideration should be given to ensure that employees, including maintenance and contract employees, receive both initial and updated training, as situations warrant. For example, if modifications are made to a process or to equipment, affected employees should be trained to use the revised operating procedures or equipment maintenance procedures and understand the affects of the changes on their job tasks. This training should be documented for each employee.

It is important that the population to be trained not be too narrowly defined. Certainly, operators and maintenance personnel require training; however, other individuals potentially exposed to process hazards require training commensurate to their degree of exposure and job function. This group could include managers, supervisors, clerks, secretaries, janitors, fork lift drivers, etc.

12.2.1. Types of Training Methods

Four types of complementary training implementation methods generally used are as follows:

- formal training (e.g., attendance and participation in a training course on-site or off-site);

- informal training:
  - on-site safety meetings, seminars, vendor equipment demonstrations and similar sessions; and
  - relatively short-term training for a few hours to a few days at one time;
- self-study and/or self-paced training modules; and
- directed on-the-job training (OJT), where the trainer performs more as a coach than a teacher.

12.2.2. Initial Training

Initial training should be given to everyone involved in operating and maintaining a process, including contractors. Refer to Appendix 12A for a listing of typical training topics. In some cases, such training is a regulatory requirement. Appendix 12B provides a partial list of OSHA regulations with training implications. These requirements should be followed where applicable.

An example of an initial operator sequence could include the following:

- overview of the process, including process safety information;
- specific operating procedures;
- basic operator knowledge;
- safety and health hazards;
- safe work practices; and
- emergency response and evacuation. (See Chapter 13, *Emergency Response*).

Training in procedures for operations and maintenance employees and contractors should emphasize the specific safety and health hazards, emergency conditions including shutdown and evacuation, and safe work practices applicable to the employee's specific job task in the process area. Training should provide effective instruction for conducting all activities safely.

Standardized lesson plans which include learning objectives or goals, written training materials (e.g., training manuals), standard methods of measuring competence, and training aids suited to the process (e.g., simulators, hands-on training and videos), provide good references for the trainer and for internal evaluations and audits of the training program.

The employer should systematically identify and record all training activities for each employee. The documentation should show how the employer verified that the employee had the requisite knowledge plus skills and abilities (e.g., years of experience, previous training or other methods).

12.2.3. Refresher Training

Refresher training should be given to those involved in operating and where required, those maintaining a process, at a frequency necessary to maintain proficiency and to meet applicable regulatory requirements. The employer may

	Management	First Line Supervisors	Control Room Operators	Field Operators	Material Handlers	Maintenance	QC Lab Technicians	Engineers	Envir. Health & Safety
Plant Safety Organization	1	1	1	1	1	1	1	1	1
Basic Fire Safety	3	1	2	1	1	2	2	3	1
Managerial Responsibilities for Safety	1	1	—	—	—	—	—	2	1
Risk Management	3	3	—	—	—	—	—	3	3
Emergency Management	3	2	3	3	3	3	3	3	2
Hazard Identification	0	2	2	2	2	2	2	1	1
Mathematics	—	—	—	—	—	—	—	3	3
Material Incompatibility	0	2	2	2	1	2	2	2	2
Accident Reporting Procedures	2	2	2	2	2	2	2	2	2
Safe Handling of Toxic Materials	0	1	3	2	1	2	2	2	2
Administrative Control and Approval Procedures for Process Safety	3	1	1	2	2	2	—	3	2

Training Frequency

0 = Initial Appreciation
1 = Annually
2 = Every Two Years
3 = Every Three Years

FIGURE 12-1. Example Refresher Training Timetable

wish to consider a refresher training timetable in the form of a matrix such as the example in Figure 12-1.

Special training may be required for employees changing jobs or processes, or for those who have not worked a job for some time.

12.2.4. Training Evaluation

Employers need to periodically evaluate their training program to see if the necessary knowledge and skills are imparted, and if procedures are properly understood and implemented by their trained employees. This is particularly important in PSM-related training where potentially dangerous incidents could result from improper operating or maintenance practices. The means or method for evaluating the training should be developed and documented along with the training program goals and objectives. Training program evaluation will help employers determine the amount of training their employees understood, and whether the desired results were obtained. If, after the evaluation, it appears that the employees are not at the level of knowledge and skills that was expected, the employer should revise the training program, provide retraining, or provide more frequent refresher sessions until the deficiency is resolved. Such remedial training should be documented.

Those who conducted the training and those who received the training should also be consulted to determine the best way of improving the training process.

Internal Evaluation. The primary purpose of internal evaluation is to determine whether the instructional development effort has accomplished what was intended. The effectiveness of the instruction can be determined through systematic analysis of written, oral, or performance-based test results. If the trainee can properly show understanding and/or perform what was taught, the instruction was successful. This does not necessarily point out whether the method used was the most effective and efficient, nor if, or how well, those trained will be able to apply the knowledge in real, sometimes stressful, job environments.

External Evaluation. The primary purpose for conducting external evaluation is to find out whether trainees who successfully complete training can perform the job for which they were trained while meeting the established job performance standards. A second purpose of external evaluation is to ensure that the job tasks, as implemented in the field, have not changed significantly since the training program was developed.

External evaluation is different from internal evaluation in two major ways. Internal evaluation is conducted as part of the actual instruction; external evaluation is conducted after the trainees have completed the instruction and have been assigned to the job. Internal evaluation is primarily concerned with whether the learning objectives have been mastered by the trainees while a well-executed external evaluation will result in finding:

- that trainees who complete the course will be able to do the job at the intended skill level;
- data that indicates trainees are not able to perform requirements of the job at the planned skill level, and on which of the job tasks they are deficient; or
- that the job requirements have changed since creation of the training course content.

12.2.5. Training Module Guidelines

This section is intended to provide employers with a model for designing, conducting, evaluating and revising training programs. The model consists of:

- identifying training program needs;
- identifying training objectives;
- developing learning activities;
- conducting the training;
- evaluating program effectiveness; and
- improving the program.

Identifying Training Program Needs. The first step in developing the program is to determine what training is needed. It is important to first identify employees working with hazardous chemicals, or who are potentially affected by these operations, in order to ensure that the training given meets the needs. It is also necessary to identify what the employee is expected to do and in what ways, if any, the employee's performance can be enhanced. This information can be obtained by conducting a job task analysis which determines the skills and knowledge an employee needs to perform a job safely. Refer to Chapter 9, *Human Factors*.

The various stages in determining training needs may include the following:

- identification of employees working with hazardous substances, considering:
  —employee's experience;
  —nature and degree of hazards;
  —statistical accident data;
  —nature of hazardous materials handled; and
  —actual tasks performed.
- identification of key tasks which require particular emphasis during training based upon:
  —tasks to be performed; and
  —use of job hazard (task or safety) analysis.
- identification of any specific regulatory requirements; and
- involvement of employees in assessing the needs of the program.

The *Training Requirements of OSHA Standards*, OSHA No. 2254 provides reference resource material which can be helpful in identifying training needs.

Identifying Training Objectives. Once the employees' training needs have been identified, employers can prepare objectives for the training. Clearly documented instructional objectives will show what employees are expected to learn as a result of the training.

Learning objectives should be well thought out before the training development begins. For an objective to be effective it should identify, as precisely as possible, what the individuals should do to demonstrate that they have learned the intended skills, or that the objective has been reached. The objectives should also describe the important conditions under which the individuals will demonstrate competence and define what constitutes acceptable performance.

Developing Learning Activities. Once employers have documented the objectives for each training program, the content of the training and specific learning activities should be identified and described. Learning activities enable employees to acquire the desired skills and knowledge. To ensure that employees transfer and apply the skills or knowledge from the learning activity to the job, the learning situation should include techniques which reasonably simulate the actual job activities. Documentation of the training modules should include an outline, training manual, visual aids, exercises and tests to verify proficiency.

Conducting the Training. Based on information generated by the tasks just outlined, the employer should be ready to begin conducting the training. The training should be presented in an unambiguous manner so that its organization and meaning are clear to the employees and should:

- provide overviews of the material to be learned;
- wherever possible, relate the new information or skills to the employees' goals, interests or experience; and
- reinforce what the employees have learned by summarizing the program's objectives and the key points of information covered.

Employers should also develop the content, structure, and format of the training. The content developed for the program, the nature of the workplace or other training site, and the resources available for training are all factors in helping employers determine the frequency and the length of the sessions, the instructional techniques, and the individuals best qualified to conduct the training.

Evaluating Program Effectiveness. An evaluation of the training is necessary to ensure the training program is accomplishing its goals. A plan for evaluating the training sessions should be developed when the course objectives and content are developed. Evaluation will assist employers in determining the amount of learning achieved and whether an employee's on-the-job performance has improved sufficiently.

However it is conducted, the documented evaluation of training should give employers the information necessary to decide whether the employees achieved the desired results, whether revisions are required and how it might be improved.

Improving the Program. If, after evaluation, it is clear that the training did not give the employees the level of knowledge and skill that was expected, it may be necessary to revise the training program or provide retraining. Use of a concise evaluation form immediately after training, or retrospectively asking questions of employees and of those who conducted the training, may be of help. The documented results of these evaluations or questionnaires will typically identify both the deficiencies and possible methods of improving the learning process.

12.3. Training Documentation

12.3.1. Training Program Documentation

Organizations implementing a training program should include in their PSM program documentation a discussion of the goals of training as well as any general procedures that outline how the training program will be implemented. The PSM training program should cover all levels of management and employees who operate, manage, and maintain the process units. Both company employees and contractor employees should be included. The program documentation may address the following topics, as pertinent to the needs of the facility or organization:

- types of training needed;
- who should be trained;
- training content;
- what levels of training are necessary;
- training and refresher training schedules;
- criteria for determining the need for unscheduled refresher training;
- how the training is given;
- who will conduct the training and who will train the trainer;
- qualifications of the trainer;
- who will evaluate the training provided and how will this be accomplished:
  —for employees and contractors;
  —for trainers; and
- training record keeping requirements.

Responsibility and accountability for the various roles in the training function should be clearly outlined in the program document to ensure that the training program is properly executed.

12.3.2. Records from Implementing the Training Element

Organizations may wish to maintain records on training module content, instructor qualifications, training programs conducted, and training status of individual employees. Each of these is discussed further below.

Records of training modules may typically include:

- course title;
- course content and objective;
- training methods (seminar, OJT, home study, or other formats);
- training aids or course materials used;
- required instructor qualifications; and
- methods of evaluation (e.g., written or oral tests, performance).

A record of each training session given may include:

- identities of the employees;
- date and location of training;
- instructor name;
- the training covered, as identified in the training manual or other training resources;
- description of training course content and objectives; and
- the means used to verify that the employee understood the training.

Standard employee training records may include:

- employee identity;
- employee work history;
- current job descriptions and work locations;
- training programs attended and dates;
- purpose of training (initial, refresher, remedial); and
- evaluation results for the training programs attended.

Records maintained for instructors may include:

- instructor identity;
- instructor work history;
- training that instructor has received; and
- list of training courses that instructor is qualified to teach.

12.4. Records Management

12.4.1. Records Management Program

In addition to the description of the training program given in Section 12.3.1, the organization's PSM documentation should define training records management practices. The records management practices should provide access to all personnel have a job need for training information (e.g., operating group supervisors). However, at the same time, some degree of access control may be appropriate considering the sensitive nature of some of the information. This topic will be developed further in this section.

12.4.2. Media and Methods

Many organizations may find that a conventional, paper-based records system is sufficient to meet their needs. It should be noted, however, that, with the possible exception of completed examinations, few training records need be kept in this hard copy format. In fact, with computer-based training and testing, even examinations can be administered via computer and records can be maintained on electronic media. Thus, some organizations may wish to pursue the potential benefits of computer-based records management systems.

12.4.3. Responsibilities and Accountability

The training program documentation should clearly specify responsibility and accountability for compiling and maintaining training program records. Records management responsibilities may be distributed; e.g., descriptions of training program content and records of instructor qualifications might be maintained by a centralized training group, while individual operator training records may be maintained by operating group supervision.

12.4.4. Records Retrieval and Access Controls

Records should be readily retrievable by those having a legitimate job need for this information. A system which identifies and sorts records by employee name, Social Security number, employee identification number, job title, date of training, type of training, etc., should be considered. Such a cross-referenced system will facilitate record retrieval. Issue and expiration dates of training-related certificates or licenses should be readily available from the system.

It should be noted, however, that some organizations regard certain training records (e.g., training examination scores) to be sensitive and, perhaps, confidential information. Appropriate access controls should be considered to prevent unauthorized access to such records. The question of records access and confidentiality may need to be addressed by legal counsel during development of the program documentation.

12.4.5. Records Retention and Purge Schedule

Training records should be maintained for as long as organizational needs, or applicable regulatory requirements mandate. Records retention and purge schedules should be developed and included in the program documentation.

12.5. Auditing

The PSM program documentation should address auditing requirements for training documentation, including contractor training documentation, whether operations, maintenance or construction. The audit procedures should determine the involvement of contractor management and the transfer of audit results from the contractor to the facility owner. The audit program documentation may address:

- the training elements to be audited;
- what is to be audited in each element;
- any applicable regulatory requirements for PSM auditing of training;
- basic content or scope required for the audit; and
- any specific documentation requirements for audit results.

The first step in auditing training documentation is to verify that a documented program exists for implementing the training element. This program would include features outlined in Sections 12.3.1 and 12.4 that are pertinent to the need of the organization or facility. The second step in auditing is to verify that the results of implementing the training element have been documented in the manner prescribed by the program.

Any deficiencies identified during the audit must be documented and promptly resolved. Further guidance can be found in Chapter 14, *Auditing* and in the CCPS book, *Guidelines for Auditing Process Safety Management Systems.*

12.6. References

The following listing offers some sources that may prove useful in developing, implementing, and documenting training programs. This is not intended as a complete reference list, but was representative of applicable literature at time of publication of this book.

AIChE–CCPS, *Guidelines for Auditing Process Safety Management,* 1993.

AIChE–CCPS, *Plant Guidelines for Technical Management of Chemical Process Safety,* 1992.

American Petroleum Institute, *Management of Process Hazards,* RP 750, Washington, D.C., 1990 .

Chemical Manufacturers Association, *Process Safety Management (Control of Acute Hazards),* Washington, DC, 1985.

Crowl, D.A., and Louvar, J.F., *Chemical Process Safety: Fundamentals with Applications,* Prentice-Hall, Inc., Englewood Cliffs, NJ, 1990.

Kelly, Barbara, *Supervisor's Safety Meeting Handbook,* Business and Legal Reports, Inc., Madison, CT, Rev. 6/93.

Kletz, T.A., *What Went Wrong? Case Histories of Process Plant Disasters,* Gulf Publishing Co., Houston, TX, 1985.

Lees, F.P., *Loss Prevention in the Process Industries,* Butterworth, London, 1986.

Mager, R.F. and Pipe, P., *Analyzing Performance Problems of 'You Really Oughta Wanna',* 2nd Edition, David S. Lake Publishers, Belmont, CA, 1984.

Mager, R.F. and Beach, K.M. Jr., *Developing Vocational Instruction,* David S. Lake Publishers, Belmont, CA, 1984.

Mager, R.F., *Preparing Instructional Objectives,* 2nd Edition, David S. Lake Publishers, Belmont, CA, 1984.

National Society for Programmed Instruction, PO Box 137, Cardinal Station, Washington, DC 20017.

OSHA, *Training Requirements of OSHA Standards,* OSHA No. 2254 U.S. Department of Labor, U.S. Government Printing Office, Washington, DC 20402. Revised 1992.

ReVelle, J.B., *Safety Training Methods,* John Wiley and Sons, New York, 1980.

Roffel, B., and Rijnsdorp, J.E., *Process Dynamics, Control and Protection,* Ann Arbor Science, Ann Arbor, MI, 1982

Appendix 12A. Training Topics

The following list illustrates the scope of training requirements which may be considered for training of operators, maintenance workers and contractors in the performance of their individual tasks. The resultant records should document the appropriate information.

Topic/Title	*Subjects/Areas for Inclusion*
Process Description/Operation	Process function, operation and control and relationship to other facility units.
Start-Up Operations	Reasons and frequency, initial start-up, start-up following a turnaround and emergency shutdown.
Normal Operations	Operating conditions, controls, limits and routine maintenance.
Shutdown Conditions	Routine, partial and emergency.
Abnormal Operations	Alternatives, emergency alarms and conditions, operating problems and deviations.
Environment, Health and Safety	Chemical properties and related hazards, controls, precautions and equipment for prevention, remediation upon exposure, quality and inventory control, special hazards and safe work practices.
Emergencies/Incidents	Reporting, evacuation and response.
Maintenance	Preventive, scheduled and abnormal, monitoring and inspection procedures and special cases (spill clean-up, etc.)
Quality Assurance & Records	Appropriate inspection, documentation development and controls.

Additional guidance regarding training may be found in Chapter 11, *Operating Procedures*, Chapter 17, *Contractors Issues* and Chapter 18, *Permit-to-Work Systems*.

Appendix 12B. OSHA Regulations

The list given on the next page identifies the titles and numbers of OSHA regulations that include references to training and is not intended to be an exhaustive list of OSHA regulations that require or imply the need for training. Furthermore, this list does not identify the nature of the training requirements. Since OSHA regulations are subject to amendment, the user should seek appropriate counsel if a definitive list of regulatory requirements is desired.

Fed. Reg. Number	*Title/Subject*
29 CFR 1910.38	Employee Emergency Plans and Fire Prevention Plans
29 CFR 1910.66	Powered Platforms for Building Maintenance
29 CFR 1910.94	Ventilation
29 CFR 1910.95	Occupational Noise Exposure
29 CFR 1910.96	Ionizing Radiation
29 CFR 1910.106	Flammable and Combustible Liquids
29 CFR 1910.109	Explosive and Blasting Agents
29 CFR 1910.110	Storage and Handling of Liquefied Petroleum Gases
29 CFR 1910.111	Storage and Handling of Anhydrous Ammonia
29 CFR 1910.119	Process Safety Management of Highly Hazardous Chemicals
29 CFR 1910.120	Hazardous Waste Operations and Emergency Response
29 CFR 1910.134	Respiratory Protection
29 CFR 1910.145	Specifications for Accident Prevention Signs and Tags
29 CFR 1910.146	Permit-Required Confined Spaces for General Industry
29 CFR 1910.151	Medical Services and First Aid
29 CFR 1910.156	Fire Brigades
29 CFR 1910.157	Portable Fire Extinguishers
29 CFR 1910.160	Fixed Extinguishing Systems, General
29 CFR 1910.165	Employee Alarm Systems
29 CFR 1910.177	Servicing Multipiece and Single Piece Rim Wheels
29 CFR 1910.178	Powered Industrial Trucks
29 CFR 1910.179	Overhead and Gantry Cranes
29 CFR 1910.180	Crawler Locomotive and Truck Cranes
29 CFR 1910.217	Mechanical Power Presses
29 CFR 1910.218	Forging Machines
29 CFR 1910.252	Welding, Cutting and Brazing—General Requirements
29 CFR 1910.261	Pulp, Paper and Paperboard Mills
29 CFR 1910.268	Telecommunications
29 CFR 1910.332	Training (Relative to Electric Shock)
29 CFR 1910.1001	Asbestos
29 CFR 1910.1003–1018	(Specific Carcinogens)
29 CFR 1910.1025	Lead
29 CFR 1910.1029	Coke Oven Emissions
29 CFR 1910.1030	Bloodborne Pathogens
29 CFR 1910.1043	Cotton Dust
29 CFR 1910.1044	1, 2-Dibromo-3-Chloropropane
29 CFR 1910.1045	Acrylonitrile
29 CFR 1910.1200	Hazard Communication

29 CFR 1910.1450	Occupational Exposure to Hazardous Chemicals in Laboratories
29 CFR 1926.20	General Safety and Health Provisions
29 CFR 1926.21	Safety Training and Education
29 CFR 1926.50	Medical Service and First Aid
29 CFR 1926.58	Asbestos, Tremolite, Anthophyllite, and Actinolite
29 CFR 1926.59	Hazard Communication
29 CFR 1926.103	Respiratory Protection
29 CFR 1926.150	Fire Protection
29 CFR 1926.302	Power-Operated Hand Tools
29 CFR 1926.350	Gas Welding and Cutting
29 CFR 1926.351	Arc Welding and Cutting
29 CFR 1926.352	Fire Prevention
29 CFR 1926.500	Guardrails, Handrails and Covers
29 CFR 1926.803	Compressed Air
29 CFR 1926.901	Blaster Qualifications
29 CFR 1926.902	Surface Transportation of Explosives
29 CFR 1926.950	Power Transmission and Distribution—General Requirements
29 CFR 1926.1060	Stairways and Ladders—Training Requirements

It is the responsibility of the employer to ensure that all regulatory requirements for training are identified and carried out.

13

Emergency Response

13.1. Overview

13.1.1. Introduction

The prevention of incidents relies on good process safety management (PSM) practices. But even with such practices in place, incidents may still occur. Whenever hazardous chemicals are handled, some risk of an incident remains, despite well-engineered equipment, detailed operating procedures and highly trained personnel.

Should an incident occur, such as an explosion or a release into the environment, its effects may be localized to the immediate vicinity or the effects may extend off-site, depending on the magnitude of the incident. An understanding of possible incident scenarios, their potential effects and, perhaps, their likelihood, serves as the basis for rational emergency response planning. Proper planning allows the facility to prepare for dealing with both the incident and its effects. These plans may involve dedicated, specialized equipment or facilities, detailed procedures and specialized training of personnel. Clear, thorough documentation is critical to this effort.

13.1.2. Goals and Benefits

The goals of the emergency response program are:

- to minimize the impacts of an incident (including health effects, costs, legal implications, loss of corporate image, and deterioration of relationships with employees and with the general public);
- to identify aspects of other programs that affect emergency preparedness, and to address these in a practical way; and
- to participate in a cooperative effort with the public, i.e., the local community, through the Local Emergency Planning Committee (LEPC) and other appropriate agencies and officials.

A wide range of benefits can result from effectively documenting an emergency response program, especially when it involves the local community and authorities. These include:

- the speed and efficiency of response in an emergency can be significantly improved, bringing lower risk to employees and members of the community;
- relations between industry and the community can be improved by the implementation of a sound, well documented and well-publicized emergency response program;
- during program preparation, changes to the facility may be identified that will reduce risk without impairing process or plant operation;
- on-site personnel will be prepared by both knowledge and training to play a proper role in emergency response; and
- the written procedures in the response program are readily amenable to auditing and updating to ensure that they remain current and effective.

13.2. Description of Emergency Response

13.2.1. Objectives

Emergency response is not just the reaction to incidents and accidents that could pose a threat to personnel, facilities or the environment. As used here, the term also includes the proactive planning that goes into preparing the response to such events. Thus, the objectives of emergency response planning, documentation, and its application are to adequately prepare for and respond to a wide range of potential emergency situations that may be encountered in operating processes and plants that utilize or produce hazardous or toxic materials. These documented preparations allow the facility to have available, and promptly apply, the right kinds of trained manpower and equipment to reduce the effects of such emergency situations.

13.2.2. Planning for Emergency Response

The appropriate time to consider how to handle an incident is *before* it happens. Sound planning begins with the identification of the hazards present in a process or facility and the evaluation of credible incident scenarios, ranging from minor incidents to realistic worst case situations. The potential incident scenarios may be evaluated using an appropriate method of process hazard analysis (PHA). See Chapter 6, *Process Hazard Analysis*.

Next, it is important to gain a perspective on the potential magnitude of the effects of each incident scenario and, perhaps, an appreciation its likelihood. It important not to underestimate potential incidents simply because they have never happened or are considered unlikely. The effects (consequences) and likelihood (probability) of incident scenarios can be evaluated using techniques described in Chapter 7, *Chemical Process Quantitative Risk Analysis*.

In some cases (e.g., major explosions, fires, or toxic gas releases) the public may potentially be affected, or the assistance of public authorities may be required. If so, the plans and procedures should be public and allow for efficient interaction with the off-site agencies that may need to be involved. An example is the need to issue an advisory to evacuate or shelter-in-place.

Note that the planning responsibility for off-site emergency response lies principally with those in the public sector. This document does not cover these details, but does reference the interfaces between a company and the public authorities necessary to achieve an overall effective response.

Facilities handling hazardous materials may wish to coordinate their emergency response plan with off-site agencies even if the scale of potential incidents is not likely to involve any public areas. Even in some small incidents, the local police and fire departments, as well as other agencies, may be involved, and proper coordination of efforts should be pre-planned. It is important to note, and factor into the organization's emergency response planning, that LEPCs have been created throughout the country to coordinate local responses to emergency situations with statewide disaster response agencies. Coordination with LEPCs should be part of an emergency response plan.

Emergency response has been extensively studied by many organizations. Section 13.7 lists a number of references containing guidance that may be beneficial for use or adaptation by other organizations as part of their emergency response planning. One such source is the *Hazardous Materials Emergency Planning Guide,* NRT-1, first issued by the National Response Team in March 1987.

Table 13-1 lists the hazardous materials planning elements detailed in the NRT document. These planning elements have been summarized in Appendix 13A, where they have been edited as necessary to reflect an on-site emergency with off-site potential. The focus of Appendix 13A is on documentation requirements. Interpretive guidance has been added, where appropriate, in italics so as to distinguish it from the original NRT-1 text. The format and element numbering convention of NRT-1 have otherwise been retained for ease of reference. Portions of the NRT-1 text that were not pertinent to the documentation issue have been purposely omitted. These omissions have been marked with a ♦ ♦ ♦.

Many aspects of emergency response are covered by federal, state, or local regulations. Applicable regulatory requirements should be considered when developing an emergency response plan for a facility.

Further guidance on emergency planning and documentation will be provided in the CCPS book *Guidelines for Technical Planning for on-Site Emergencies*. *Guidelines for Technical Planning for on-Site Emergencies* may be more comprehensive than NRT-1 and may suggest documentation in addition to that discussed here. However, the documentation practices described in this book are applicable to any documentation needs identified elsewhere.

TABLE 13-1
NRT-1 Hazardous Material Planning Elements

This list is drawn from Chapter 5 of the *Hazardous Materials Emergency Planning Guide,* NRT-1, March 1987. Relevant aspects of these elements are addressed in Appendix 13A of this book.

ELEMENT A—INTRODUCTION

A.1 Incident Information Summary
A.2 Promulgation Document
A.3 Legal Authority and Responsibility for Responding
A.4 Table of Contents
A.5 Abbreviations and Definitions
A.6 Assumptions/Planning Factors
A.7 Concept of Operations
A.8 Instructions on Plan Use
A.9 Record of Amendments

ELEMENT B—EMERGENCY ASSISTANCE TELEPHONE ROSTER

ELEMENT C—RESPONSE FUNCTIONS

C.1 Initial Notification of Response Action
C.2 Direction and Control
C.3 Communications (among Responders)
C.4 Warning Systems and Emergency Public Notification
C.5 Public Information/Community Relations
C.6 Resource Management
C.7 Health and Medical
C.8 Response Personnel Safety
C.9 Personal Protection of Citizens
C.10 Fire and Rescue
C.11 Law Enforcement
C.12 Ongoing Incident Assessment
C.13 Human Services
C.14 Public Works
C.15 Others

ELEMENT D—CONTAINMENT AND CLEAN UP

D.1 Techniques for Spill Containment and Clean Up
D.2 Resources for Clean Up and Disposal

ELEMENT E—DOCUMENTATION AND INVESTIGATIVE FOLLOW UP

ELEMENT F—PROCEDURES FOR TESTING AND UPDATING PLAN

F.1 Testing the Plan
F.2 Updating the Plan

ELEMENT G—HAZARDS ANALYSIS (SUMMARY)

ELEMENT H–REFERENCES

H.1 Laboratory, Consultant, and other Technical Support Resources
H.2 Technical Library

13.3. Emergency Response Documentation

13.3.1. Emergency Response Program Documentation

Organizations implementing an emergency response program should document, as part of their PSM program documentation, the goals for the emergency response program, as well as any detailed procedures required to implement emergency response. Such program guidance would address, but not be limited to, the topics discussed below.

Responsibilities and Accountability for Emergency Response Planning. The organization's program documentation should clearly articulate responsibilities and accountability for this function. For example, at large sites, an individual (perhaps a member of the safety and security staff) might be assigned overall responsibility for the preparation of the facility emergency response plan, and be assisted by a team assembled from different areas of the facility.

During plan preparation, the team leader or his designate would act as liaison with the various off-site groups (e.g., other industrial facilities and public agencies in the vicinity of the facility). Assistance from knowledgeable personnel at such facilities and agencies ensures the availability of needed information, builds communication channels and trust between mutually dependent parties and, thus, can help ensure cohesion of the program.

Smaller organizations or facilities may require external assistance when developing the plan; however, the responsibility for the plan and its application remains with the facility.

The responsible person should ensure that the plan is periodically audited and updated, as required. Considerations that might prompt the need for updating the plan include:

- changes in the nature or magnitude of process hazards;
- significant changes in organization (either within the facility or in an external organization) that may affect emergency response capabilities;
- changes in communication systems (including phone numbers and radio frequencies);
- changes in external population distributions;
- changes in regulatory requirements; and
- recommendations made as a result of an audit, incident, or exercise.

Plan Content. The program should outline general requirements for the content of the emergency response plan. The responsible person at a site, with management approval, must then decide which specific elements (such as described in Section 13.2) should be included in the facility plan. Before making this determination, it is necessary to identify and evaluate a range of credible incident scenarios, in order to assess the extent of involvement of off-site agencies.

Plan Preparation and Timing. The complete documentation for emergency preparedness must be available before a new facility handles hazardous materials for the first time. This means that plan preparation must begin well ahead of receipt of chemicals and subsequent start-up. Depending on the project schedule, plan preparation could begin as early as during the design phase. The plan should be subsequently updated in a timely manner to reflect significant modifications to the facility, processes, or personnel.

Often, local communities will have an emergency response plan in place (e.g., prepared by the LEPC). If so, coordination with the facility plan must be established. If there is a potential requirement for a higher standard of external emergency response and/or aid than currently exists, the facility organization may need to take a lead role in assisting external organizations in upgrading their plans. Documentation will be a key aspect in defining these needs and supporting this effort.

Plan Review and Authorization. The program documentation should establish procedures for the review, and define authority levels for the approval, of the emergency response plan. While off-site organizations would typically not be involved in the approval of the plan, the issuing organization may find it beneficial to solicit their opinions during the review process.

Exercises and Testing. Experience has demonstrated that emergency response does not work satisfactorily unless it is properly planned, documented, and rehearsed. Training and familiarization of personnel in the contents of the plan and what is expected of them should be given regularly so they can respond rapidly and effectively when actual emergencies arise. Additionally, some form of periodic exercise or validation is important, to ensure that the system works as intended and that deficiencies are corrected as they are found.

The program for emergency response should contain requirements for performing and recording the results of such exercises. A procedure should be established for incorporating lessons learned into revisions of the plan.

13.3.2. Records from Implementing an Emergency Response Program

Section 13.2 and Appendix 13A give a very broad summary of the documentation needed to define and record the implementation of an emergency response program. Not all of this information will be required in every instance. For example, some facilities may have demonstrated, by calculations and examination of their processes, that they do not have the potential to cause harm off-site. These plants may have documentation that is limited to their own facility with minimal external interface.

The following paragraphs supplement, or elaborate on, the documentation requirements described in Appendix 13A.

Organization Charts. Among the documents that are a necessary part of an emergency response plan are organization charts for bóth on-site management and external organizations (e.g., surrounding industrial facilities and public emergency response agencies). These charts should be prepared early in the development of a response plan since they: can improve the quality of the planning by highlighting levels of responsibility and authority for participating organizations; and may uncover required functions inadvertently left unassigned or unrecognized.

The chart for the on-site organization should show the chain of command from the manager to the lead member of each team on-site and should include the chain of command for any specialist personnel (e.g. security staff, safety engineers, fire department, etc.) involved. The chart should highlight the names and phone numbers of personnel who should be contacted during an emergency, as well as those of their alternates.

The off-site organization document should include the organizations that could be involved in an emergency and the links between them.

Documentation of Incident Reporting. Speed and accuracy are essential in initial documentation efforts; extremely detailed information is not necessary and could inhibit or delay reports. For example, one company that trains all its personnel how to report incidents efficiently requires only the following information:

- name and location of the person reporting the incident;
- location of the incident (giving the unit, tank or vessel number); and
- nature of the incident (spill, fire, explosion, etc.).

This organization's preplanned emergency procedures are set up to deal with this type of brief report, based on the tank or vessel named. More knowledgeable personnel are notified with the information collected in the first stage of reporting; the second stage is, of necessity, more comprehensive.

Refer to Chapter 15, *Incident Investigation,* for further discussion on documentation.

Real-Time Emergency Response Computers. Some facilities, typically installations storing or handling toxic or flammable gases, have installed computer systems to predict and/or monitor the travel of toxic gas clouds. These systems can be used in a *real-time* mode to assist in evaluating emergency response options. Where such a system exists, the emergency response plan documentation should include a sufficiently detailed description of the system, its capabilities and use, its inputs and outputs, backup capabilities, etc.

Such systems often monitor and record information from meteorological data towers and/or remote gas sensors. This information, along with the results of calculations made by the system, should be protected and documented as part of the incident record.

Some systems of this type include:

- CHARM (Radian Corp., Austin, TX);
- DAISY (Dow Chemical Co., Midland, MI);
- HASTE (ERT, Concord, MA);
- MIDAS (Pickard, Lowe and Garrick Inc., Newport Beach, CA);
- PLANT SAFE (Pacer Systems Inc, Horsham, PA); and
- SAFER (DuPont Safer Systems, Westlake Village, CA).

13.4. Records Management

The organization's program documentation should establish practices to be followed for managing records related to the emergency response effort. Responsibilities and accountability for records management should be clearly defined. This section addresses pertinent records management issues. The reader may wish to refer to Chapter 4, *Records Management*, for more general guidance on practices and procedures.

13.4.1. Where Are Records Maintained?

One important aspect of the records management task for an emergency response program is the diverse nature of the program documentation and the requirement that it be maintained in a variety of locations, including some that might be off-site and beyond the control of the issuing organization. Locations that may require some portion, or perhaps all, of the documentation include:

- facility control room;
- training center;
- site emergency control center;
- backup control center (required if there are credible scenarios that could prompt evacuation from the site);
- site fire department;
- site management offices;
- corporate office; and
- local municipal and/or emergency response agencies (city and/or county administration, fire, police, hospitals, and emergency mutual aid organizations).

The facility issuing the documentation should determine, in conjunction with other involved organizations, the appropriate locations, and decide which documents are required at each location to meet the needs of the group or function at that location. Particular care should be given to making these determinations. During an incident, there is seldom time to look for information; such information must be readily available to allow decisive, effective response.

Appropriate records control procedures should be implemented to ensure that updates and revisions are properly distributed, and that obsolete documentation is purged.

13.4.2. What Media Should Be Used?

A significant portion of the documentation will be in the form of hard copy documents. These would include procedures manuals, charts or maps, reference books and other similar records.

As previously noted, some facilities utilize computer modeling systems to monitor and/or predict the dispersion of toxic or flammable releases. These systems can be used to manage data such as:

- a list of substances that are held on-site and that can be modeled;
- physical and toxicological properties of these substances;
- local meteorological data (based on historical records and/or real-time input); and
- map of the facility and surrounding regions (including key buildings and roadways, population distributions, etc.).

Conventional computer-based data management systems can be used to store and quickly retrieve other pertinent data, such as:

- pertinent process information, such as discussed in Chapter 5, *Process Knowledge*:
- emergency response procedures and information about evacuation needs; and
- other information required in emergencies (e.g., organization charts and telephone numbers).

Back-up is required for computer-based systems and may include either hard copy duplicates of the computer databases or redundant computer capabilities (e.g., a second database on a battery-powered laptop computer, to serve in the event of a power failure).

13.5. Auditing

The PSM program documentation should establish auditing requirements for the emergency response program. The first step in auditing emergency response is to verify that a documented program, such as described in Section 13.3.1 and Appendix 13A, exists for implementing emergency response.

The second step is to verify that the results of implementing emergency response have been documented in the manner described by the program. Particular emphasis should be placed on the documentation of the resolution of recommendations resulting from incidents and drills.

Any deficiencies identified during the audit must be documented and resolved in a timely manner. Further details on auditing procedures and methods can be found in Chapter 14, *Auditing*, and Chapter 12, Appendix A of the CCPS book, *Guidelines for Auditing Process Safety Management Systems.*

Emergency response procedures can quickly become out of date as facilities, personnel, processes, and requirements change. Periodic audits can help identify where such changes have occurred and, thus, assist in keeping the emergency response program current.

13.6. Examples

13.6.1. Lack of Preparedness

Lack of preparedness has led to massive environmental pollution from firewater runoff. Examples include herbicides from a warehouse fire in England in February 1982, and agricultural chemicals (including insecticides and mercury compounds) from a fire in Switzerland in November 1986.

Such problems are foreseeable and require pre-planning, including documentation, to mitigate or prevent the massive pollution of lakes, rivers, and water supplies.

13.6.2. Effective Evacuation Planning

One of the best known examples of large-scale evacuation took place at Mississauga, Ontario, in 1979. A train derailed, several rail tank cars of flammable liquid BLEVE'd, and a chlorine rail tank car suffered a large burn hole in the fire. Most of the immediate chlorine release was drawn aloft in the fireballs, but the residual chlorine in the tank car remained for several days, slowly boiling off. The Borough of Peel had a comprehensive emergency preparedness program in place, and it was put into effect. In seven stages, nearly a quarter of a million people were evacuated from their homes, some for several days. The initial stages involved small numbers of people, but this quickly changed as the potential impact of the incident was evaluated. Some later stages required evacuation from different downwind directions as the wind shifted. No fatalities or injuries occurred; the key to the successful evacuation was the well-prepared and documented plan, an efficient core team to manage the situation, and good communications.

13.7. References

The documents listed below include many that provide useful information to assist in preparing emergency plans for hazardous material handling sites, where the hazard can extend off-site into public areas. While many of these documents have been written for public emergency response to a wide variety of specific hazards, from hurricanes to nuclear attack, much of the guidance they contain is applicable for any hazard.

The publication dates shown on the documents below are those available at the time of publication of these *Guidelines*. Where later editions are published, these should be used.

AIChE-CCPS, *Guidelines for Auditing Process Safety Management Systems,* 1993

AIChE-CCPS, *Guidelines for Technical Planning for On-Site Emergencies*, work-in-progress

Chemical Manufacturers' Association, *Community Emergency Response Guidebook,* 1992

Chlorine Institute, *Chlorine Institute Public Information Handbook for Emergency Planning and Community Right-to-Know,* 1992

Federal Emergency Management Administration, *Disaster Planning Guide for Business and Industry,* FEMA 141, August, 1987 (many useful checklists)

Federal Emergency Management Administration, *Guide for Development of State and Local Emergency Operations Plans,* FEMA CPG1-8, October, 1985

National Association of Manufacturers, *Preparing for Emergency Planning*

National Institute for Chemical Studies/University of Charleston (WV), *Protecting the Public in a Hazardous Material Emergency,* September, 1988

National Response Team, *Hazardous Materials Emergency Planning Guide,* NRT-1, March, 1987

Occupational Safety and Health Administration, *HAZWOPER Compliance Directive CPL2-2.59,* October 25, 1993; 29 CFR 1910.120(q), Emergency Response to Hazardous Substance Releases

Appendix 13A. NRT-1 Hazardous Material Planning Elements

The portions of NRT-1 included in this appendix are from Chapter 5: HAZARDOUS MATERIALS PLANNING ELEMENTS, Sub-Chapter 5.2 (pp. 40–66). The following text has been edited and supplemented as described in Section 13.2.2.

Planning Element A: Introduction

Planning Element A.1: Incident Information Summary

❑ Develop a format or logsheet for recording (in the emergency control center) essential information about the incident:

- Date and time
- Name of person receiving call
- Location
- Nature (e.g., leak, explosion, spill, fire)
- Time of release
- Possible health effects/medical emergency information

- Name of material(s) released, if known
  —Manifest/shipping invoice/billing label
  —Shipper/manufacturer identification
  —Container type (e.g., *tank, vessel,* truck, rail car, pipeline, drum)
  —Railcar/truck 4-digit identification numbers
  —Placard/label information *(HAZCOM labels)*
- Characteristics of material (e.g., color, smell, physical effects), if readily detectable
- Present physical state of the material (i.e., gas, liquid, solid)
- Total amount of material that may be released
- Other hazardous materials in area
- Amount of material released so far/duration of release
- Whether significant amounts of the material appear to be entering the atmosphere, nearby water, storm drains, or soil
- Direction, height, color, odor of any vapor clouds or plumes
- Personnel at the scene
- Number of casualties, or negative report

♦ ♦ ♦

This initial emergency response notification section should be:

BRIEF—never more than one page in length.

EASILY ACCESSIBLE—usually located on the cover or first page of the plan. It should also be repeated at least once inside the plan, in case the cover is torn off. *Blank forms should be readily available in the Emergency Response Center for immediate use.*

SIMPLE—Reporting information and emergency telephone numbers should be kept to a minimum.

♦ ♦ ♦

Much of the information will not come with the initial report but should be filled in as soon as possible.

Planning Element A.2: Promulgation Document

❑ Statement of plan authority

♦ ♦ ♦

This document should be signed by the most senior manager and should list all parties and individuals responsible for any part of the emergency preparedness plan, particularly who is to be in charge and who is authorized to take any necessary action. A 24-hour basis should be assumed.

Planning Element A.3: Legal Authority and Responsibility for Responding

The main document may list the following for reference:

❑ Authorizing legislation and regulations

♦ ♦ ♦

❑ Mandated agency responsibilities
❑ Letters of agreement

♦ ♦ ♦

Site Management—The following personnel or those who carry out these roles:

- site manager/plant manager;
- plant superintendent;
- emergency coordinator (shift);
- safety and security officer;
- site fire chief;
- environmental officer; and
- site medical staff.

Corporate Management—Those who may not normally be present at this site:

- corporate vice president responsible for safety;
- corporate safety manager;
- corporate counsel;
- corporate personnel manager;
- corporate public relations manager; and
- corporate line management responsible for site.

Planning Element A.4: Table of Contents
Comment: All sections of the plan should be listed here and should be clearly labeled with a tab for easy access.

Planning Element A.5: Abbreviations and Definitions
Comment: Frequently used abbreviations, acronyms, and definitions should be gathered here for easy reference.

Planning Element A.6: Assumptions/Planning Factors
❑ Geography—*Discuss on-site and off-site separately*

- Sensitive environmental areas
- Land use (actual and potential, in accordance with local development codes)
- Water supplies
- Public transportation network (roads, trains, buses)
- Population density
- Particularly sensitive institutions (e.g., schools, hospitals, homes for the aged)
- *Staging area or off-site assembly points*

❑ Climate/weather statistics

- ❑ Time variables (e.g., rush hour, vacation season)
- ❑ Particular characteristics of each facility and the transportation routes for which the plan is intended.
    - Onsite details
    - Neighboring population
    - Surrounding terrain
    - Known impediments (tunnels, bridges)
    - Other areas at risk
- ❑ Assumptions—*Include advance judgments concerning what would happen in the case of an accidental spill or release*

♦ ♦ ♦

Appropriate maps may be included in this section. Maps should show air intakes, water intakes, environmentally sensitive areas, major chemical manufacturing or storage facilities, population centers, and the location of response resources.

Planning Element A.7: Concept of Operations

Planning Element A.7a: Governing Principles
Comment: The plan should include brief statements of precisely what is expected to be accomplished if an incident should occur.

Planning Element A.7b: Organization Roles and Responsibilities—*Internal*

♦ ♦ ♦

Comment: This section lists all organizations and officials responsible for planning and/or executing the **pre-response** (planning and prevention), **response** (implementing the plan during an incident), and **post-response** (cleanup and restoration) activities to a hazardous materials incident. **One organization should be given command** and control responsibility for each of these three phases of the emergency response. The role of each organization/official should be clearly described. The plan should clearly designate who is in charge and should anticipate the potential involvement of state and federal agencies and other response organizations.

♦ ♦ ♦

Reference should be made to any arrangements made with neighboring facilities *(e.g., a Mutual Aid System)*. Details of how such a system operates must be kept with the documentation held at the Emergency Control Center.

Planning Element A.7c: Organization Roles and Responsibilities—External

♦ ♦ ♦

This section should include information about municipal, county, state, and federal organizations, agencies, and officials who may be involved in any external emergency plan. These details can be obtained from the local area plan.

Planning Element A.8: Instructions on Plan Use

Planning Element A.8a: Purpose
Comment: This section should be a clear and succinct statement of when and how the plan is meant to be used.

♦ ♦ ♦

Planning Element A.8b: Plan Distribution
❑ List of organizations/persons receiving plan.

♦ ♦ ♦

Planning Element A.9: Record of Amendments

❑ Change record sheet—*include columns for:*
- Date of change
- Recording signature
- Page numbers of changes made

♦ ♦ ♦

Planning Element B: Emergency Assistance Telephone Roster

❑ List of telephone numbers for:
- Participating agencies
- Technical and response personnel
- CHEMTREC
- Public and private sector support groups
- National Response Center

Comment: An accurate and up-to-date emergency telephone roster is essential. The name of a contact person (and alternate) and the telephone number should be listed. Briefly indicate the types of expertise, services, or equipment that each agency or group can provide. Indicate the times of day when the number will be answered; *note all 24-hour telephone numbers. Include FAX numbers if possible.* All phone numbers and names of personnel should be verified regularly. When alternate numbers are available, these should be listed. The list should be dated to aid identification of obsolete listings. This section of the plan should stand alone so that copies can be carried by emergency response personnel and others. Examples of organizations for possible inclusion in a telephone roster are as follows:
❑ **Telephone Roster**—*On-Site/Corporate*
- Site manager

- All plant superintendents
- Safety and security officer
- Medical staff
- Fire chief and officers
- Corporate officials

❑ **Telephone Roster**—*External*

The list should specifically include 24-hour and/or 800 numbers of agencies that could provide emergency assistance and/or advice, for example,

- *Local officials named in the local emergency response plan*
- CHEMTREC
- CHEMNET
- CHLOREP
- NACA Pesticide Safety Team
- Association of American Railroads
- *Federal Aviation Administration (to close air space if necessary).*
- *LEPC (Local Emergency Planning Committee)*
- *U.S. Coast Guard*

Planning Element C: Response Functions

Comment: Each function should be clearly marked with a tab so that it can be located quickly. When revising and updating a plan, companies might decide to add, delete, or combine individual functions.

Each response "function" usually includes several response activities. Some communities prepare a matrix that lists all response agencies down the left side of the page and all response activities across the top of the page. Planners can then easily determine which response activities need interagency coordination and which, if any, activities are not adequately provided for in the plan.

Function C.1: Initial Notification of Response Agencies

❑ 24-hour emergency response hot line telephone numbers

- Local number to notify area public officials and response personnel
- Number to notify state authorities
- *Number of the* National Response Center (800-424-8802; or 202-267-2675 in Washington, DC area)

❑ Other agencies (with telephone numbers) to notify immediately (e.g., hospitals, health department, Red Cross)

Comment: The local 24-hour emergency response hotline should be called first and therefore should have a prominent place in the plan. Provision should be made for notifying nearby municipalities and counties that could be affected by a vapor cloud or liquid plumes in a water supply.

♦ ♦ ♦

The plan should indicate how volunteer and off-duty personnel will be summoned. Similarly, there should be a method to notify special facilities (e.g., school districts, private schools, nursing homes, day care centers, industries, detention centers), according to the severity of the incident.

Function C.2: Direction and Control

- ❑ Name of on-scene authority
- ❑ Chain of command (illustrated in a block diagram)
- ❑ Criteria for activating emergency operating center
- ❑ Method for establishing on-scene command post and communications net work for response team(s)
- ❑ Method for activating emergency response teams
- ❑ List of priorities for response actions
- ❑ Levels of response based on incident severity

♦ ♦ ♦

Responsibilities:

Work out, in advance, the following:

1. Who will be in charge (lead organization)
2. What will be the chain of command
3. Who will activate the emergency operating center, if required
4. Who will maintain the on-scene command post and keep it secure
5. Who will have advisory roles (and what their precise roles are)
6. Who will make the technical recommendations on response actions to the lead agency
7. Who (if anyone) will have veto power
8. Who is responsible for requesting assistance from outside the community

This chain of command should be clearly illustrated in a block diagram.

Response action checklists are a way of condensing much useful information. They are helpful for a quick assessment of the response operation. If checklists are used, they should be prepared in sufficient detail to ensure that all crucial activities are included.

Planners should consider whether to have categories of response actions based on severity. The severity of an incident influences decisions on the level (or degree) of response to be made. This will determine how much equipment and how many personnel will be called, the extent of evacuation, and other factors.

The following chart summarizes who and what are involved in three typical emergency conditions. Information about the three response levels should be provided to special facilities (e.g., school districts, private schools, day care centers, hospitals, nursing homes, industries, detention centers).

Chart of Response Levels

Response Level	Description	Contact
I. Potential Emergency Condition	An incident or threat of a release that can be controlled by the first response agencies and does not require evacuation of other than the involved structure or the immediate outdoor area. The incident is confined to a small area and does not pose an immediate threat to life or property.	Fire Department Emergency Medical Services Police Department Partial EOC Staff Public Information Office CHEMTREC National Response Center *Local Emergency Planning Committee (LEPC)*
II. Limited Emergency Condition	An incident involving a greater hazard or larger area that poses a potential threat to life or property and that may require a limited evacuation of the surrounding area.	All Agencies in Level I HAZMAT Teams EOC Staff Public Works Department Health Department Red Cross County Emergency Management Agency State Police Public Utilities Media
III. Full Emergency Condition	An incident involving a severe hazard or a large area that poses an extreme threat to life and property and will probably require a large-scale evacuation; or an incident requiring the expertise or resources of county, state, federal, or private agencies/organizations.	All Level I and II Agencies plus the following as needed: Mutual Aid Fire, Police. Emergency Medical State Emergency Management Agency State Department of Environmental Resources State Department of Health EPA USCG ATSDR FEMA OSC/RRT

Function C.3: Communications (among Responders)

❑ Any form(s) of exchanging information or ideas for emergency response with other entities, either internal or external to the existing organizational structure.

♦ ♦ ♦

The plan should refer to the radio frequencies used by potentially involved emergency responders since such frequencies are seldom common. It is vital that communications are not hindered by this problem.

Function C.4: Warning Systems and Emergency Public Notification

Two distinct phases exist: on-site and off-site.

The primary plan should cover on-site warning systems. These might include:

- *visible (flashing light alarms);*
- *audible alarms (siren, warbler, whistle);*
- *public address (loudspeaker broadcast); and*
- *dedicated telephone system (one-on-one to control centers).*

Coverage should be included for fire, toxic gas, and evacuation. Testing should be required to determine that the alarm can seen or heard in the required coverage areas. Potentially affected personnel should be trained on what the different signals mean and the appropriate response to each signal. Visitors to a facility having such systems should either have such training before entry or be escorted while on the facility.

Off-site plans should include the following, to be activated according to the overall plan:

❑ Method for alerting the public.

- Title and telephone number of person responsible for alerting the public **as soon as word of the incident is received.**
- List of essential data to be passed on (e.g., health hazards, precautions for personal protection, evacuation routes and shelters, hospitals to be used) *or a prescribed plan contained in the documentation.*

Comment: This section should contain precise information about how any sirens or other signals will be used to alert the public in case of an emergency. This should include information on what the different signals mean, how to coordinate the use of sirens, and the geographic area covered by each siren. (If possible, a backup procedure should be identified.) While a siren alerts those who hear it, an emergency broadcast is necessary as backup to provide detailed information about the emergency and what people should do.

Sample Emergency Broadcast System (EBS) messages should be prepared with blank spaces that can be filled in with precise information about the accident. One sample message should provide fundamental information about the incident and urge citizens to remain calm and await further information and instructions. Another sample message should be for an evacuation. Another should describe any necessary school evacuations so that parents will know where their children are. A sample message should be prepared to tell citizens to take shelter and inform them of other precautions they may take to protect themselves. The message should clearly identify those areas in which protective actions are recommended, using familiar boundaries. Messages might be developed in languages other than English, if customarily spoken in the area.

The plan must document how and to whom warnings are to be provided in cases where life-threatening materials are released. In such cases, speed of action may be critical and the documentation should address this problem to ensure rapid and efficient implementation.

Function C.5: Public Information/Community Relations

Good public relations with the local community involves emergency preparedness and can be compromised if the plan is not comprehensive. This section should include:

- ❑ a method to educate the public for possible emergencies.
- ❑ a method for keeping the public informed.

◆ ◆ ◆

The emergency response plan should identify one person as spokesperson, responsible for public relations and for dealing with the media. This individual must be adequately briefed on the company's operations, and also on the incident occurring. It is important that this system be established and adequately documented. It is useful to identify a location where the media can assemble and be briefed.

◆ ◆ ◆

Function C.6: Resource Management

- ❑ List of personnel needed for emergency response.
- ❑ Training programs, including schedules for training of local emergency response and medical personnel.
- ❑ List of vehicles needed for emergency response.
- ❑ List of equipment (both heavy equipment and personal protective equipment) needed for emergency response.
- ❑ *Mutual aid resources.*
- ❑ *List of contractors.*

Comment: This section should list the resources that may be needed and state where the people, equipment, and vehicles are located or can be obtained. A major task in the planning process is to identify what resources are already available and what must still be provided. Information should include:

◆ ◆ ◆

Function C.7: Health and Medical

◆ ◆ ◆

Comment: This section should document how medical personnel and emergency medical services can be summoned. It should integrate with the established public EMS *(Emergency Medical Service)*. *Documentation should include on-site medical resources and activities before, during, and after an emergency.* Medical personnel must be made aware of significant chemical hazards in order to train properly and prepare for possible incidents. Emergency medical teams and hospital personnel must be trained in proper methods for decontaminating and treating persons exposed to hazardous chemicals that are handled onsite. Protective action recommendations for sanitation, water supplies, recovery, and reentry should be addressed in this section.

Function C.8: Response Personnel Safety

This section should document the policies and procedures for the following:

- ❑ Standard procedure for entering and leaving the site. *(This is critically important if releases of toxic gas are involved. In this case, more than one access is required and the way to identify the correct one to use must be stated.)*
- ❑ Accountability for personnel entering and leaving the sites *(and for personnel required to muster at an assembly point.)*
- ❑ Decontamination procedures.
- ❑ Recommended safety and health equipment.
- ❑ Personal safety precautions—*List personnel protective equipment normally used or available in an emergency.*

♦ ♦ ♦

Plant owner/operators should consider the requirements of federal regulations when preparing instructions for mutual aid teams, fire brigades, spill control teams, etc. One example is ***OSHA 1910.120*** *the Hazardous Waste Operations and Emergency Response standard* ***(HAZWOPER.)***

Function C.9 : Personal Protection of Citizens

Function C.9a: Indoor Protection

❑ Hazard-specific personal protection.

Comment: The plan should clearly indicate what protective action should be taken in especially hazardous situations. **Evacuation is sometimes, but not always, necessary** (see Function 9b). For some hazardous materials it is safer to keep citizens inside with doors and windows closed rather than to evacuate.

♦ ♦ ♦

The shelter-in-place policy should be clearly documented where relevant. To be effective, planning and preparedness activities should provide:

- an emergency management system and decision-making criteria for determining when an indoor protection strategy should be used;
- a system for warning and advising the public;
- a system for determining when a *vapor* cloud has cleared a particular area;
- a system for advising people to leave a building at an appropriate time; and

- public education on the value of indoor protection and on expedient means to reduce ventilation.

In facilities with potential for vapor cloud explosions (VCEs), reinforced buildings may exist for protection of personnel from blast waves. The plan should clearly indicate the action to be taken by personnel (e.g., seek shelter or evacuate) immediately following a flammable vapor release.

Function C.9b: Evacuation Procedures

- ❑ Title of person and alternate(s) who can order/recommend an evacuation.
- ❑ Vulnerable zones where evacuation could be necessary and a method for notifying these places.
- ❑ Provisions for a precautionary evacuation.
- ❑ Methods for controlling traffic flow and providing alternate traffic routes.
- ❑ Shelter locations and other provisions for evacuations (e.g., special assistance for hospitals, *nursing homes, homes for the elderly, and those who are not ambulatory*).
- ❑ Agreements with nearby jurisdictions to receive evacuees.
- ❑ Agreements with hospitals outside the local jurisdictions.
- ❑ Protective shelter for relocated populations.
- ❑ Reception and care of evacuees.
- ❑ Reentry procedures.

Comment: Evacuation is the most sweeping response to an accidental release. The plan should clearly document the circumstances in which evacuation would be appropriate and necessary *and how the extent of the evacuated area is to be determined. Since an evacuation is the most stressful response, strong consideration needs to be given to shelter-in-place as the alternative.*

♦ ♦ ♦

Maps (drawn to the same scale) with evacuation routes and alternatives clearly identified should be prepared for each risk zone in the area. Maps should indicate precise routes to another location where special populations (e.g., from schools, hospitals, nursing homes, homes for the physically or mentally disabled) can be taken during an emergency evacuation, and the methods of transportation during the evacuation.

Consideration of when and how evacuees will return to their homes should be part of this section.

This section on evacuation should include a description of how other agencies will coordinate with the medical community.

Copies of evacuation procedures should be provided to all appropriate agencies and organizations ♦ ♦ ♦ and could periodically be published in the local newspaper(s).

Function C.9c: Other Public Protection Strategies

Where any of the following could be relevant, appropriate details should be documented.

- ❑ Relocation.
- ❑ Water supply protection.
- ❑ Sewage system protection.

♦ ♦ ♦

Function C.10: Fire and Rescue

Firefighters should be trained in proper safety procedures for approaching a hazardous materials incident. Such procedures need to be documented and issued to appropriate organizations for training. Other items to be listed include:

- ❑ Chain of command among firefighters (especially the link between onsite or mutual aid systems and public fire departments).
- ❑ List of available support systems.
- ❑ List of all tasks for firefighters.

♦ ♦ ♦

Function C.11: Law Enforcement

The section should document the following:

- ❑ Chain of command for law enforcement officials as it would affect the facility.
- ❑ List of all *facility-related* tasks to be performed by law enforcement personnel.
- ❑ *Special responsibilities following all evacuations.*

♦ ♦ ♦

Comment: This section should include maps that indicate control points where police officers should be stationed in order to expedite the movement of responders toward the scene and of evacuees away from the scene, to restrict unnecessary traffic from entering the scene, and to control the possible spread of contamination.

Function C.12: Ongoing Incident Assessment

This section should document who is responsible to monitor the size, concentration, and movement of leaks, spills, and releases, and how they will do their work.

- ❑ Field monitoring teams
- ❑ Provision for environmental assessment, biological monitoring, and contamination surveys.
- ❑ Flood/water controls.

Comment: After the notification that a release has occurred, it is crucial to monitor the release and assess its impact, both on and off site. A detailed log of all sampling results should be maintained. Health officials should be kept informed of the situation.

♦ ♦ ♦

The location of the best sampling and test equipment for this purpose should be recorded.

Function C.13: Human Services

This section can document the link with organizations, often voluntary bodies, that may provide support.

- ❑ List of agencies providing human services.
- ❑ List of human services tasks.

♦ ♦ ♦

Function C.14: Public Works

In some instances, public works personnel may be required. The document should list appropriate tasks for public works personnel.

♦ ♦ ♦

Function C.15: Others

Comment: If the preceding list of functions does not adequately cover the various tasks to be performed during emergency responses, additional response functions can be developed, and the necessary documentation prepared describing them and the way they will perform.

Planning Element D: Containment and Cleanup

Planning Element D.1: Techniques for Spill Containment and Cleanup

- ❑ Containment and mitigation actions.
- ❑ Cleanup methods.
- ❑ Restoration of the surrounding environment.

♦ ♦ ♦

A clear and succinct list of appropriate containment and cleanup countermeasures should be prepared for each hazardous material present onsite in significant quantities. This section should be coordinated with the section on "Response Personnel Safety" so that response teams are subjected to minimal danger. Planners should concentrate on the techniques that are applicable to the hazardous materials and terrain of their area. It may be helpful to include sketches and details on how cleanup should occur for certain areas where spills are more likely.

♦ ♦ ♦

Documentation should describe:

- *the circumstances under which a fire should be extinguished or allowed to burn;*
- *containment and disposal of contaminated fire water;*
- *materials that are water-reactive and pose a greater hazard when in contact with water;*
- *the appropriate use of water and/or foam; and*
- *dilution and disposal.*

Planning Element D.2: Resources for Cleanup and Disposal

The section should list and document the following:

- ❑ Cleanup/disposal contractors and services provided.
- ❑ Cleanup material and equipment.
- ❑ Communications equipment.
- ❑ Provision for long-term site control during extended cleanups.
- ❑ Emergency transportation (e.g., aircraft, four-wheel-drive vehicles, boats).
- ❑ Cleanup personnel.
- ❑ Personal protective equipment.
- ❑ Approved disposal sites.

♦ ♦ ♦

Both CERCLA and RCRA regulate waste disposal and it is important that this section reflect the requirements of these regulations and those of states for on-site disposal, transportation, and off-site disposal.

♦ ♦ ♦

Planning Element E: Documentation and Investigative Follow-Up

The documentation of emergency preparedness should extend to the procedure for setting up a post-incident investigation. Further details are given in Chapter 15, but the following is a summary of the basic documentation needs.

- ❑ List of required reports.
- ❑ Reasons for requiring the reports.
- ❑ Format for reports.
- ❑ Methods for determining whether the response mechanism worked properly.
- ❑ Provision for cost recovery.

Comment: This section indicates what information should be gathered about the release and the response operation. Key response personnel could be instructed to maintain an accurate log of their activities. Actual response costs should be documented in order to facilitate cost recovery.

♦ ♦ ♦

Planning Element F: Procedures for Testing and Updating Plan

Planning Element F.1: Testing the Plan

♦ ♦ ♦

Comment: Exercises or drills are important tools in keeping a plan functionally up-to-date. These are simulated accidental releases where emergency response personnel act out their duties. The exercises can be table-top and/or they can be realistic enough so that equipment is deployed, communication gear is tested, and "victims" are sent to hospitals with simulated injuries.

♦ ♦ ♦

This section should specify:

1. the organization in charge of the exercise;
2. the types of exercises, and how they should be arranged;
3. the frequency of exercises; and
4. a procedure for evaluating performance, making changes to plans, and correcting identified deficiencies in response capabilities as necessary.

Planning Element F.2: Updating the Plan

Documented emergency procedures are likely to become progressively out-of-date. To overcome this problem, the procedures must be updated regularly. The policy for updating the plan should be documented with the following:

❑ Title and organization of responsible person(s).
❑ Change notification procedures.
❑ How often the plan should be audited and what mechanisms will be used to change the plan.

Comment: Responsibility should be delegated to someone to make sure that the plan is updated frequently and that all plan holders are informed of the changes. Notification of changes should be by written memorandum or letter; the changes should be recorded in the RECORD OF AMENDMENTS page at the front of the completed plan. Changes should be consecutively numbered for ease of tracking and accounting.

Following are examples of information that should regularly be checked for accuracy:

1. identity and phone numbers of response personnel;
2. name, quantity, properties, and location of hazardous materials onsite. (If new hazardous materials are made, used, or stored, or are transported in the community, revise the plan as needed);
3. facility maps;
4. transportation routes;
5. emergency services available; and
6. resource availability.

Planning Element G: Hazards Analysis (Summary)

Comment: This analysis is a crucial aspect of the planning process. It consists of determining where hazards are likely to exist, what places would most likely be adversely affected, what hazardous materials could be involved and what conditions might exist during a spill or release.

♦ ♦ ♦

Most of this information can be developed following identification in Process Hazard Analysis (see Chapter 6). The documentation should cover the following:

- ❑ Identification of hazards.
- ❑ Analysis of vulnerability.
- ❑ Analysis of risk.

Planning Element H: References

Planning Element H.1: Laboratory, Consultant, and Other Technical Support Resources

- ❑ Telephone directory of technical support services
  - laboratories (environmental and public health)
  - private consultants
  - colleges or universities (chemistry departments and special courses)
  - local chemical plants

Comment: This section should identify the various groups capable of providing technical support and the specific person to be contacted. Note should be made about the ability of these laboratories to provide rapid analysis *and whether they are staffed 24 hours a day.* These technical experts can provide advice during a disaster and also be of great service during the development of this plan. For this reason, one of the first planning steps should be gathering and recording of information for this section.

Planning Element H.2: Technical Library

The documentation should include a list of references, their location, and their availability.

Normally, these reference documents should be available in the Emergency Control Center.

- general planning references;
- specific references for hazardous materials;
- *calculation methods for assessing the impact of leaks or spills;*
- technical references and methods for using national databases; and
- maps.

♦ ♦ ♦

14

Auditing

14.1. Overview

14.1.1. Introduction

Auditing is a fundamental part of an effective process safety management (PSM) program; through auditing organizations can confirm the existence of process safety systems and verify their effectiveness. This subject has previously been addressed in the CCPS book *Guidelines for Auditing Process Safety Management Systems*, and the reader may wish to refer to this book to gain further perspective on the audit function as it is generally applied to PSM. This chapter presents guidance on documentation of the audit function. It supplements the individual chapters in this book, which contain specific guidance for auditing the documentation of the various PSM elements or PSM-related topics.

14.1.2. Goals and Benefits

The primary goals of auditing PSM systems are to:

- enhance safety awareness among facility personnel; and
- determine whether the PSM procedures are timely, complete, up-to-date and applied in compliance with company policies, applicable government regulations and good process safety practices.

There are a very wide variety of PSM elements or PSM-related topics (e.g., Training, Emergency Response, Management of Change) that can benefit from some form of formal audit to meet the goals outlined above. Some benefits from auditing PSM systems are that management can:

- measure effectiveness of PSM elements individually and as a whole;
- gain feedback for corrective action and improvements; and
- have increased confidence that PSM works.

Collectively, PSM audits provide assurance that organizations, and their individual employees, are working effectively toward the goal of providing the highest possible level of process safety.

14.2. Description of Auditing

14.2.1. Objectives

An audit is a systematic, independent review to verify conformance with established guidelines or standards. Organizations conduct audits to determine the status of existing systems, to identify what needs improvement, and to avoid potential incidents. Auditing helps establish management control of the PSM elements and helps meet the following objectives:

- improves the performance of the function being audited;
- measures results and permits trend analysis;
- reinforces accountability, based on audit findings;
- facilitates compliance with internal guidelines and governmental regulations; and
- communicates the results to affected audiences.

There are several types of audit that will be described below.

14.2.2. Types of Audits

The objectives of the various audits can be different, and sometimes require different approaches. Three principal types are briefly described in the following paragraphs.

PSM Systems Audit. This type of audit consists of a careful review of the systems involved in the management of the PSM elements. Basically, it is a confirmation check that procedures and practices exist. Verification activities should include checking to see that:

- documentation is adequate and used effectively (a sampling approach may be necessary);
- resources (personnel and otherwise) are sufficient;
- audit schedules are being met; and
- responses to problems are timely.

Process Safety Audits. This audit type is generally more detailed and specific than a PSM systems audit. Process safety audits seek to verify that the actual practices are in accordance with established procedures. Each PSM element may have a specific audit procedure or focus. For large facilities, it may be necessary to divide the documentation system into more manageable parts to ensure comprehensive coverage of the various PSM elements and their documentation. See Appendixes 13B, C, and D in the CCPS book, *Plant Guidelines for Technical Management of Chemical Process Safety*, for specific examples.

Sampling for verification is an important aspect of performing audits. In addition, efforts should be concentrated on areas subject to change. This includes both documented change involving change request and authorization forms and hard-to-recognize incremental changes that may occur in a facility over long periods of time.

Compliance Reviews. The Environmental Protection Agency (EPA) and the Occupational Safety and Health Administration (OSHA) each have programs for auditing facilities for compliance with their regulations. There are two documents that can provide useful information on items that are likely to be audited by government inspectors. The documents are:

- EPA, *Chemical Safety Audit Program* (1989); and
- OSHA, *Compliance Guidelines and Enforcement Procedures*,

OSHA Instruction CPL 2-2.45A for 29 CFR 1910.119 (1992).
Organizations wishing to conduct self-audits against the same criteria that an inspector from a regulatory organization might use may find it helpful to refer to publications such as those cited above. Such internal audits can be conducted according to a set schedule, which is related to regulatory requirements. Alternatively, audits can be performed as needed, perhaps after some documentation file has been updated. Such audits are useful to confirm that the PSM program element and its documentation meet regulatory requirements.

14.2.3. Audit Methodology References

Information in the CCPS book, *Guidelines on Auditing Process Safety Management Systems*, can be used when preparing for or performing PSM systems audits. Other useful information appears in Appendix 13A of another CCPS book, *Plant Guidelines for Technical Management of Chemical Process Safety*, which provides suitable checklists for auditing the implementation of the twelve PSM elements listed in Table 1-2. Recommendations for auditing the documentation of each PSM element as well as other PSM-related topics are included in each of the relevant chapters in this book.

In the book, *Plant Guidelines for the Technical Management of Chemical Process Safety,* CCPS has developed a set of questions to be raised in safety audits of the technical programs. These cover the following areas:

- process hazard review program;
- process hazards review;
- process hazards report;

- Management of Change (technology):
  —test authorization; and
  —test conclusion.
- operating instructions:
  —general;
  —safety and occupational health; and
  —operating steps.
- spill and chemical release control;
- unusual incident investigation/reporting/communication;
- PSM system auditing (performance versus process hazards management guidelines);
- equipment tests and inspections;
- pre-start-up safety checklists;
- Management of Change (of facilities):
  —general;
  —projects; and
  —field changes.
- new manufacturing facilities;
- quality assurance:
  —communication of desired criteria;
  —fabrication of critical service equipment; and
  —assembly and installation.
- process safety audit:
  —process change control;
  —safety of plant design;
  —maintenance and operation of process safety equipment;
  —plant safety, as built;
  —training and behaviors; and
  —continuous process safety improvement.

14.3. Audit Documentation

14.3.1. Audit Program Documentation

Organizations implementing an audit program should include in their PSM program documentation a discussion of the goals of auditing as well as any general procedures required for implementation of the function. Such program guidance would generally address the following items:

Audit Program Responsibility. Functional responsibility and accountability for the audit function should be clearly defined in the program documentation.

Audit Team Composition. Since there are a wide variety of PSM systems in place, independent teams with appropriate qualifications should perform these auditing functions on behalf of management. The number and experience of those conducting the audit will vary according to the unique circumstances of the activity being audited. Auditing techniques need to be learned and practiced, and organizations may wish to consider establishing minimum standards of training for those individuals filling key roles on the audit team.

Audit team leaders are typically personnel independent from the systems being audited. Generally, one member of the team should be responsible for the collection of the various inputs to the audit and for recording the audit. Where complex technology is involved, an expert in the technology is needed, at least as an advisor. Smaller audits may be performed by only one person; however, auditors could benefit from the assistance of an independent person or persons to cover the following topics:

- chemistry;
- process engineering;
- maintenance;
- operations;
- safety (fire protection, electrical, materials handling); and
- auditing (technique).

Some organizations have established the practice of interchanging employee auditors during major systems audits. In cases where operations are of a similar nature and qualifications are comparable, this practice may be appropriate. This exchange of ideas among colleagues can lead to improved audits. Trading employees temporarily for audits helps foster improved knowledge and, by bringing in impartial observers, helps maintain the objectivity of the evaluation process.

Audit Scope. Within the context of the documentation task, the audit should be focused on verification that a documentation program exists and that it is being followed as described. For example, in auditing Process Hazard Analysis (PHA), the team should first check that the program for implementing PHA defines the documentation requirements for PHA. The team should next check that the records produced when implementing PHA are documented in a manner that complies with the requirements of the program.

It should be remembered that, when auditing PSM documentation, the intent is not to confirm the technical accuracy or the merit of the contents of the documentation. Rather, the intent is to confirm that the documentation practices conform to the established program requirements.

Auditing may, however, reveal areas of potential improvement in the implementation of the PSM element itself. Such findings, while beyond the scope of the documentation audit, could be separately documented and forwarded to the responsible parties for consideration.

Audit Frequency. The frequency of audits will vary, as dictated by the needs of the organization or facility, or according to regulatory requirements. In defining a schedule for audits, consideration should also be given to:

- degree of risk in the facility;
- seriousness of findings in the last audit;
- age of the installation;
- interval since the last audit;
- extent of changes since the last audit; and
- incident history since the last audit.

Conduct of the Audit. PSM system audits can be conducted in a variety of ways. To enhance the value of the audit, the method chosen should be structured and appropriate to the scope and focus of the audit. In the process safety field, it is helpful if the audit methodology can be reproducibly applied for subsequent audits at the same location and at other sites. Such reproducibility allows for trend analysis and valid comparisons of audit results. Using a checklist approach can be one such effective method.

When using the checklist methodology, each question posed in the checklist typically requires an answer: *YES/NO*. In a few cases some degree of performance achievement is sought, either as a percentage or in terms of *always/sometimes/never*, or their appropriate equivalent. In some instances, additional space is required on the checklist for an answer, for example where a brief description of an alternate being followed is needed.

Some sources of PSM audit checklists were listed in Section 14.2.3.

14.3.2. Records from Implementing the Auditing Element

Audits are typically documented via a formal summary report. This permits dissemination of the audit results to those facility and management personnel having a need to know and establishes a historical record that permits future trend analysis of the performance of the PSM function.

Audit report contents. The report should identify the facilities audited plus the area or section of the documentation that was covered in the audit. There may be a few large audit reports, each covering a large number of PSM elements, or numerous small reports each addressing one or more elements.

The report of the audit team findings would typically include an summary assessing how the documentation of each element is being performed, as well as any recommendations for improving the documentation of the PSM policies and procedures. Recommendations should be expressed in clear statements defining, and justifying, the need for the recommendation; i.e., what is to be done and what it is intended to achieve. The report should clearly define who is responsible for carrying out the recommendation, and the date by which completion or a progress report is required.

The audit report would typically contain the following:

- introduction, outlining the method of approach;
- scope and objectives;

- areas covered;
- team members and their experience;
- dates of audits;
- main findings and conclusions; and
- recommendations, with responsibilities and schedules for their completion.

Any resulting action lists should be monitored by management through to final resolution of all open items.

Dissemination of Results. The organization's PSM program documentation should address who will receive the audit reports. Since the audit is a report to management on the effectiveness of one or more facility PSM elements, the report would typically be addressed to the responsible site manager, who will have overall responsibility, and to appropriate members of the organization's management. Other copies, or as a minimum, relevant extracts, should be provided to each person who is assigned responsibility for resolution of recommendations contained in the report.

14.3.3. Resolution of Audit Recommendations

It is common, as a result of an audit, to formulate recommendations (e.g., to correct a documentation deficiency identified during the audit). It is important to promptly and fully resolve each recommendation, just as was the case for PHA recommendations in Chapter 6. Reference should be made to Section 6.3.3 for a discussion of documentation of recommendations and their resolution. The resolution of recommendations may be documented within the audit report itself, within an addendum to the report, or in a separate follow-up report issued at a later date.

14.4. Records Management

In addition to the description of the audit program given in Section 14.3.1, the organization's PSM program documentation should define pertinent audit program records management practices. There are no records management considerations that are unique to the audit function, and the reader may wish to refer to Chapter 4, *Records Management*, for general guidance.

14.4.1. Storage Locations

Audit records should be accessible to those parties responsible for implementing PSM within a facility or organization, especially those charged with the responsibility for resolving audit findings or recommendations. Additionally, organizations may find it helpful to have a centralized compilation of audit records to facilitate trend analysis or to aid in demonstrating compliance with applicable regulatory requirements.

14.4.2. Media and Methods

There are no considerations, unique to the audit function, that would dictate the choice of one media over another. However, large facilities or organizations having a centralized audit function might find a computer-based records management system helpful for tracking the resolution of audit recommendations.

14.4.3. Records Retention and Purge Schedules

The organization's PSM program documentation should establishing guidelines for the retention and purging of audit records. It should be noted that current regulatory requirements dictate for covered processes that documentation, including any corresponding follow-up action reports, be maintained for the two most recent audits.

14.5. References

AIChE–CCPS, *Guidelines for Auditing Process Safety Management Systems,* 1993.

AIChE–CCPS, *Plant Guidelines for Technical Management of Chemical Process Safety,* 1989.

Environmental Protection Agency, *Chemical Safety Audit Program,* Washington, D.C., 1989.

Greeno, J.L., et al., *Environmental Auditing—Fundamentals and Techniques,* 2nd Edition, Arthur D. Little, Cambridge, MA, 1987.

Krivan, S.P., "Avoiding Catastrophic Loss: Technical Safety Audit and Process Safety Review." *Professional Safety,* Feb., 1986, pp. 21–26.

Occupational Safety and Health Administration, *Compliance Guidelines and Enforcement Procedures,* Instruction CPL 2-2.45A, Washington, D.C., 1992.

Occupational Safety and Health Administration, *Process Safety Management of Highly Hazardous Chemicals,* 29 CFR 1910.119, Washington, D.C., 1992.

15

Incident Investigation

15.1. Overview

15.1.1. Introduction

An incident is an unplanned event with the potential for undesirable consequences. This definition is meant to include both accidents, and situations in which there were *near misses*. Thus, the definition covers all cases where there was, or could have been: injury; loss of, or damage to, property; or the release of hazardous or toxic material to the environment.

Organizations typically investigate incidents to determine their causes and to seek ways of preventing their recurrence. The process of incident investigation generally utilizes a team-based approach involving: gathering and analyzing the evidence; drawing conclusions as to the causes of the incident; generating corrective or preventive actions; and summarizing, documenting, and disseminating the findings. Proper documentation of incident investigation results provides the basis for corrective actions leading to improvements in the organization's process safety management (PSM) program and, accordingly, should enhance plant safety. This chapter discusses PSM documentation requirements for incident investigation.

A description of many of the available methods of incident investigation techniques and how they can be applied is given in the CCPS book, *Guidelines for Investigating Chemical Process Incidents.*

15.1.2. Goals and Benefits

The goal of an incident investigation is to define the facts related to the event, determine the causes, and develop remedial actions to control or eliminate recurrence. Each of these steps should be properly documented.

An effective incident reporting and investigation program has many benefits, including:

- assurance that incidents will be reported and investigated;
- confidence that the underlying or root causes of incidents have been discovered;
- identification of appropriate corrective actions for management review and implementation;
- development of improved knowledge, techniques, and facilities that should reduce recurrence of similar incidents;
- enhancement of process safety within the affected facility;
- heightened awareness of operational safety requirements fostered by the dissemination of reports and recommendations;
- identification of needed PSM program improvements by analysis of *lessons learned*;
- compliance with applicable regulatory requirements;
- development of detailed, credible information for litigation support; and
- reduction in downtime and operating costs by control of accidental losses.

15.2. Description of Incident Investigation

15.2.1. Objectives

The primary objective of an incident investigation is to determine and document the root causes and contributing causes of an incident. Recommendations and corrective action plans are then developed with the intent of preventing such incidents in the future.

15.2.2. Conducting Incident Investigations

Incidents cannot be investigated if they are not reported. To ensure reporting, an incident should be viewed as any event that is unusual or out of the ordinary, whether or not harm resulted from its occurrence. The initial reporting of such events should not be limited to those perceived to be serious or potentially serious incidents, since the seriousness of an event often cannot be determined without analysis. It is important that all incidents be reported so that corrective action can be established before a similar occurrence results in a serious accident. Therefore, barriers to incident reporting (e.g., fear of management reprisal) must be removed to promote timely reporting by employees at any level of the organization.

The investigation of an incident involves four stages of activity: gathering evidence, analyzing the evidence, drawing conclusions, and formulating recommendations. Each of these stages are discussed further below.

Gathering Evidence. Investigation of an incident should commence as soon as possible since the quality and quantity of evidence begin to decline immediately after the incident. This is the case with the memories of witnesses and can also be true for certain types of physical evidence. For maximum effectiveness, an incident investigation should start within 48 hours; this is a requirement for processes covered by current regulatory requirements.

Evidence gathering activities include:

- interviewing witnesses;
- taking photographs (or making video tapes) of the incident scene;
- identifying and securing physical evidence (e.g., pieces of a damaged vessel);
- collecting paper or computer records (e.g., procedures, instrument charts, log books); and
- sampling key process streams and arranging for analysis.

Although the gathering of evidence should begin as soon as possible, completion of this task may require a significant interval of time; e.g., if time-consuming laboratory analyses are required.

Analyzing the Evidence. The typical incident investigation will amass a significant amount of evidence, only a portion of which will be pertinent to explaining the incident. It is often helpful to arrange this data based on a chronological sequence, attempting to establish a time line for the events leading up to, and perhaps following, the incident. This allows the identification of conflicting and supporting evidence, and may point to information gaps that require additional evidence.

Drawing Conclusions. The goal of an incident investigation is to identify the root causes of the event. The identification of root causes is not always obvious and sometimes requires special analytical techniques. Selection of the method of analysis depends on the unique circumstances of the case at hand; a description of many of the methods of incident analysis techniques and how they can be applied is given in the CCPS book, *Guidelines for Investigating Chemical Process Incidents.*

Formulating Recommendations. Once the investigation team has analyzed the evidence from the incident investigation and agreed upon the probable causes, recommendations for corrective or remedial actions should be formulated. Generally, recommendations fall into one of two classes:

- those for correcting the specific identified problems (causes) involved in the incident; and
- those for correcting systemic problems (causes) identified during the investigation.

At least one recommendation for corrective action should be developed for each causal finding reported. In evaluating recommendations, the incident investigation team should consider the following:

- evaluation of the safety of the relevant operation;
- identification of objectives and priorities for the corrective action;

- limiting criteria (i.e., limits of performance and/or cost and collateral impacts);
- practical alternatives to correct the deficiency;
- testing of possible alternative recommendations against the objectives, limiting criteria and evaluation of possible adverse effects; and
- implementation procedures, controls, responsibilities, and feedback communication.

There are other benefits that may be derived from incident investigations, in addition to achieving the immediate goal of preventing the recurrence of a particular incident. These are discussed below.

Improved Operational Safety. Documentation of incident investigations provides a historical record that can be used to search for similarities in past events. Use of such records can lead to improvements in future designs and operational procedures and can help to identify trends more clearly. Appropriate categories or classifications of incident information can be established in the documentation so that accident analysis can focus on possible links. Examples of categories for classifying incident data may include:

- root cause;
- contributing (multiple) causes;
- related PSM elements (e.g., process knowledge, training, process equipment integrity); and
- hazard source (e.g., toxicity, flammability, ignition sources, corrosive nature).

Improved PSM Program. The incident investigation program may identify areas for potential improvement in either the overall PSM program, or in individual PSM elements. These areas of improvement should be documented. Some examples are:

- updating of purging requirements as a result of an explosion attributed to residual volatile organics in a vessel prepared for maintenance (Operating Procedures);
- necessity for retraining of plant operators based on the revised operating instructions (Training); and
- increasing the frequency of periodic maintenance on critical safety equipment (Process Equipment Integrity).

Satisfaction of Legal Requirements. The procedures and documentation involved in incident investigation must satisfy applicable federal, state, and local regulatory requirements. Insurers may also have specific requirements that should be incorporated into the procedures.

15.3. Incident Investigation Documentation

15.3.1. Incident Investigation Program Documentation

An organization implementing an incident investigation program should establish a documented program defining the goals and requirements of incident investigation and providing detailed procedures outlining how incident investigations will be performed. Such preplanning is an important part of a well designed incident investigation program and should be clearly documented. The program documentation should also clearly establish responsibilities and accountability for the incident investigation element.

The program documentation would typically address the following:

- basic statement of policy;
- which types of incidents are to be investigated;
- the importance of promptly initiating the investigation;
- who is to initiate an investigation;
- who is to lead the investigation;
- responsibilities of the leader, who will:
  —determine the scope of team activities;
  —call and preside over team meetings;
  —establish the time table and assign tasks;
  —ensure that all pertinent data has been collected;
  —advise management as to the progress of the investigation;
  —bring the investigation process to a timely conclusion; and
  —issue a timely report to management.
- team selection;
- objectives of the investigation;
- selection of the investigation method(s) to be used;
- documentation requirements for incident investigation;
- resolution of all recommendations;
- any special policies regarding incidents that may be subject to litigation;
- interactions with regulatory authorities;
- records control, storage and issue, retention; and
- auditing of incident reports.

The program documentation should also address requirements and procedures for communicating the conclusions of the investigation to management and to affected employees and contractors.

15.3.2. Documentation of Incident Investigation Results

Incident investigations are typically documented via a formal summary report containing the following information:

- a summary of the incident;
- a statement of who performed the investigation and when it was conducted;
- a discussion, where pertinent, of the incident analysis technique used;
- an analysis of the evidence, citing pertinent events leading up to the incident and identifying its causes, including root causes;
- recommendations proposed to prevent a repeat of the incident; and
- a summary of supporting data.

Some organizations use a standardized form for the incident investigation report. An example of one such form is provided in Figure 15-1 (pages 260–263).

The supporting data for an investigation report can be quite voluminous, and may include the following:

- documents received, examined, analyzed or consulted during the investigation (the actual documents may be included, or they may be referenced);
- transcripts or audio tapes of interviews;
- visual evidence (e.g., photographs, video, etc.);
- maps or sketches of the incident scene;
- calculations;
- inventories of physical evidence, indicating storage location; and
- physical and/or chemical analysis or test reports.

Appendixes can be very helpful in organizing this supporting data. An appendix index and/or table of contents would be advisable, in addition to cross referencing in the main body of the report. Alternatively, some organizations may chose to document the more voluminous portions of this information, not in appendixes to the report, but as a separate report with limited distribution.

Regulatory requirements may dictate that report findings be communicated to affected personnel. If this is the case, confirmation of such communications should also be documented.

15.3.3. Resolution of Incident Report Recommendations

Follow-up is one of the most important aspects of incident investigation. Unless causes are identified and ways to prevent recurrence are implemented, repetition of the incident remains possible. Resolution of recommendations is, therefore, vital to achieve PSM program goals. The following documentation practices for recommendations should be followed:

- each recommended action should be clearly and unambiguously stated in the investigation report (including the intent underlying the recommendation);
- the report should identify:
  —the person responsible implementing the resolution; and
  —a target date for implementation, or for presenting a report on the recommendation to the responsible manager.

- the final resolution of each recommendation should be documented showing that either:
  —the recommendation has been implemented as proposed;
  —an equivalent alternative has been adopted; or
  —upon further consideration, the recommendation has been deemed to be impractical or unnecessary. In this case, the rationale for the decision should also be documented.

The documentation of resolution of recommendations may be included in the original incident investigation report, in an addendum to the report, or in a separate follow up report.

15.4. Records Management

In addition to the description of the incident investigation program given in Section 15.3.1, the organization's PSM documentation should define pertinent records management practices. Some records management considerations are discussed in this section. The reader may also wish to refer to Chapter 4, *Records Management*, for more general guidance.

15.4.1. Responsibilities and Accountability

The responsibility for collection, storage, and management of incident investigation records should be clearly defined. A member of the safety organization, because of routine involvement in the investigation program, might be an appropriate choice.

15.4.2. Records Control

Records media will likely vary. Frequently, the evidence examined during an investigation will consists of paper documents (including logbooks, logsheets, and instrument recorder charts), photographs, magnetic tape, etc. In a broader sense, however, the concept of incident investigation records should be expanded to include physical evidence such as hardware and pieces of equipment. All of these various types of records must be controlled in accordance with the organization's records management procedures. These procedures should address not only the needs of the organization, but also applicable legal, regulatory, and insurance requirements.

As pointed out in the CCPS book, *Guidelines for Investigating Chemical Process Incidents,* the potential for litigation associated with incidents may dictate unique records control practices. For example, records associated with pending litigation may require a *chain-of-custody* record. Organizations may wish to enlist the assistance of legal counsel when promulgating incident investigation records control practices, particularly with regard to retention and purge schedules.

Accident/Incident Investigation Form

Name Last First Initial	Incident Date	Date Reported	Division Code
Site	Plant	Incident Location	Incident Time
Job Title	Supervisor	Occupational ☐ Non-occupational ☐	Undetermined ☐ Precautionary ☐

How did incident occur?

Classification:

☐ 1. First Aid	☐ 8. Overexposure
☐ 2. Medical	☐ 9. Illness
☐ 3. Restricted Duty	☐ 10. Spill
☐ 4. Days Away From Work	☐ 11. Release
	☐ 12. Property Loss
☐ 5. Near Miss	☐ 13. Permit Excursion
☐ 6. Fire	☐ 14. Off-Site
☐ 7. Contractor	☐ 15. Other

Complete Reverse Side.

Chemical or substance involved:	Exposure above I.H. Limit Value:
Amount discharged to air:	CERCLA or SARA reportable quantity exceeded?
Amount discharged to land:	Agencies notified:
Amount discharged to sewer:	Disposition of material:

Describe injury/property loss:

Estimated cost of property loss:	Estimated remedial cost:

Why did it happen?

Immediate corrective action(s):

Corrective action(s) to prevent recurrence:	Responsibility	Target Date

Investigator/Employee Date	Supervisor Date
Safety & Loss Prevention Date	Department Head Date
Ecology Date	Plant/Site Manager Date

Rev. 3/92

FIGURE 15-1. Typical Accident/Incident Investigation Form (Sheet 1 of 2)

Instructions

Complete all sections on the front page. The Accident/Incident Investigation Report is not complete until the appropriate signatures are obtained.

Copies of all injury and illness investigations, except first aids, must be sent to Corporate Safety and Loss Prevention and environmental incident investigations

This investigation is "open" until corrective action has been completed.

The following is for the purpose of maintaining computerized Statistics. Complete each block "A" through "V" using the appropriate code number assigned for each entry from the data below.

When applicable, complete the below section for number of restricted days and/or number of days away from work.

Industrial Injuries Code

A.	☐☐☐☐☐☐☐☐☐ Employee Soc. Sec. No.			
B.	☐☐☐☐ SIC Code	M.	☐☐	Time in that job
C.	☐☐☐☐ Division	N.	☐☐	Accident type
D.	☐☐☐☐ Site	O.	☐☐	Accident agency
E.	☐☐☐☐ Plant	P	☐☐	Nature of
F.	☐☐ Age	Q.	☐☐	Body part affected
G.	☐ Sex	R.	☐☐	Primary cause
H.	☐☐ Hours Worked	S.	☐☐	Contrib. cause
I.	☐ Overtime	T.	☐☐	Contrib. cause
J.	☐ Employee Status	U	☐☐	Contrib. cause
K.	☐ Occupation	V.	☐	Unsafe act/condition
L.	☐☐ Time Employed			

Number of restricted days: ____________ Final ☐ Est. ☐

Number of days away from work: ________ Final ☐ Est. ☐

A. Employee Social Security Number
B. SIC code

C. Division

O. Accident agency/involved equipment
- 01-Fired vessels-boiler, incin., etc.
- 02-Reactors, columns,vessels, etc. >15 psig
- 03-Process eqpt. tanks, bins <15 psig
- 04-Gas or liquid handling (into)
- 05-Solids handling (into)
- 06-Mechanical power transmission-gears, couplings, belts, pulleys
- 07-Portable eqpt. machinery
- 08-Hoists, cranes, etc.
- 09-Over-the-road automobiles, and trucks, incl. tank trucks
- 10-Industrial trucks, forklifts, end loaders, tractors, bicycles
- 11-Railroad, rolling stock, incl. tank cars
- 12-Piping, hoses, valves and fittings
- 13-Containers-drums, boxes, pails, cylinders etc.
- 14-Ladders, scaffold
- 15-Floors, working/walking surfaces
- 16-Tools-hand (wrenches, etc.)
- 17-Knives, scissors
- 18-Tools-powered (elec., air, etc.)
- 19-Electrical distrib. sys./apparatus
- 20-Office equipment
- 21-Laboratory equipment
- 22-Fabrication, assembly or machine shop equipment
- 23-Chemicals
- 24-Hot liquids/gas
- 88-Other
- 99-Unknown

P. Nature of injury/illness
- 01-Amputation, avulsion
- 02-Fracture, dislocate, crush
- 03-Cut, scrape, puncture, sting bite
- 04-Bruise, contusion
- 05-Irritation
- 06-Hernia, rupture
- 07-Sprain, strain
- 08-Burn-chemical
- 09-Burn-thermal or electrical
- 10-Heat stress, exhaustion, sunstroke,
- 11-Suffocate, drown, asphyxiate (lack of oxygen)
- 12-Concussion, unconscious
- 13-Poisoning-acute
- 14-Other
- 21-Skin disease or disorder
- 22-Dust disease of the lung
- 23-Respiratory-toxic agents
- 24-Poisoning-chronic
- 25-Physical agents-radiation, etc.
- 26-Repeated trauma-noise, etc.
- 29-Other illness, heart cont., etc.
- 00-No injury

Q. Body part affected
- 10-Head
  - 11-Eye(s)
  - 12-Ear(s)
  - 13-Face
  - 14-Neck
- 30-Upper Extremeties
  - 31-Upper arm
  - 32-Elbow
  - 33-Forearm
  - 34-Wrist
  - 35-Hand
  - 36-Finger(s)
- 50-Body Systems
  - 51-Circulatory
  - 52-Respiratory
  - 53-Neurological
  - 54-Reproductive
  - 00-No body part injured

D. Site Site Code

E. Plant Plant Code

F. Age
Age in years at time of incident

G. Sex
M-Male F-Female

H. Hours worked before incident

01-1	04-4	07-7	10-10	13->12
02-2	05-5	08-8	11-11	
03-3	06-6	09-9	12-12	

I. Overtime
Y - Yes N - No

J. Employee status
1-Reg. full-time 3-Temporary
2-Reg. part-time 4-Non-employee

K. Occupation
1-Warehouse, shipping & receiving
2-Production/utilities worker
3-Maintenance/construction worker
4-Plant services, janitors, guards
5-Vehicle driver
6-Foreman/Supervisor
7-Lab-QC, R&D & pilot plant
8-Sales, marketing, techincal services
9-Administrative, clerical

L. Time employed, years
01-less than 1 04-10–20
02-1 to 5 05->20
03-5 to 10

M. Time in that job, years
01-less than 1 04-10 to 20
02-1 to 5 05->20
03-5 to 10

N. Accident type
01-Fall from elevation
02-Fall, same level
03-Slip or trip without fall
04-Struck against object
05-Struck by object
06-Caught in, under, between
07-Overexertion, strain
08-Public or contracted transportation
09-Motor vehicle (employee's or company's)
10-Drowning, buried
11-Explosion, implosion
Contact by:
12-Chemicals
13-Electricity
14-Temperature extremes
15-Noise
16-Radiation
17-Other physical agents
18-Animal, insect, plant
88-Other

20 Trunk, Torso
21-Shoulder
22-Chest
23-Back, spine
24-Abdomen, groin
25-Hip

40 Lower Extremities
41-Thigh
42-Knee
43-Shin, calf
44-Ankle
45-Foot
46-Toe(s)

R.-U. Causal factors (primary and contributory causes)

Supervision
10-Incorrect/incomplete procedures, instructions
11-Rules, procedures, work methods not enforced
12-Inadequate training of employee(s)
13-Proper tools, equipment not provided
14-Deficient storage/material handling practices
15-Inadequate housekeeping, area inspections
16-Too much rush on job by supervisor

Employee
20-Physical limitation
21-Deficient in skill or ability
22-Influence of drugs or alcohol
23-Lack of alertness
30-Failure to follow written procedures or rule
31-Confined space entry procedure
32-Hot work procedure
33-Line breaking procedure
34-Lockout/tagout procedure
35-Maintenance, adjustment or cleaning on moving/pressurized equipment/line
40-Failure to follow oral instructions
50-Failure to use personal protective equipment
51-Operating without authority
52-Taking an unsafe position
53-Unsafe speed, haste, short cut
54-Improper use of tool, equipment, material
55-Use of incorrect tool/equipment/material
56-Improper manual material handling

Equipment
60-Defective equipment, tool material
61-Inadequate or missing guards
62-Inadequate or bypassed safety devices
63-Inadequate maintenance, equipment inspections
64-Inadequate lighting
65-Inadequate ventilation
66-Inadequate design/layout (congestion)
67-Inadequate fabrication/installation

Environment
70-Horseplay/distraction by fellow employee
71-Error by fellow employee
72-Unsafe eqpt./matl's./actions of 3rd party
73-Upset conditions-fire/explosion/spill, etc.
74-Exposure to chem./phys./biological agents
75-Weather-rain, snow, ice, wind, etc.
99-No other causes

V. Unsafe act/condition
1-Unsafe act 2-Unsafe condition

FIGURE 15-1. Typical Accident/Incident Investigation Form (Sheet 2 of 2)

15.5. Auditing

The PSM program documentation should establish auditing requirements for the incident investigation element. The first step in auditing the element is to verify that a documented program, such as described in Section 15.3.1, exists for implementing incident investigation.

The second step is to verify that the results of implementing the incident investigation element have been documented in the manner described by the program. Particular emphasis should be placed on the documentation of the resolution of recommendations. It should be possible to trace the disposition of each recommendation made in the incident investigation reports.

Deficiencies identified during the audit must be documented and resolved in a timely manner. Further details on auditing procedures and methods can be found in Chapter 14, *Auditing* and the CCPS book, *Guidelines for Auditing Process Safety Management Systems.*

15.6. Examples

This section gives examples that illustrate the importance of thorough incident investigation, documentation, and adequate follow-up of recommendations, to ensure that incidents do not recur. For additional examples see *What Went Wrong? Case Histories of Process Plant Disasters,* Second Edition, by Trevor Kletz.

15.6.1. Inadequate Follow-Up to Incident Investigation Causes Fire

A change was made in the operating practices for taking a reactor off-line for catalyst change, because a small fire had occurred during a prior catalyst change. The change was recorded in the log book and in a handwritten note at the plant, but apparently was not formally documented nor followed-up after investigation of the incident. The note was eventually lost and an operator resumed using the original operating procedures, contained in the standard operating procedures manual. Another fire occurred.

Appropriate documentation of the results of the incident investigation and recommended actions and a subsequent audit of recommendation follow-up could have prevented the second fire.

15.6.2. Proper Follow-Up Prevents Recurrent Equipment Damage

A refinery experienced significant, costly damage to the internals of a heavy gas/oil fractionator on numerous occasions during start-up operations. Damage was thought to be from water (condensate) left in the system after steaming, which flashed upon contact with the hot oil introduced during the dry out/warm up period. These incidents and conditions were extensively investigated and documented. It was eventually determined that circulation for not less than 18 hours at an oil temperature not exceeding 275°F was needed to safely and completely dry out the accumulated condensate in the lower portion of the fractionator and its associated piping. This recommendation was documented within the incident investigation report.

Facility personnel incorporated the appropriate instructions into the start-up procedure for the fractionator. The revised procedure prevented flashing and damage to vessel internals, and avoided costs previously incurred for downtime and repairs.

15.7. References

AIChE–CCPS, *Guidelines for Auditing Process Safety Management Systems,* 1993.

AIChE–CCPS, *Guidelines for Investigating Chemical Process Accidents,* 1992.

Kletz, T.A., *Lessons from Disaster: How Organizations Have No Memory and Accidents Recur,* Institution of Chemical Engineers, UK, 1993.

Kletz, T.A., *What Went Wrong? Case Histories of Process Plant Disasters,* Gulf Publishing Co., Houston, 1985.

Lees, F.P., *Loss Prevention in the Process Industries,* Butterworth, London, England, 1986.

Sanders, Roy, E., *Management of Change in Chemical Plants: Learning from Case Histories,* Butterworth, London, England, 1993.

Occupational Safety and Health Administration, *Process Safety Management of Highly Hazardous Chemicals,* 29 CFR 1910.119, Washington, D.C., 1992.

16

Standards, Codes, and Regulations

16.1. Overview

16.1.1. Introduction

Many aspects of process safety management (PSM) are based on the cumulative knowledge and experience reflected in company standards, industry codes, and regulations from all levels of government. Adherence to the practices outlined in these documents is sometimes referred to as the implementation of *good engineering practice*. Knowledge of the existence and proper management and use of these documents provides essential information for those involved in planning, designing, constructing, operating, and maintaining chemical, oil and gas, and pharmaceutical facilities. It is incumbent on such personnel to be aware of this knowledge resource and to understand how it should be utilized in their work.

This chapter describes some of the more common types and sources of this information and suggests guidelines on how to manage the relevant documents that communicate this knowledge.

16.1.2. Goals and Benefits

There are a variety of organizations involved in the promulgation and use of standards, codes, and regulations. Each have unique goals that they wish to achieve as a consequence of their efforts.

Industry organizations that operate chemical plants and other facilities that handle hazardous materials have PSM goals that are met through the use of company standards, external standards, codes, and regulations. These goals are to:

- communicate the organization's intentions regarding minimum essential practices in support of safe operations; and
- assure that operating locations within the organization share a common approach, where appropriate, to process safety.

Industry associations such as the American Society of Mechanical Engineers (ASME) and the Chlorine Institute (CI) actively record, review, analyze, and disseminate to their members and make available to the public, specific standards and codes that reflect the best available knowledge of safe practices. The goals of these organizations are to:

- communicate the best available practices to all who have a need; and
- continually improve the current practices through sharing of problems and solutions.

State, local, and federal governmental organizations issue regulations regarding process safety to:

- protect workers; and
- protect the community and the environment.

Operating companies and associated organizations such as design engineering firms and equipment manufacturers use these standards, codes, and regulations to carry out their work.

Some of the benefits derived are:

- assurance of the application of good engineering practice;
- reduced accident potential;
- increased efficiency in the workplace through standardization;
- simplified record keeping by use of references to applicable codes or standards;
- an improved image of the chemical process industries (CPI) as a result of enhanced safety in the workplace; and
- reduced legal liability.

16.2. Description of Standards, Codes, and Regulations

16.2.1. Objectives

Some objectives of the agencies and organizations that write standards, codes, and regulations are to:

- provide reasonably certain protection of life and property; and
- provide a safety margin to give a reasonably long, safe period of usefulness.

These objectives are quoted from the foreword of the ASME Boiler and Pressure Vessel Code but apply equally well to the entire body of knowledge that is encompassed by standards, codes, and regulations.

16.2.2. Differentiation among Standards, Codes, and Regulations

There frequently exists a degree of ambiguity as to what is meant by the terms *standards*, *codes*, and *regulations*, and organizations and individuals often use conflicting definitions. The following provides an understanding of the usage of these terms, as applied in this book.

Standards. These can be either internal (company generated) or external (created by an outside organization).

Owners of a manufacturing facility need to be cognizant of practices recommended by trade associations or general purpose technical standards organizations. One example is the Chlorine Institute (CI). This trade organization forms technical committees that produce design standards for items such as valves and vaporizers used in chlorine service. Plant designers and operators handling chlorine generally follow the recommended practices of the CI. Another example of external standards are those issued by the National Fire Protection Association (NFPA).

Internal (or company) standards are prepared to record best practices to be used by design, operations, and maintenance personnel so that a level of safety and consistency is established and maintained within the company. Large companies may have hundreds of design, construction, maintenance, and operating standards. Some examples include standards:

- specifying maximum safe velocity for sulfuric acid in carbon steel pipe;
- defining minimum line size for ethylene service;
- specifying that all heat exchangers shall be TEMA rated;
- defining the shape of a plug valve symbol on a P&ID; and
- specifying gasket material for liquid chlorine service.

In many cases internal standards are derived from actual facility operating experience. For example, the sulfuric acid velocity standard mentioned above was created to address pipe wall erosion problems that resulted when acid flowed at high velocities.

Internal standards may cite external standards that are to be used to supplement and clarify company requirements. For example, heat exchanger specifications written by a chemical manufacturer might incorporate, by reference, the requirements of external standards promulgated by the Tubular Exchanger Manufacturers Association (TEMA).

Standards, whether they are internal or external, are used to communicate best practices intended to enhance operational safety. While the use of such standards typically is not mandatory, it is generally recognized as *good engineering practice* to do so by insurers and regulators.

Codes. These are standard practices, typically issued by national organizations, which in some cases have a higher degree of legal implication than do standards. For example, pressure vessel design practice usually follows the ASME-recommended Boiler and Pressure Vessel Code. A committee of the ASME has established safety standards governing the design, fabrication, and inspection of such equipment. This collection of standard requirements are so widely accepted that they have become a *code*. Further, some states and municipalities have adopted sections of this code as a legal requirement for the design, construction, and operation of pressure vessels.

Designers, operators, and contractors have a responsibility to know of and use applicable codes. In many situations, organizations may be judged to be negligent for failing to do so.

Regulations. Regulations make standards and codes mandatory, with penalties for failure to comply. These regulations may be issued by federal, state or local governments. Many regulations are process safety related and have been mandated to protect workers in facilities as well as the public in the surrounding community.

An example at the state level is in Louisiana. The state fire marshall's office issues permits to construct industrial plants after reviewing applications and design drawings submitted by state-licensed architects and engineers. Instead of promulgating unique standards, the state refers to and establishes NFPA and American Society for Testing and Materials (ASTM) standards as mandatory requirements. For example, LAC 55:303D, a section of the state architectural code or law, requires fire extinguishers to be located in buildings according to the recommendations in NFPA 10.

An example at the federal level is the Occupational Safety and Health Administration (OSHA), which issues regulations to administer legislative mandates. One such recent rule is OSHA 29 CFR 1910.119 which covers *Process Safety Management of Highly Hazardous Chemicals*. This regulation, issued in 1992, requires the utilization of *good engineering practice.* While the regulation leaves the definition of *good engineering practice* subject to individual interpretation, many organizations interpret this to mean that the use of generally accepted standards, codes, and practices is now mandated by OSHA.

16.2.3. Sources of Standards, Codes, and Regulations

Organizations must apply or comply with a large variety of standards, codes, and regulations depending on the size, complexity, physical makeup, and process hazard level of their facility operations. These documents cover plant design, construction, operation, and maintenance and they can range from *good engineering practice* to legal requirements subject to rigid enforcement. For this reason, plant technical and operating personnel, as well as management, must develop a working knowledge of the major standards, codes, and regulations affecting their daily operations.

Appendix 16A provides a list of organizations that issue nationally applicable standards, codes, and regulations. Such organizations typically maintain publications lists that can be used to identify documents pertinent to the needs of the facility or organization.

16.3. Documentation of Standards, Codes, and Regulations

16.3.1. Program Documentation for Standards, Codes, and Regulations

The PSM program documentation should establish the goals intended from, and provide the necessary instructions required for, implementing the standards, codes, and regulations element. A written program is helpful in providing a uniform and adequate approach, defining responsibilities, and ensuring that proper documentation exists. Such a program should cover applicable records management practices, and define a system of audits to verify compliance with the documentation program:

The program documentation should communicate the organizations's policies on the application of standards, codes, and regulations. For example, an organization might allow facility management to waive compliance with internal standards, provided that an alternate means of achieving equivalent protection can be documented.

Some organizations may create and maintain their own internal standards. The mechanisms for the preparation, approval, issue, distribution, and revision of these standards, as well as the responsibilities for this program, should be clearly articulated in the PSM program documentation.

The body of knowledge comprised by standards, codes, and regulations is in a constant state of change as organizations create new documents and revise or withdraw existing ones. The responsibilities and accountability for monitoring standards, codes, and regulations for changes in recommendations and requirements should be clearly defined.

The frequency with which change occurs within this knowledge base suggests the need for an audit system to confirm that documents in use are up to date. Such audit requirements should be described in the program documentation.

16.3.2. Records from Implementing the Standards, Codes, and Regulations Element

The *records* that come from implementing this element are the files of standards, codes, and regulations that have been compiled for use in designing, constructing, operating, and maintaining the facility. Generally, these are the most recent revisions of the applicable documents; however, as discussed further in the next section, there may be instances where it is appropriate to maintain, for historical purposes, past revisions of standards, codes, or regulations.

16.4. Records Management

The organization's PSM program documentation should define applicable records management practices for the standards, codes, and regulations element. While many of the general considerations outlined in Chapter 4, *Records Management*, are applicable, this element is somewhat unique due to the breadth of the information, and documentation, that it encompasses and the frequency with which this information undergoes modification. This section addresses some of these unique considerations.

16.4.1. Where Are Records Maintained?

The information embodied in codes, standards, and regulations has broad applicability and, as might be expected, the documentation can be equally broadly distributed. Many large facilities or organizations may have one or more centralized records centers where comprehensive inventories of applicable standards, codes, and regulations are maintained. Table 16-1, shows one example plan for creating such a network of records centers. However, the necessity of frequent, perhaps daily, reference to certain types of information dictates that many documents will be maintained by individuals within the organization. Thus, the responsibility for complying with applicable records management practices may be shared among many individuals.

16.4.2. Records Procurement

As previously defined, internal standards are those generated by the organization itself. The programs for creating, reviewing, approving, distributing, and revising such internal standards should address the specific needs, and capabilities, of the organization. A detailed discussion of such programs is beyond the scope of this chapter. However, there are close parallels to the create, review, approve, and revise cycle described for procedures in Chapter 11.

External standards, codes, and regulations are obtained from outside organizations. Many documents in these categories undergo periodic review and revision, and the task of monitoring to identify revised (or new) standards, codes, and regulations, as they come available, can be demanding. A knowledge of organizations, services, and agencies that issue these documents is essential.

Federal regulations such as those from OSHA and the Environmental Protection Agency (EPA) can be systematically obtained by subscribing to the Federal Register which is published Monday through Friday. All public regulations and legal notices issued by federal agencies are published in this document. Daily reading of the entire Federal Register would be too time consuming for most people; fortunately, there are indexes and other specialist publications that identify, summarize, and report new or revised regulatory requirements. Some of these include:

- the *Federal Register Index,* a monthly cumulative publication;
- *List of CFR Sections Affected,* (LSA), a monthly compilation of updates;
- the *Environment Reporter,* which covers federal and state regulations plus related activities, published by the Bureau of National Affairs (BNA) Inc.;

TABLE 16-1
Sample Plan for Repository of External Codes/Regulations

Codes/Regulations Information Category	*Corporate Environmental/ Risk Dept.*	*Business/ Technology Center*	*Plant Site*	
			Technical	*Safety*
Equipment & Piping Design	—	C	S	—
Fire & Safety System Design Testing/Inspection Practices	S	C	C	S
Federal Government				
Air/Water Pollution	C	S	S	S
Hazardous Waste/Toxic Substance	C	S	S	S
Occupational Health/ Safety	C	S	S	C
Product Safety	C	—	—	S
Transportation Safety	C	C	S	S
State Government				
Air/Water Pollution	C	S		C
Hazardous Waste/Toxic Sub	C	S		C
Fire Codes	C	C		C
Local Government				
Municipal Waste Treatment	—	S		C
Building Codes	—	S		C
CAFR	C	—		C
Fire Codes	—	S		C

Notes:
C = complete copy of codes/regulations
S = selected codes/regulations or applicable sections

Source: CCPS *Guidelines for Technical Management of Chemical Process Safety*, Table 12-2

- the *Occupational Safety and Health Reporter,* which is similar (BNA); and
- the *Chemical Process Safety Report* (published by Thompson Publishing Group).

The last three publications have a newsletter format. There are similar publications in certain states, for example the *Louisiana Engineers.* Professional and trade association journals can also be helpful in keeping aware of recent regulatory activities and applicable legal rulings.

Standards and codes are best obtained directly from the issuing organizations. Such organizations typically offer indexes or catalogs identifying titles and current revision numbers and/or issue dates of their publications. Catalogs should be regularly replaced to ensure that this information remains current.

Single copies of individual standards or codes can usually be purchased, and this may meet the needs of the incidental user. Additionally, some promulgating organizations (such as NFPA) offer a subscription service where revised copies of standards and codes are forwarded automatically to subscribers. Such services may better meet the needs of a library or central records center, or of an individual specialist needing frequent access to a broad selection of codes and standards.

16.4.3. Media and Methods

Standards, codes, and regulations have traditionally been available as paper copy. Binding conventions vary from one organization to another; some standards are issued in loose leaf form and others may be published as hard or soft cover books or pamphlets. Standards organizations offering subscription services usually provide ring binders to facilitate storage and updating. Documents are also available in other hard copy formats such as microfiche or microfilm.

Standards, codes, and regulations are increasingly available on magnetic or optical media such as diskette or CD-ROM. The latter, in particular, can efficiently manage enormous amounts of information in a relatively small space, and may be particularly appropriate for centralized storage facilities that must handle a broad spectrum of such information. Subscription services reissue the CD-ROM disks periodically so they can continue to be current.

Finally, many standards, codes, and regulations are available via computer on-line services. For example, OSHA standards are accessible via modem from the Department of Labor in Washington, D.C.

The reader may wish to refer to Chapter 4 for additional guidance on media selection. It is likely that a large facility or organization will use a diverse selection of media, appropriate to individual applications.

16.4.4. Records Retention and Purge Procedures

It is important to ensure that standards, codes, and regulations in use by an organization are up to date so that users can be confident that their efforts are guided by, and in compliance with, the most recent recommendations and requirements. Most publicly issued standards, codes, and regulations are identified with the following information:

- title and/or identification number;
- issuing organization;
- revision or issue number; and
- date of issue or revision.

By comparing this information with that provided in current indexes or publications catalogs, it can be verified whether or not the document represents the most recent revision. As mentioned previously, this is a responsibility that is shared among document custodians and users. New or revised standards, codes, and regulations should be reviewed to identify those changes and additions which could affect existing installations. It should be noted that organizations sometimes withdraw standards, codes, or regulations and do not replace them. Copies of documents that have been superseded or withdrawn are often discarded; however, there are some circumstances where it may be appropriate to retain a copy for historical purposes.

For example, currently installed equipment was probably designed to the standards and codes existing at the time of purchase. It may be necessary, or advisable, to retain a copy of pertinent standards for the remaining life of the equipment in order to document the basis for equipment selection or fabrication or to aid in equipment maintenance. Refer to Chapter 8, *Process Equipment Integrity*.

Similarly, complications can occur when replacing existing equipment, if significant modifications have been made in associated standards. Equipment that meets the original standards may no longer be available and equipment built to current standards may not be compatible with or interchangeable with the original equipment. In these circumstances it may be necessary to fabricate equipment according to the old standards and, to do this, the purchasing organization may have to provide copies of such standards. In such circumstances, it is incumbent on the purchasing organization to demonstrate that use of older standards revisions will not compromise the safety of the equipment. The justification for utilizing the older standards should be documented.

Obsolete revisions of codes, standards, or regulations retained for historical purposes should be clearly identified as such to ensure against their inadvertent use. Some form of more secure storage may be warranted.

16.5. Auditing

The PSM program documentation should establish the auditing requirements for this element. As with other elements, the first step in auditing would be to verify that a documented program exists for implementing the standards, codes, and regulations element. This program would include features outlined in Sections 16.3.1 and 16.4 that are pertinent to the needs of the facility or organization.

The second step in auditing would be to verify that the documentation program is being implemented as described. Particular emphasis should be applied to auditing of files to ensure that the standards, codes, and regulations in use are up to date. This could be accomplished by checking the documents themselves, perhaps augmented by checking recent design and purchasing documents to determine if the referenced standards, codes, and regulations are current.

Any deficiencies identified during the audit must be documented and resolved in a timely manner. Further guidance on auditing procedures, methods, and applications can be found in Chapter 14, *Auditing,* and in the CCPS book *Guidelines for Auditing Process Safety Management Systems.*

16.6. References

AIChE–CCPS, *Guidelines for Auditing Process Safety Management Systems,* 1993

AIChE–CCPS, *Guidelines for Technical Management of Chemical Process Safety*, 1989

The Bureau of National Affairs Inc., *Environment Reporter,* a weekly publication, BNA, 1231 25th Street NW, Washington, D.C. 20037

The Bureau of National Affairs Inc., *Occupational Safety and Health Reporter,* a weekly publication, BNA, 1231 25th Street NW, Washington, D.C. 20037

National Fire Protection Association, *Portable Fire Extinguishers*, NFPA 10, Quincy, MA

Occupational Safety and Health Administration, *Process Safety Management of Highly Hazardous Chemicals,* 29 CFR 1910.119, Washington, D.C., 1992

Thompson Publishing Group, *Clean Air Permits*, a monthly publication, 1725 K Street NW, Suite 200, Washington, D.C. 20006

Thompson Publishing Group, *Chemical Process Safety Report*, 1725 K Street NW, Suite 200, Washington, D.C. 20006

Appendix 16A

Organizations in the United States That Write Codes, Standards, and Regulations

A. FEDERAL REGULATIONS

These appear in the various volumes of the Code of Federal Regulations (CFR). Bound volumes of each chapter are reissued annually, with amendments and additions published, when required, in the Federal Register (FR), which is issued daily.

Principal chapters of the CFR that are important to PSM and should be reviewed include the following:

- 29 CFR Department of Labor/Occupational Safety and Health Administration;
- 40 CFR Environmental Protection Agency; and
- 49 CFR Department of Transportation/National Transportation Safety Board.

B. STATE REGULATIONS

These vary from state to state, and require local scrutiny to establish their requirements. Their requirements may be in excess of those based on Federal Law. Appropriate state regulatory agencies should be consulted.

C. LOCAL CODES AND ORDINANCES

In addition to federal and state regulations, there are often local codes, issued by cities, towns or counties, which may apply. These often affect building codes.

D. NATIONAL STANDARDS AND CODES

There are a number of organizations that issue nationally applicable standards and codes. A list of organizations that commonly issue standards and codes having potential PSM relevance is given at the end of this appendix. This list is not intended to be comprehensive.

The American National Standards Institute (ANSI) publishes a wide variety of standards, many of them sponsored directly by ANSI, and some adopted from other organizations. In the latter case, it is common to find references such as ANSI/NFPA 70, the *National Electrical Code Handbook*, which is written by NFPA and adopted by ANSI. A date added to the number indicates the date of ANSI approval. Not all the standards of the individual organizations get adopted by ANSI.

PRINCIPAL STANDARDS AND CODES ISSUING ORGANIZATIONS

AABC	Associated Air Balance Council 1518 K Street N.W., Washington, D.C. 20005 (202) 737-0202
ABMA	American Boiler Manufacturer's Association 160,950 N. Glebe Rd., Arlington, VA 22203 (703) 522-7350
ACGIH	American Conference of Governmental Industrial Hygienists 6500 Glenway Ave., Bldg. D-7, Cincinnati, OH 45211-4438 (513) 661-7881
ACI	American Concrete Institute P.O. Box 19150, Detroit, MI 48219-0150 (313) 532-2600
ACS	American Chemical Society 1155 16th Street, N.W., Washington, DC 20036 (202) 872-4600
ADC	Air Diffusion Council One Illinois Center 111 E. Wacker Drive, Suite #200 Chicago, IL 60601-4298
AGA	American Gas Association 1515 Wilson Boulevard, Arlington, VA 22209 (703) 841-8400
AIS	American Insurance Services Group 85 John Street, New York, NY 10038 (212) 669-0400
AMCA	Air Movement and Control Association 30 West University Dr., Arlington Heights, IL 60004 (708) 394-0404

ANSI	American National Standards Institute 1430 Broadway, New York, NY 10018 (212) 642-4900
APCA	Air Pollution Control Association P.O. Box 2861, Pittsburgh, PA 15230 (412) 232-3444
API	American Petroleum Institute 1220 L Street, N.W., #900, Washington, D.C. 20005 (202) 682-8000
ARI	Air-Conditioning and Refrigeration Institute 1501 Wilson Blvd. 6th Flr., Arlington, VA 22209 (703) 524-8800
ASA	Acoustical Society of America 120 Wall Street, 32nd Floor New York, NY 10005-3993
ASCE	American Society of Civil Engineers 345 East 47th Street, New York, N.Y. 10017 (212) 705-7496
ASHRAE	American Society of Heating, Refrigerating and Air Conditioning Engineers 1791 Tullie Circle N.E., Atlantic, GA 30329-2305 (404) 636-8400
ASME	American Society of Mechanical Engineers 345 East 47th Street, New York, NY 10017 (212) 705-7746
ASNT	American Society for Nondestructive Testing 4153 Arlingate Plaza, Columbus, OH 43228-0518 (614) 274-6003
ASSE	American Society of Safety Engineers 1800 E. Oakton Street, Des Plaines, IL 60018 (708) 692-4121
ASTM	American Society for Testing and Materials 1916 Race Street, Philadelphia, PA 19103 (215) 299-5400
AWS	American Welding Society 550 N.W. Lejune Road, P.O. Box 351040, Miami, FLA 33135 1 (800) 443-9353
AWWA	American Water Works Association 6666 West Quincy Avenue Denver, CO 80235 (303) 794-7711

BOCA	Building Officials and Code Administration Intl. Publications Order Department 4051 West Flossmoor Rd., Country Club Hills, IL 60478-5795 (708) 799-2300
CGA	Compressed Gas Association 1235 Jefferson Davis Highway, Arlington, VA 22202 (703) 979-0900
CI	Chlorine Institute 2001 L. Street, N.W., Washington, D.C. 20036 (202) 775-2790
CII	Construction Industry Institute 3208 Red River St. Suite 300, Austin, TX 78705-2650 (512) 471-1793
CMA	Chemical Manufacturers Association 2501 M Street, N.W., Washington, D.C. 20037 (202) 887-1100
CTI	Cooling Tower Institute P.O. Box 73383, Houston, TX 77273 (713) 583-4087
EJMA	Expansion Joint Manufacturers Association 25 North Broadway, Tarrytown, NY 10591 (914) 332-0040
EPA	Environmental Protection Agency 401 M Street, S.W., Washington, D.C. 20460 (202) 260-2090
FEMA	Federal Emergency Management Agency 500 C Street S.W., Washington, D.C. 20472 (202) 646-3923
FMED	Factory Mutual Engineering Division P.O. Box 9102, Norwood, MA 02062 (617) 762-4300
HEI	Heat Exchanger Institute 122 E 42nd Street, New York, NY 10017
HI	Hydraulic Institute 9 Sylvan Way Ste. 360, Parsippany, NJ 07054-3802 (201) 267-9055
HMTB	Hazardous Material Transportation Board 400 7th Street, S.W., Washington, D.C. 20590 (202) 366-0656

ICBO	International Conference of Building Officials 5360 South Workman Mill Road, Whittier, CA 90601 (213) 699-0541
IEEE	Institute of Electrical and Electronics Engineers 345 East 47th Street, New York, NY 10017 (212) 705-7900
IRI	Industrial Risk Insurers 85 Woodland St., Hartford, CT 06105-1226
ISA	Instrument Society of America 67 Alexander Drive, P.O. Box 12277, Research Triangle Park, NC 27709 (919) 549-8411
ISO	International Organization for Standardization Geneva, Switzerland
MAB	Materials Advisory Board Washington, D.C. (202) 334-3505
MSS	Manufacturers Standardization Society of the Valve and Fittings Industry, Inc. 127 Park Street, N.E., Vienna, VA 22180 (703) 281-6613
NACE	National Association of Corrosion Engineers P.O. Box 218340, Houston, TX 77218 (713) 492-0535
NAS	National Academy of Sciences 2101 Constitution Avenue, N.W., Washington, DC 20418 (202) 334-2000
NEMA	National Electrical Manufacturers Association NEMA Publications Distribution Center P.O. Box 338, Annapolis Junction, MD 20701-0338 (301) 604-8002
NFPA	National Fire Protection Association 1 Batterymarch Park, P.O. Box 9101, Quincy, MA 02269 (617) 770-3000
NIOSH	National Institute for Occupational Safety & Health 4676 Columbia Parkway Cincinnati, OH 45226
NRC	National Research Council 2101 Constitution Avenue, N.W., Washington, DC 20418 (202) 334-2000

NTIS	National Technical Information Service Springfield, VA 22161 1-800-553-6847
NTSB	National Transportation Safety Board Department of Transportation Washington, D.C. 20590 (202) 366-4000
OSHA	Occupational Safety and Health Administration 200 Constitution Avenue, N.W., Washington, D.C. 20210 (202) 523-8148
SAMA	Scientific Apparatus Makers Association 225 Reinekers Lane #625, Alexandria, VA 22314 (703) 836-1360
SMACNA	Sheet Metal and Air Conditioning Contractors National Association 4201 Lafayette Center Drive, Chantilly, VA 22021 (703) 803-2989
TEMA	Tubular Exchanger Manufacturers Association 25 North Broadway, Tarrytown, NY 10591 (914) 332-0040
TIMA	Thermal Insulation Manufacturer's Association 29 Bank Street Stamford, CT 06901 (203) 324-7533
UL	Underwriters Laboratories, Inc. 333 Pfingsten Road, Northbrook, IL 60062 (708) 272-8800
USBM	United States Bureau of Mines 2401 E Street, N.W., Washington, D.C. 20241 (202) 634-1300
USCG	United States Coast Guard 2100 2nd Street, S.W., Washington, D.C. 20590 (202) 267-1748

17

Contractor Issues

17.1. Overview

17.1.1. Introduction

Contractor employees are defined as those who work on a site but who are not directly employed by the owner. There are commonly three areas in which contractor employees may be utilized. First, they can be used to construct or modify process facilities. Second, contract employees may perform maintenance work on a facility. Third, contractors are sometimes hired to operate a facility. In the latter two roles, the contractors may be employed to address a short term, specific need or they may have a long term, continuing role at the facility.

In any case, the safety standards, including controls and documentation, required of the contract organization should be just as rigorous as those expected of the owner's employees, were they to perform the same work. The basic responsibility for the performance and safety of contractor employees lies with the contractor management. However, the site owner has the responsibility to inform the contractor of potential hazards associated with the work and of appropriate facility health and safety work practices, procedures, and regulations.

This chapter provides guidance on defining and documenting the individual and joint responsibilities and procedures required of contractor and employer organizations to ensure the continued safe operation of the facility.

17.1.2. Goals and Benefits

Contractor organizations should develop PSM programs that support and interpret the owner's PSM program, in a way that is meaningful to the contractor employees. This PSM program should emphasize the responsibility of contractors to contribute toward achieving the overall facility goal of maintaining an excellent safety performance regardless of who performs the work.

These PSM guidelines for contractors will help owners, operators, and contractors to work together to perform their tasks in a safe manner with a sound knowledge of the process hazards that may be involved. A properly operated and maintained process, where safety is emphasized as a normal part of the work environment, will benefit all who work in the facility.

The major benefits to be realized are a reduced injury rate, and the reduction in the likelihood of potentially catastrophic incidents. There should also be benefits resulting from a reduction in time lost in the contractor activities. Other less tangible, but significant, benefits may be an enhanced corporate image and a reduced potential for litigation that can be associated with incidents.

17.2. Description of Contractor PSM Programs

17.2.1. Objectives

The principle objective of addressing contractor issues is to ensure that contractor employees can perform their job tasks without jeopardizing themselves or those around them. To do this, they must have, in addition to their inherent job skills, a proper knowledge of basic PSM principles, the unique process safety hazards present at the facility, and their individual responsibilities in support of process safety.

Where there is joint owner/contractor responsibility, the program should ensure that roles and responsibilities are understood and interfaces are well-defined.

17.2.2. Owner Responsibilities

In order for the owner organization to demonstrate and define its responsibilities, it is important that the PSM program includes well-defined overall objectives; stipulates their applicability to both the facility and contractor organizations; and specifies how the objectives are to be accomplished.

A comprehensive facility PSM program should include elements that influence directly or indirectly the safety performance of all contractors. The program should address and document, among other issues, the following:

- specific guidelines for selection and management of contractors;
- facility owner management and assignment of responsibilities that affect the contractor;
- contractor management and assignment of responsibilities that affect the owner;
- sound operating and maintenance discipline;
- appropriate training; and
- job-site auditing.

In some areas, such as site emergency preparedness and response procedures and plans, the owner is responsible for all personnel (owner and contractor) who may be present at the facility. Other basic owner responsibilities include:

- obtaining and evaluating a contractor's safety performance records and safe work practice program prior to selection;
- providing contractors pertinent information describing known hazards, applicable regulations and the site emergency action plan;
- verifying contractor's knowledge of the types of work involved;
- developing and implementing safe work practices to control entrance, presence, and exit of the contractor's employees;
- developing medical testing guidelines, if applicable; and
- periodically auditing the contractor's safety program performance.

17.2.3. Contractor Responsibilities

Contractors should define and document a project or facility-specific PSM program. This program may be an adaptation and/or revision of the contractor's existing PSM program, or a new PSM program specifically tailored to the job or site requirements.

The contractor's facility-specific PSM program should be applicable to its employees and subcontractors. It should address the pertinent aspects of the owner's PSM program, providing interpretive details and specific instructions for implementation, as required. The program documentation should relate contractor activities to the facility PSM goals by ensuring that:

- each employee has been trained to perform tasks safely;
- each employee has been instructed in known hazards and emergency response procedures;
- contractor employee training records are complete (including dates of training and the means used to verify understanding);
- contractor employees comply with facility safety rules;
- contractor employees advise the owner in a timely manner of any unique or unexpected hazards encountered during the performance of their job; and
- appropriate lines of communication exist between the contractor and the owner.

The contractor's program should be reviewed jointly by facility and contractor management. Facility management should examine the program for quality and application; since contractors may fill a wide variety of roles, it is essential that the program and its associated documentation are sufficiently comprehensive to match the actual scope of work. Standards, responsibilities and accountability, permits, training, emergency procedures, and applicable regulatory requirements should be reviewed. Such a preplanned and documented program should assist in resolving difficulties that may arise both before and during actual work performance.

Contractor employees may perform a variety of tasks in and around process facilities. The nature of these tasks differ widely and may require unique guidance within the PSM program. These tasks may include:

New Construction. This work is often, but not always, located on a dedicated part of a site, which may not be near an operating process. In these cases, the contractor should be required by the owner to adopt a PSM program that covers the contractor's specific work activities. In addition, the contractor should incorporate the relevant parts of the owner's program covering access to the site, emergency procedures, security aspects, compliance with legal requirements, and other pertinent requirements.

In the case of construction within an operating area, or for modification to or upgrading of an existing plant, the program should include additional elements that may be relevant to the owner's plant operations, such as restrictions on lifting heavy loads over or near operating equipment or personnel, hot work, or preparing existing equipment for modification or process tie-ins.

Maintenance. Many facilities hire contractors to perform all or part of their regular maintenance activities. In this situation, the maintenance contractor should incorporate in its PSM program pertinent requirements from the owner's program governing the performance of maintenance work on that operating facility.

Turnarounds. Many facilities plan a periodic shutdown, during which major preventive maintenance tasks, unit overhauls, cleanouts, catalyst changes, equipment modifications, and similar activities are performed. Normally, only minor preventive and scheduled maintenance and, as necessary, immediately needed repairs occur outside of this period.

These turnarounds require extensive preplanning involving representatives of the owner and all contractors. Each participating organization should incorporate in their PSM documentation any unique requirements identified during the planning.

For example, the owner's program should include provision for the preparation and availability of a safe work area, which is effectively isolated, sufficiently decontaminated, and ready for the scheduled tasks to proceed. Such preparations supplement the safe work practices used for normal operations.

The owner's PSM program and the programs of each of the turnaround contractors should address the performance of tasks on a shutdown facility that may still contain hazardous material. Special requirements relating to such situations should be identified and documented. These requirements may be more stringent than those for an operating facility since potential for incidents may be increased by the nature of the work being performed.

Another unique aspect of a turnaround is that a large number of contract employees, who will be working in the facility for a relatively brief period of time, may not have recently (or ever) worked in the facility. This requires inclusion of a significant training or retraining element in the PSM program.

Specialty Work. Occasionally there may be some special tasks performed, either on a separate construction site or an operating facility, which could pose major hazards to the facility and its personnel. Some examples include heavy lifts, use of high reach cranes, and radioactive testing. Procedures for such tasks should be included in the owner's PSM program documentation, and appropriate elements of the owner's program should be included in the specialty contractor's program.

17.3. Description of Documentation

The preceding description of the responsibilities of the owner and the contractor clearly indicates that there should be two complementary PSM programs. Often, the contractor already has such a program, although it may need to be supplemented or otherwise modified to be consistent with the owner's PSM program. It is important that any differences between the two programs be identified and resolved by the two parties before the contractor's activities commence.

17.3.1. Owner's Program Documentation

Use of contractors will require that the owner's PSM program documentation be supplemented. As a minimum, the owner's PSM program documentation should set objectives and standards for a contractor's program and include provisions to verify that these will be met.

Employers who use contractors to perform work in and around processes that involve highly hazardous chemicals should establish a screening process to identify qualified contractors whose capabilities foster confidence that can support the PSM objectives of the facility. This screening process is particularly important when evaluating contractors whose performance on previous jobs of a similar nature is not known to the hiring employer. The employer should establish a set of objective criteria for evaluating contractors. The employer should first verify that the contractor has the appropriate job skills, knowledge, and certifications for the work being performed. The selection process should include verification of the potential contractor's past safety performance, and a review of the contractor's current PSM program.

The owner's PSM program documentation should set out in detail the overall requirements and any specific procedures for administering contractor issues. This may include, but is not limited to, the following documents and related activities:

- policy statement regarding administration of contractor issues;
- responsibilities and accountability for administration;
- safety performance questionnaire as a prerequisite for contractors to be placed on request-for-bid list;
- pre-bid contractor qualification process using safety performance and PSM program information;
- contractor bid review, qualification, and final selection procedures;
- postselection contract finalization steps including pre-job meeting;
- contractor work rules on the owner's property;

- method of monitoring contractor personnel's access to and location within the facilities;
- work monitoring activities;
- postjob contractor evaluation; and
- audit requirements for compliance, as well as for program review and improvement.

17.3.2. Contractor's Program Documentation

The contractor's PSM program documentation should be based on the owners's program, supplying interpretive guidance and specific instructions for implementation. Some contractors may have a general PSM program that they modify specific to the needs of the current facility and the type of work they perform. The following are typical of the items the program may address:

- policy statement;
- responsibilities and accountability for administration;
- work rules to be followed on the owner's property;
- training;
- permit-to-work systems;
- hazard communication;
- work monitoring activities;
- accident reporting and first aid;
- incident reporting and emergency preparedness; and
- audit program for review of the PSM program.

The reader may wish to refer to other chapters in this book (e.g., see Chapter 18, *Permit-to-Work Systems)* for additional details on some of the above topics.

Due to the specific, and sometimes unique, nature of their work activities, contractors may have safety program requirements that go beyond those in the owner's PSM program. There should, however, not be conflicting requirements in the two programs. Where facility employees and contractor employees perform similar operating or maintenance activities in an operating facility, there should be no significant difference in the PSM procedures being followed or the documentation being used by the two groups.

17.3.3. Records from Addressing Contractor Issues

In addition to the relevant PSM program documentation as just described, the records generated when addressing contractor issues may include the following special records.

Contract Employees Injury/Illness Records. Owners may require that contractors (and all subcontractors) provide them with the following documents:

- the log and summary of occupational injuries and illnesses;

- accident/incident investigation reports; and
- any other injury/illness information that the owner requires.

Contractor Employee Training Records. The contractor should document that their employees, by name and date trained, have received and understood training given in:

- work practices necessary to perform their job safely;
- known potential fire, explosion, or toxic release hazards related to their jobs and the process; and
- applicable provisions of the emergency response and evacuation plan.

The training topics should be listed and traceable to a training lesson plan and/or training materials used in the instruction process. See Chapter 12, *Training,* for a more complete discussion on training documentation.

Audit Records. The owner may require that the contractor share the results of PSM program audits. See Section 17.5 below.

17.4. Records Management

Both the owner and the contractor should be responsible for maintaining current and accurate records. Generally, there are few records management considerations unique to the topic of contractor issues. Refer to Chapter 4, *Records Management*, for a discussion of basic records management practices.

17.4.1. Where Are Records Maintained?

Some records may need to be maintained at the facility by both owner and contractor to ensure that they are readily available for reference, inspection, and audit purposes. Additionally, some records may need to be kept at the point of use in the process area (e.g., hot work and confined-space entry permits).

Attention should be given to identifying those records generated by the contract organization that are to be turned over to the owner at the completion of the job. These may include critical equipment test records required as part of the process equipment integrity program. Refer to Chapter 8, *Process Equipment Integrity*.

17.4.2. Retention/Purge Schedule

The owner and contractor should develop procedures for records retention and purging to fulfill their specific PSM requirements, to meet regulatory requirements and meet other needs that may be identified by the owner. The owner should monitor and approve the adequacy of contractor plans in this area.

17.4.3. Access Controls

The owner and the contractor, as part of their own PSM programs, should determine if any special access controls for records or other documentation are warranted. This activity may need to consider proprietary or trade secret information as well as privacy rights and other legal aspects.

17.5. Auditing

This chapter has described, for both the owner and the contractor, PSM program requirements related to the use of contract employees for a variety of functions such as construction, operations, and maintenance. It was recommended that each contract organization should have a documented PSM program that is derived from, and implements, applicable portions of the owner's PSM program. Further, it was recommended that the owner's PSM program should be broadened to include policies and procedures governing the selection and oversight of contractor organizations.

Both the owner and the contractor organizations have a responsibility to audit the content and implementation of their respective PSM programs. The detailed audit requirements for owners and for contractors will differ according to the contents of their PSM programs, but both should follow the general approach described in this book for auditing PSM elements.

The first step in auditing an element in a PSM program is to verify that a documented program exists for its implementation. The second step in auditing is to verify that the results of implementing the element have been documented in the manner prescribed by the program.

Each organization should document the results of their audits, and any deficiencies identified during the audits should be promptly resolved. General guidance on auditing can be found in Chapter 14, *Auditing,* and in the CCPS book, *Guidelines for Auditing Process Safety Management Systems.*

17.5.1. Auditing by the Contractor

The contractor's PSM program would incorporate pertinent features outlined in Section 17.3.2. Many of these features are analogous to elements of the owner's PSM program and audit requirements would be similar. The reader may wish to refer to the applicable chapters of this book for more detailed guidance.

The scope of the audit may include, but not be limited to, the following:

- permit system requirements;
- written procedures to parallel those of the owner;
- that the owner's emergency preparedness procedures are available to the contractor;
- training records;

- Management of Change records;
- incident investigation and documentation procedures; and
- that the contractor's PSM documentation includes requirements for subcontractors, where appropriate.

17.5.2. Auditing by the Owner

The owner organization should audit its program for administering contractors issues; i.e., that a documented program exists, that it addresses pertinent features discussed in Section 17.3.1, and that the results of its implementation are documented as prescribed in the program.

The owner organization should also, as part of its oversight roll, audit the contractor's PSM program and its documentation to ensure that both are consistent with the facility requirements.

17.6. Examples

In 1989 there was a vapor cloud explosion (VCE) at a plant in Texas. Many of the 23 workers that were killed were contract workers. There appeared to be inadequacies in vehicle and hot-work permits, and in respiratory protection and hazard communications programs. It appears the owner did not require the contractor to have a PSM program and supporting documentation consistent with the owner's; did not audit the contractor's PSM program (if one existed); and possibly had an inadequate, or inadequately audited owner PSM program.

In 1990 there was a fire and explosion at a Texas plant where the owner employed several different contractors on-site and maintained control over them and their working conditions. This is apparently an instance of the owner, who was responsible for overall worker safety, not insisting that the various contractors have and enforce their own documented PSM systems.

Four contractor employees were among the six killed in an explosion and fire at a Louisiana refinery in 1991. Improvements in contractor training were recommended as one result of the investigation.

Contractor employees may be subject to hazardous situations for which the owner is liable. Proactive owner/contractor joint PSM programs, which are subjected to periodic audits, can help reduce both the frequency and severity of incidents.

17.7. References

American Petroleum Institute, *Cleaning Mobile Tanks Used for Transportation of Flammable Liquids* in *Accident Prevention Manual,* No 13, Washington, D.C., 1958.

American Petroleum Institute, *Cleaning Petroleum Storage Tanks,* RP 2015, Washington, D.C.

American Petroleum Institute, *Gas and Electric Cutting and Welding,* RP 2009, Washington, D.C.

American Petroleum Institute, *Owners and Contractors Responsibility for Safety,* RP 2250, Washington, D.C.

American National Standards Institute, *Lockout/Tagout of Energy Sources, Safety Requirements,* ANSI-Z244.1, New York, NY.

American National Standards Institute, *Safety in Welding and Cutting,* ANSI-Z49.1, New York, NY.

American National Standards Institute, *Safety Requirements for Working in Tanks and Other Confined Spaces,* ANSI-Z117.1, New York, NY.

Construction Safety Association of Ontario, *Confined Space in Construction,* Toronto, Ontario.

National Fire Protection Association, *Cleaning or Safeguarding Small Tanks and Containers Without Entry,* NFPA 327, Quincy, MA.

National Institute of Safety and Health, Department of Health and Human Services, *A Guide to Safety in Confined Spaces,* No. 87-113, Washington, D.C.

Occupational Safety and Health Administration, *Permit-Required Confined Spaces for General Industry,* 29 CFR 1910.146.

Occupational Safety and Health Administration, *The Control of Hazardous Energy (Lockout/Tagout),* 29 CFR 1910.147.

The National Safety Council, *Accident Prevention Manual for Industrial Operations,* Itasca, IL.

The National Safety Council, *Personnel Safety in Chemical and Allied Industries,* Itasca, IL.

18

Permit-to-Work Systems

18.1. Overview

18.1.1. Introduction

Routine operations in process plants normally take place in a work environment designed to limit the potential hazards. For instance, in processes involving flammable materials, potential ignition sources such as open flames, spark-producing equipment and electrical equipment are typically excluded by a combination of design features and administrative procedures. In situations where it becomes necessary to temporarily bypass these protections, special permit-to-work systems are established to control and monitor job or work-site conditions to ensure that an equivalent measure of safety is maintained.

Such intermittent or nonroutine work conducted in process areas should be consistently controlled through the use of documented procedures to ensure that the work is accomplished safely. The hazards associated with this work should be communicated not only to those doing the work, but also to other personnel whose work is affected, either directly or indirectly. Safe work practices and permit-to-work systems are part of an organization's PSM program for identifying, assessing, and controlling potentially hazardous nonroutine tasks.

The following list includes many of the safe work practice and permit-to-work systems found in process plants:

- confined space entry;
- equipment isolation (e.g., lockout, tagout);
- excavation and shoring;
- fire equipment impairment;
- hot work;
- pipeline breaking; and
- other practices and permits controlling high-risk work of a unique nature (e.g., working with ionizing radiation).

Some organizations also use a permit system for work not involving the other specialized permits. This is known by various titles such as a cold-work permit or clearance certificate. Note that some organizations use different terminology when defining and describing the above practices and permits. Also, be aware that a task may require the use of more than one type of permit in its scope of work.

Some permit-to-work systems apply to process operations with specific hazards; however, most apply to maintenance, modification, and other similar activities.

18.1.2. Goals and Benefits

The goal of a permit-to-work system is to ensure that affected parties (e.g., the employer, employees, and contractors) are made aware that nonroutine work is being performed in or near a process area and to ensure that appropriate safety precautions are taken prior to and during the course of this work. Permit systems are intended to keep the employer and employees informed of work activities, provide better coordination of such activities, and enhance management control over the work being performed in the process area. The benefits to be gained from implementing a permit-to-work system include higher standards of safety which should result in fewer incidents, and lower costs as a result of reduced production downtime, fewer injuries, and less damage to equipment.

Federal, state, and local regulations may require the adoption of safe-work permit systems. Comprehensive and clear documentation is necessary to ensure that such systems are in place and effective.

18.2. Description of Permit-to-Work Systems

18.2.1. Objectives

Under normal process operating conditions, personnel do not come into direct contact with hazardous chemicals or equipment. This is because hazardous chemicals are usually contained within plant equipment, and hazardous equipment is fitted with protective devices to prevent personal contact or exposure. However, during many maintenance operations and some intermittent process operations, the potential for exposure to such hazards may increase, since the containment system is opened and/or protective devices on equipment are removed.

Safe work practices are required to ensure that such situations cannot occur indiscriminately, and that potential hazardous exposures are minimized and adequately controlled. Safe work practices include a permit-to-work system, which is basically administrative and may be related to specific protective systems (e.g., personal protection) or procedures.

While safe work practices are the guidelines for properly and safely performing specific work assignments, permit-to-work systems are used to:

- document responsibility for authorization of work;
- ensure that affected personnel are aware that the work is being done;
- identify the facility and/or equipment on which the work is to be done;
- ensure that appropriate safety precautions are taken prior to beginning the work;
- document that the work has been satisfactorily completed and that all equipment has been returned to an operable condition; or
- where necessary, document that the work has not been completed on return of the permit, and/or that the equipment is still not operable.

Experience indicates that without such permit-to-work systems in place, accidents are more likely to occur. These accidents are considered preventable in PSM programs, and accident prevention is the primary objective of safe work practices and permit systems.

18.2.2. Implementing Permit-to-Work Systems

The procedures for implementing permit-to-work systems vary significantly from one organization to the next; therefore, it is not feasible to provide definitive "how-to" guidance here. Each organization, after understanding the intent of permit-to-work systems, should define a program that meets their goals and needs.

Not all areas of a facility will require the use of a permit-to-work system. Permit-free work areas are areas which, after thorough review, have been deemed to be free of hazards that would require implementation of the permit-to-work system. An example would be areas of new construction specifically declared permit-free by the proper authority.

Some general rules and responsibilities for permit-to-work systems are suggested in Appendix 18A.

18.3. Permit-to-Work Systems Documentation

18.3.1. Permit-to-Work Systems Program Documentation

Organizations implementing a permit-to-work system should include, in their PSM program documentation, guidance on the goals of the system as well as any procedures require for its implementation. The program documentation might cover:

- purpose;
- definitions (e.g., hot work, cold work, confined spaces, etc.);
- authorization of persons responsible for issuing permits, and definition of requirements for their training;

- permit responsibilities;
- permit application and format;
- requirements for display of permits;
- requirements for monitoring of work performed under a permit;
- minimum requirements (e.g., atmosphere testing, employee training, operating conditions, etc.);
- record keeping requirements; and
- audit requirements.

Some of these topics are elaborated on below:

Purpose. A system of work permits provides for orderly and safe job preparation and performance; ensures adequate communication with all concerned personnel; and assigns responsibilities. Proper permit systems reduce the likelihood of trapping employees in moving machinery or confined spaces; of fires and explosions from hot work; of inadvertent release of highly hazardous materials; and of other potential hazards that might result from the performance of maintenance or other special tasks.

Permit Responsibilities. The issuance of a permit-to-work implies the acceptance of responsibility for the safe performance of the work. The permit should clearly identify who is responsible for the various aspects of the work, including:

- the prime responsibility for coordination and approval of permits-to-work should be established at a supervisory level within the organization having custodial responsibility for the equipment or area;
- the supervisor responsible for overseeing the work (e.g., a maintenance foreman) should also be assigned responsibility to ensure that the requirements of the permit-to-work are fulfilled during the performance of the work;
- additional responsibility should be placed on the supervisor having day-to-day operational responsibility for the equipment or area to ensure that permit-to-work requirements are fulfilled; and
- the delegation of responsibility to line personnel for ensuring that detailed permit-to-work requirements are enforced should neither lessen nor relieve authorized persons of final responsibility.

Refer to Appendix 18A for a more detailed discussion of permit responsibilities.

Permit Application and Format. To avoid any confusion and to assist in providing clear information to those involved, an organization should consider the use of a consistent format, using standardized forms for each type of permit. The following basic considerations should apply:

- a permit-to-work should be required for the performance of any work involving maintenance, repairs, alterations, or installations above and beyond normal procedures (also in certain normal situations where hazardous materials are exposed);

- a permit-to-work should be govern all affected personnel;
- subsidiary permits (e.g., lockout, confined space entry, line breaking, hot work) and necessary approvals should be obtained before the principal permit-to-work is issued;
- as a minimum requirement, the permit-to-work should contain the following information:
  —date and time of issuance;
  —effective dates and times of the permit;
  —location and outline of work to be performed;
  —approvals and clearances (involved departments, sections, units, etc.);
  —precautions to be taken; and
  —emergency procedures.
- the permit-to-work document should cover the use of any special equipment, materials or procedures that could cause personnel injury or property damage, such as radioactive materials, laser beams, high pressure equipment, explosives, high ultraviolet and infrared light intensities, flammable materials, extreme temperatures, etc;
- safeguards should be identified to protect against materials, equipment or job site hazards; and
- the work permit may contain a log section or attachment to denote job progress, shifts and trades involved, and any reassignments made during the course of the job.

Examples of typical permit forms are contained in the appendixes at the end of this chapter. These examples illustrate how the basic requirements previously described can be set out in a permit format. Other formats can also be developed to meet the basic requirements, as well as any specific needs. The contents of the appendixes are as follows; permit titles may vary from one organization to another.

APPENDIX 18A	**Typical Rules and Responsibilities for Permit-to-Work Systems**
APPENDIX 18B	**Hot-Work Permit**
APPENDIX 18C	**General Permit-to-Work**
APPENDIX 18D	**Lockout/Tagout Permit**
APPENDIX 18E	**Pipeline Breaking Permit**
APPENDIX 18F	**Confined Space Entry Permit**
APPENDIX 18G	**Other Permit Systems**

Requirements for Display of Permits. In order to facilitate providing information to those involved, the program should require that:

- issued permits should be conspicuously displayed at the worksite for the duration of the work; and

- permits should be protected from damage and other causes that could result in degraded legibility.

18.3.2. Records from Implementing a Permit-to-Work System

The records generated when implementing a permit-to-work system consist of the completed permit forms, as well as any subsidiary special procedures or field test results (e.g., atmosphere testing). Organizations typically prepare two copies of each permit, one for posting at the job site and another to be retained by the issuer. Additional copies may be necessary to fulfill unique records management requirements established by the organization.

18.4. Records Management

18.4.1. Records Management Program

The organization's PSM program documentation should establish practices to be followed for managing records for the permit-to-work systems. Responsibilities for the management of the documents should be clearly designated (by organization and title). This section addresses pertinent records management issues. Refer to Chapter 4, *Records Management*, for more general guidance.

18.4.2. Where Are Records Maintained?

Normally, current permits are kept in two places:

- one copy is maintained at or near the work site by the user during the period the work is taking place; and
- a second copy is typically retained by the issuer. Often, this location is a control room or a facility supervisor's office.

The work site copy is typically returned to the issuer at the completion of the job, or upon expiration of the permit. This copy typically indicates the status of the job, documents any special testing performed (such as atmosphere testing, as described in Appendix 18A), and identifies those parties that worked on the job. Many organizations may choose to file this copy with, or in lieu of, the issuer's copy as a historical record (subject to auditing) of the job.

18.4.3. What Media Should Be Used?

Traditionally, most permit-to-work documents have been on paper forms. To facilitate identification, many organizations use carbon or carbonless copy systems, with the different copies numbered and/or color coded. After the form is filled out and authorized, one copy is given to the user and the other is retained by the issuer.

The mechanism for generating the permit form could be computer-based, with the issuer filling out the form using a blank form template and then printing the permit for signature. Copies could then be distributed to the issuer and user, as described above.

18.4.4. Retention/Purge Schedule for Records

The PSM program documentation should define a records retention and purge schedule for completed work permits. Consideration should be given to the potential need to retain a permit for a job where an incident or injury has occurred and the need to retain records until applicable audit requirements have been met.

18.4.5. Revision Controls for Permit System

As the permit system is used and regularly audited the need for changes may be identified; e.g., to improve clarity of the forms and enhance the understanding for those involved in the work, or to implement more stringent requirements deemed necessary to ensure adequate protection of personnel. The program documentation should establish how these changes are to be made and approved.

Organizations may wish to implement a controlled revision procedure similar to that described for procedures in Chapter 11, *Operating Procedures*, reasoning that the permit-to-work form is, essentially, a procedure for preparing the job site prior to work.

After such a revision is approved, the person(s) responsible for the permit-to-work system should ensure that permit issuers are notified of the changes and the reasons for them; that obsolete versions of permit forms are destroyed; and that retraining, where necessary, is given.

18.5. Auditing

The auditing of the permit-to-work system is an important step toward retaining a high level of PSM achievement. Audits can be conducted by persons involved in the operation, company internal auditors, or external auditors. The program documentation should establish applicable audit requirements.

The first step in auditing is to verify that a documented program for implementing the permit-to-work system exists. The program would include features outlined in Sections 18.3.1 and 18.4 that are pertinent to the needs of the organization or facility. The second step in auditing is to verify that the results of implementing the program have been documented in the manner prescribed by the program documentation.

Any deficiencies identified during the audit must be documented and promptly resolved. See Chapter 14, *Auditing*, the CCPS book, *Guidelines for Auditing Process Safety Management Systems* for additional guidance on auditing.

18.6. Examples

This section provides some illustrations of where inadequate permit-to-work systems have failed to prevent accidents.

18.6.1. Inadequate Implementation of Permit-to-Work System

In 1988, an offshore platform in the North Sea was totally destroyed in a series of explosions, taking the lives of 165 of the 225 persons on board. Two rescuers in a life craft also lost their lives. A full report on this accident has been published which identified unsafe work practices as one of the contributory causes. Improved PSM documentation could have helped prevent this incident. In the report, 106 recommendations were made, including 7 related to permits. These included:

- the operator's permit-to-work system should be part of its PSM program;
- all permit-to-work systems should incorporate a mechanical isolation procedure which involves the physical locking closed and tagging of isolation valves; and
- copies of all issued permits should be displayed at a convenient location and in a systematic arrangement such that process operating staff can readily see and check which equipment is under maintenance and not available for operation.

18.6.2. Inadequate Documentation

In January 1973, a caustic soda liquor tank exploded, resulting in a fatality. A hot-work permit, requiring the work area to be free of flammables, had been issued for welding high above the tank. However, the tank had also contained a hydrogen–air mixture resulting from degassing of the liquor. Falling sparks ignited the hydrogen–air mixture issuing from the tank's grade level overflow (which also served as the vent). When the hot-work permit was filled out, the issuer had not recognized that falling sparks would, effectively, extend the work area to grade level.

18.6.3. Application of Permit-to-Work System Not Comprehensive

In October 1989, a major release of process fluids formed a vapor cloud which exploded, damaging or destroying much of a chemical complex in Texas. In the OSHA report of their investigation, two aspects of poor safe-work permit performance were highlighted. Certain process operations had been excluded from the permitting system, and the application of permits to employees and to contractor employees during maintenance operations was not consistent.

18.6.4. Ambiguous Information Entered on Work Permit

Atmosphere testing had been completed for a vessel prior to the start of a job requiring a vessel entry. The notation *none found* had been entered on the work permit by the technician performing the test. An employee interpreted this to mean that *no hydrocarbons* were found, when it actually meant that *no oxygen* was found. Fortunately, the error was detected before a serious injury or death resulted. Field test results should be recorded in an unambiguous manner using the appropriate units.

18.7. References

AIChE–CCPS, *Guidelines for Auditing Process Safety Management Systems,* 1993.

AIChE–CCPS, *Guidelines for Investigating Chemical Process Accidents*, 1992.

AIChE–CCPS, *Guidelines for Preventing Human Error in Chemical Process Safety*, 1994.

ANSI, *Lockout/Tagout of Energy Sources, Safety Regulations,* ANSI-Z244.1.

Department of Energy (U.K.), *The Public Inquiry into the Piper Alpha Disaster,* HMSO, London, 1990.

Kletz, T.A., *Lessons form Disaster: How Organizations Have No Memory and Accidents Recur*, Institution of Chemical Engineers, 1993.

Occupational Safety and Health Administration, *Permit-Required Confined Spaces for General Industry*, 29 CFR 1910.146.

Appendix 18A. Typical Rules and Responsibilities for Permit-to-Work Systems

This appendix suggests rules and responsibilities for a hypothetical permit-to-work system. Note that the following may not be applicable to all situations; organizations should develop and document a program that addresses their unique needs and circumstances.

18A.1. Typical Permit Rules

Permit-to-work systems should be based on firm rules for implementation, such as those listed below:

- a work permit must be obtained before the job is started, if the work to be accomplished falls within the scope of a permit-to-work system;
- the work permit should be issued by a representative of the group having custodial responsibility for the equipment or work area;

- some organizations allow the permit to be approved by the issuer; other organizations require a higher level of authorization;
- when work is to be performed on equipment where custodial responsibility is not clearly defined, the permit should be issued by the next level of authority or that person's designated representative;
- if responsibility for the equipment or area involved in the work permit is shared by two or more operating units or departments, the permit should be signed by representatives of each group;
- normally, the employee who performs the work receives the permit and signs it. In some cases, the employee's supervisor receives and signs the permit;
- a copy of the permit must be kept at the work site;
- permits should be valid for a specific period of time (e.g., from time of issuance until a set time has elapsed, or until the end of the shift);
- if work is not started, or if it is stopped because the conditions, controls, or procedures required by the permit could not be established or maintained, the permit should be voided and a new permit issued, after reevaluation, before work is resumed;
- if work is stopped or delayed for other reasons, the appropriate facility representative should be notified, and approval to resume work must be obtained;
- if the work extends beyond the period authorized by the permit, work must be halted until, after reevaluation, the existing permit is extended or a new permit is issued; and
- when the work is completed, or at the end of the workday, the permit should be returned to the issuer, marked to indicate the status of the work, and signed by the employee.

Special consideration is required in some cases. Blanket work permits could be issued during turnaround or construction jobs, but usually only after representatives of the operations, maintenance, and safety departments have decided on the coverage to be provided and have determined that unit isolation has been completed and equipment has been properly prepared for the work to be done. Upon receipt of the work permit, it is the responsibility of the supervisors to thoroughly explain to the workers the conditions and coverage for which the blanket permit was issued.

18A.2. Typical Permit Responsibilities

The responsibility for the safety of the work is shared by all persons involved with the job covered by the permit. The issuing party has the primary responsibility and should:

- ensure that the permit is accurately and completely filled out, and properly approved;

- ensure that the equipment or area has been properly prepared by making an on-site inspection before issuing the permit;
- inspect the work site just prior to the start of the job;
- clearly indicate on the permit whether the site is inspected and ready for work;
- arrange for the atmosphere to be tested by a qualified employee to determine the concentration of oxygen and/or combustible gas, hydrogen sulfide, carbon monoxide or any other harmful gas suspected of being present, if required by the permit (e.g., for vessel entry or hot work);
- enter the results of atmospheric testing on or attach to the permit;
- ensure that the person receiving the permit understands the requirements of the permit and any unusual conditions connected with the job;
- stop the work if unusual operating conditions affect the safety of the work;
- pass on information about work in progress to relief persons;
- inspect the worksite for housekeeping at the completion of the job; and
- retain the completed copies of permits as required by the records management program.

The employee receiving the permit is responsible for the following:

- inspecting the worksite for safety conditions prior to the start of the job;
- performing only the specific work called for on the permit;
- complying with all conditions stated on the permit;
- keeping the employee's copy of the permit available at the jobsite;
- notifying the issuer of any change in the condition of the job or area;
- arranging for permit extension or re-issuance (if necessary);
- signing and returning the permit to the issuer at the end of the job or end of the workday with the status of the job clearly stated (e.g., completed, not complete, etc.); and
- cleaning up the work area.

Appendix 18B. Hot-Work Permit

This appendix gives the requirements for a typical hot-work permit, i.e., where the temperature achieved in the performance of the work could ignite flammable material that may be present, whether solid, liquid or vapor. Hot work includes welding, burning, grinding, use of power tools, tar pot operation, resistance soldering, pouring of lead, use of salamanders, sandblasting, and similar work involving open flames, high temperatures, or sparks.

The use motor vehicles and other internal-combustion-driven equipment within operating unit processing areas can also pose an unintended ignition hazard; a related permit is the Permit to Operate Motorized Equipment.

The hot-work permit should address the following issues, as appropriate:

- free areas for hot work (e.g., the areas in which hot-work permits are not required);

- any atmospheric testing for flammable vapors before start of, and periodically during, the work;
- identification, for sealing or covering, of openings, cracks, and holes into which sparks may drop to combustible materials;
- fire extinguishers and/or hoses required to be present at the job site;
- a fire watcher, whenever welding is performed in any locations where other than a minor fire might develop;
- written authorization by the person responsible for all welding and cutting operations;
- site inspection and designation of appropriate precautions prior to written authorization;
- description of precautions associated with combustible materials on floors, walls, partitions and ceilings, or roofs of combustible construction;
- prohibition of welding in unauthorized areas:
  —in sprinklered areas while such protection is impaired;
  —in the presence of potentially explosive atmospheres; and
  —in storage areas for large quantities of readily ignitable materials.
- relocation of combustibles, where practical, or covering with fireproofed covers where not practical;
- identification, for shutdown, of any ducts or conveyor systems that may convey sparks to distant combustibles;
- precautions when welding on components (e.g., steel members, pipes, etc.) that could transmit heat by radiation or conduction to unobserved combustibles;
- identification of hazards associated with welding on walls, partitions, ceilings or roofs with combustible coverings, or welding on walls or panels of sandwich-type construction; and
- emergency procedures.

Figure 18-1 shows a typical hot-work permit form.

Appendix 18C. General Permit-to-Work

The general permit-to-work is used for a variety of activities that are not performed in a permit-free area and are not covered by one of the specific work permits, such as hot work or confined space entry. This permit goes under a wide variety of names, such as cold-work permit, clearance certificate, etc. Examples of jobs requiring such permits-to-work are as follows:

- work on a vessel or lines that have contained flammable liquids, gases or toxic and corrosive material (some organizations use a special line breaking permit for this work, see Appendix 18E); and
- work on process equipment associated with an operating unit or facility.

Figure 18-2 shows a typical cold-work permit form.

HOT WORK PERMIT

REQUIRED FOR WELDING, CUTTING, BURNING OR OTHER HOT WORK
IN ANY LOCATION OTHER THAN ESTABLISHED SHOPS.
THIS PERMIT TO BE ISSUED ONLY AFTER WORK SITE HAS BEEN INSPECTED.
THIS PERMIT IS SUSPENDED IN THE EVENT OF PLANT ALERT OR EVACUATION.

DATE	FROM A.M. P.M.	TO A.M. P.M.	WORK AREA

JOB DESCRIPTION ______________________________

____

SITE PREPARATION

1. Equipment Preparation
   Steamed [ ] Washed [ ] Purged with ______________
2. Has equipment been checked for linings, deposits, or pockets that could be flammable, corrosive, or toxic? [ ] Yes [ ] No
3. Explosimeter check performed?
   If yes:
   Time area was checked ______________
   Name of person who made check ______________
4. Have hazards of nearby areas been checked?
   Other Floor Levels [ ] Yes [ ] No
   Neighboring Bldgs. [ ] Yes [ ] No
   Sewers [ ] Yes [ ] No
   Other Equipment [ ] Yes [ ] No
   Radiation [ ] Yes [ ] No
   Welding Machines Properly Grounded [ ] Yes [ ] No

SAFETY EQUIPMENT

5. Protective equipment needed
   Fire Protection [ ] Clothing [ ] Ventilation [ ]
   Respirator [ ] Ear [ ] Other ______________

WORK PROCEDURES

6. Will an operating representative or fire watch be present? [ ] Yes [ ] No
7. Is a Vessel Entry Permit required? [ ] Yes [ ] No
8. Special procedures ______________

____

____

ADDITIONAL SPACE ON BACK

SIGNATURE OF PERSON AUTHORIZING PERMIT	
DEPARTMENT OR GROUP ASSIGNED JOB	
SIGNATURE OF EMPLOYEES ASSIGNED TO JOB Space for additional signatures on back	

SAFETY STANDARD

FIGURE 18-1. Example Hot-Work Permit

SAFE WORK PERMIT/CHECKLIST

☐ SAFEWORK ☐ VEHICLE ENTRY ☐ HOT WORK

DATE ______ TIME WORK STARTS ______ TIME WORK STOPS ______

DEPARTMENT ______ ISSUED TO ______ COMPANY ______

LOCATION/EQUIPMENT ______ NO. OF WORKERS ASSIGNED ______

WORK TO BE DONE ______

	YES	NO	N/A
PIPING/EQUIPMENT			
Are all connections blinded, disconnected, or double blocked and bled?	☐	☐	☐
Has the equipment been depressured and drained?	☐	☐	☐
Has the equipment been flushed and cleaned for this job?	☐	☐	☐
Were prior contents corrosive?	☐	☐	☐
Were prior contents combustible?	☐	☐	☐
Were prior contents toxic?	☐	☐	☐
HAZARD COMMUNICATIONS			
Have workers been advised of chemicals used in this area related to this job and the location of MSD sheets? (Note the Chemicals) ______	☐		☐
FIRE/EXPLOSION PREVENTION			
Has the combustible gas meter been calibrated and does it work properly?	☐		☐
Does the combustible gas test show safe working/vehicle entry conditions for the area and equipment?	☐	☐	☐
Do workers know fire extinguisher location?	☐		☐
LOCKOUT/TAGOUT			
Is lockout/tagout required?	☐	☐	
Has owner performed lockout/tagout?	☐	☐	☐
Is a Master Red Tag Installed?	☐	☐	
If Yes, Location: ______ Number(s) ______			
Has craft performed lockout/tagout?	☐		☐
METHOD: ☐ At each Blocking Device ☐ Sign Master Red Tag ☐ Lockout Electrical			
Are motor starters off or fuses removed?	☐		☐
Is a stand-by man needed?	☐		☐
WORK ENVIRONMENT/PERSONAL PROT. EQUIP.			
Is the atmosphere ok for breathing?	☐	☐	
Is purging or ventilation needed during work?	☐	☐	
Is the area where work is to be performed safe for this job?	☐		
Is any extra personal safety equipment necessary to do this job? (If yes, note them)	☐	☐	

☐ Monogoggles ☐ Acid Suit ☐ SCBA
☐ Full Face Shield ☐ Flash Coat & Hood ☐ Paper Coveralls
☐ Rubber Gloves ☐ Linemans Gloves ☐ Cotton Gloves
☐ Rubber Boots ☐ Rubber Mats ☐ Dust Masks
☐ Slicker Suit ☐ Respirator ☐ Safety Harness
☐ Acid Hood ☐ Hot Stick Elect. ☐ Ear Protection

	YES	NO
Are other special protective measures or instructions needed? (If yes, note them) ______	☐	☐

	YES	NO	N/A
MATERIAL HANDLING			
Does the operator of the mobile equipment have a valid license?	☐		☐
Does the lift equipment have current inspection tag?	☐		☐
Is the lift area barricaded as per Safety Policy #1?	☐		☐
Note restricted lift areas within the boom reach of the equipment being used. ______			
PERSONS INVOLVED			
Do all Contractors in the crew have Block Specific Orientation?	☐		
Do workers know safety shower/eye bath location(s)?	☐		
Does the person accepting the permit have the proper attire and safety equipment?	☐		
Do workers know seek shelter location(s) listed?	☐		
Do workers know block emergency assembly points? ______	☐		
The above items are correct. ______ Signature Crew Leader/Foreman			
OTHER REQUIREMENTS			
Are other required permits, approvals, and procedures completed, attached, and in compliance with safety rules and standards? (If yes, note them)	☐	☐	☐

Confined Space Entry ☐ Radiography ☐
Excavation ☐ Personnel Lift ☐
Electrical (Over 600V) ☐ Hot Work Procedures ☐
Critical Lift ☐ Isolation Procedures ☐
Other ☐ ______

SIGNATURES/NAMES

Process person notified ______

Stand-by-person ______ Company ______

Firewatch (1) ______ (2) ______

We have personally inspected for compliance with all of the above requirements.

Issued by ______

Accepted by ______ Company ______

Co-signed by ______ Company ______

CLOSE OUT

Time ______

Job Comp. ______ (Yes/No) Tags & locks removed ______

Master Red Tag Signed Release ______

Area Cleaned ______

Returned by ______

FIGURE 18-2. Example Cold-Work Permit

Appendix 18D. Lockout/Tagout Permit

Lockout and/or tagout procedures are used to protect workers from exposure to hazardous materials and/or sources of energy that might still be present in the work area. An example might be isolating an equipment system from a source of corrosive chemicals. Another example might be where the accidental restarting of electrically, mechanically, or hydraulically driven equipment or equipment containing stored energy could cause a hazardous situation.

Situations in which such hazardous exposures might occur should be identifiable in a safety review and addressed in work permits. A lockout/tagout permit provides a means of specifically identifying the isolations points (e.g., electrical circuit breakers to be opened or block valves to be closed) and means of isolation required to safely prepare the job site.

Lockouts require the installation of a lock to physically prevent the operation of equipment (e.g., closing a circuit breaker or opening a valve) and typically include installation of an information tag explaining the purpose of the lockout, identifying the person installing the lock, and indicating the date that the lock was installed. Tagouts (installation of the tag, but without the lock) can be used when it is impractical or undesirable to install a lock.

Many organizations prefer a lockout to a tagout because of the physical barrier to equipment operation that the lock provides. The organization's PSM documentation should describe any acceptable alternatives to lockout and describe how they are to be performed and documented. Alternatives should be adequately rigorous to ensure the safety of those performing work that relies on effective isolation for its safe execution.

The lockout/tagout permit may be issued independently, or in association with another permit (e.g., confined space entry).

Figure 18-3 shows a typical lockout permit form.

CONFINED SPACE ENTRY PERMIT

SPACE LOCATION/ID: ______________________________

PURPOSE OF ENTRY: ______________________________

AUTHORIZED ENTRANTS/ATTENDANTS:	

RESCUE/EMERGENCY SERVICES

NAME	NOTIFIED	NAME	NOTIFIED

MEANS OF COMMUNICATION: (Entrant to Attendant)

___ VERBAL/VISUAL ___ RADIO

___ OTHER ______________

INSTRUCTIONS TO SUMMON RESCUE: ______________________________

____

TESTING RECORD		**TIME**				**REVALIDATION TESTING**			
		: AM/PM	: AM/PM	: AM/PM	: AM/PM	: AM/PM	: AM/PM	: AM/PM	: AM/PM
Parameter	**Acceptable Conditions**	**RESULTS**							
Oxygen	19.5% - 21%								
Flammability	0% LEL								
Toxics (Specify)									
Tester Initials									

Comb Gas/02 Monitor Make/#: Span Gas Test Date:

Tube/Instrument Used for Toxics: Span Gas Test Date:

EQUIPMENT REQUIRED FOR ENTRY

FIRE EXTINGUISHER	LADDER	TWO-WAY RADIO FOR ATTENDANT	OTHER(List) ______
SAFETY HARNESS & LINE	LIGHTING	AIR MOVER (E.G. BLOWER)	______
COMB. GAS/O_2 MONITOR	GFCI	PRESSURE RESPIRATORY EQUIPMENT	______
		NON ENTRY RESCUE EQUIPMENT	______

HAZARDS OF CONFINED SPACE: ______

EXCEPTIONS AND/OR SPECIAL INSTRUCTIONS: ______

APPROVAL SIGNATURES ______ OPERATIONS ______ MAINTENANCE

ADDITIONAL PERMITS ISSUED
___ HOT WORK ___ LOCK, TAG, TRY ___ OTHER ______

I HAVE INSPECTED EACH REQUIREMENT ON THIS PERMIT AND STATE THAT THIS WORK CAN BE DONE IN COMPLIANCE WITH THE SAFE ENTRY PROCEDURE AND LOCK, TAG AND TRY

SIGNATURES ______ OWNING DEPARTMENT AUTHORIZATION ______ DATE/TIME OF ISSUANCE ______ SERVICING DEPT. AUTHORIZATION

PROBLEMS ENCOUNTERED DURING ENTRY ______

PERMIT TERMINATION DATE / TIME: ______

THIS PERMIT IS VALID FOR A MAXIMUM OF 16 HOURS

FIGURE 18-5. Example Confined Space Entry Permit

EXCAVATION PERMIT C0001

REQUESTED BY:	DATE NEEDED:	FOR:

LOCATION AND DIMENSIONS

INSTRUCTIONS AND PRECAUTIONS

THIS PERMIT IS VOID UNLESS ACCOMPANIED BY THE USUAL HOT OR COLD WORK PERMITS ISSUED BY THE AUTHORIZED PERSON RESPONSIBLE FOR THE AREA IN WHICH THE ABOVE WORK IS TO BE PERFORMED AND UNTIL THE SIGNATURES REQUIRED BELOW ARE OBTAINED.

SIGNATURES REQUIRED		RESPONSIBLE FOR LOCATION OF THESE LINES
UTILITIES REPRESENTATIVE	DATE	ELECTRICAL DISTRIBUTION, RAW WATER SALT WATER, ELECTRIC LINES, STEAM LINES, CITY WATER, NITROGEN, AND AIR MAINS.
FIRE DEPT. REPRESENTATIVE	DATE	FIRE LINES
CONSERVATION SUPERVISOR	DATE	PROCESS AND CONDENSER SEWERS
BUILDINGS REPRESENTATIVE	DATE	SANITARY SEWERS
UNIT AREA SUPERVISOR	DATE	LOCAL ELECTRICAL INSTRUMENT, PROCESS, OR ANY OTHER LINES NOT ADDRESSED ABOVE

I HAVE READ THE ABOVE PERMIT AND UNDERSTAND THE WORK TO BE DONE AND THE PRECAUTIONS REQUIRED.

____
PRINT NAME OF PERSON RESPONSIBLE FOR EXCAVATION

____
SIGNATURE

☐ EMPLOYEE ☐ CONTRACTOR ________________________ PRINT CONTRACTOR COMPANY NAME

AUTHORIZED PERSON (Person Issuing Permit)	DATE ISSUED
CONTRACTOR COORDINATOR	PERMIT EXTENSION 60 CALENDAR DAYS
UNIT AREA REPRESENTATIVE	FROM __________ DATE TO __________ DATE

THIS PERMIT IS VOID 60 CALENDAR DAYS AFTER DATE ISSUED, EXCEPT WHEN EXTENSION REQUEST IS COMPLETED. WORK MUST BEGIN WITHIN 2 CALENDAR WEEKS OF ISSUE DATE. DISCOVERY OF UNIDENTIFIED CABLES, PIPES, OR RED CEMENT SHOULD BE REPORTED AND THEN WORK STOPPED UNTIL FURTHER APPROVAL IS RECEIVED.

FIGURE 18-6. Example Excavation Permit

prevent injury to personnel or interruption of critical services that might result from the damage of underground piping and cabling.

Procedures should be established to ensure that proper reviews are conducted to identify the location and depth of underground piping, power lines, telephone cables, instrument signal cables, and utility lines. Although it should be issued only after maps and reference materials are checked, the permit does not guarantee that the area is free of live cables and pipes. The possibility of such hazards may need to be stated on the permit; in any case, caution should be exercised on all excavations. Any cable, pipes or red cement (used by some organizations to identify and protect electrical cabling) found during excavation should be reported and assumed to be in service until evidence proves otherwise.

Precautions required to prevent collapse of the excavation should be detailed in the permit. Also, if the depth is significant, applicable federal, state, and local regulations should be followed and these requirements identified in the permit.

Some organizations require a similar permit for chipping or core drilling through concrete within buildings. Figure 18-6 shows a typical excavation permit form.

19

Control Software Documentation

19.1. Overview

19.1.1. Introduction

Many operations in modern process plants are digitally controlled by a class of devices known generically as programmable electronic systems (PES). These devices include the distributed control system (DCS) and the programmable logic controller (PLC). As the PES evolved, so did new documentation requirements. This new class of documentation is required to accurately describe the design, function, operation, and maintenance of control systems, and covers both software and hardware elements.

This documentation is vitally important for a number of reasons. Process upsets due to inadequate control software documentation could result in a major incident resulting in serious injury, loss of life, or damage to property or the environment. In addition, the ability to efficiently modify a control program to enhance or change a process operation can be hampered by inaccurate or out-of-date software documentation or control system drawings. Thus, the control system documentation needs to be flexible to allow incorporation of any necessary changes, while maintaining strict control of configuration, operating characteristics and procedures to ensure process safety and efficiency.

The approach to control software documentation presented in this chapter is a distillation of practices used by a number of companies and is based on concepts that have evolved over the last 20 to 30 years. As will be seen, the approach is very prescriptive and is intended to address those situations where particular attention to detail is warranted; e.g., due to the size or scope of the installation or due to stringent performance standards required by the particular application. Less documentation may be required in simpler situations. Combining document features to fit the situation can be beneficial and more cost effective; however, the requirement remains to maintain documentation adequate to ensure system integrity and handle the needs for controlled system changes.

The concepts, terminology, and definitions used in this chapter are based on IEEE standards.

19.1.2. Goals and Benefits

The goals of process control software documentation are to accurately and completely define the control system and to provide a basis for the review and authorization of control system changes before they are made. The benefits to be gained from such a documentation system include:

- fewer incidents caused by an incorrect software applications;
- better understanding by employees of control schemes, alarm points and interlocks; and
- facilitated examination and evaluation of the software design for process hazards.

19.2. Control Software Documentation

19.2.1. Objectives

The purpose of using formalized procedures for control software development and documentation is to provide visibility to what is, by its nature, an *invisible* product, and to provide a framework in which design and operating personnel may contribute constructively and consistently throughout the software life cycle. Good design and development documentation is essential to the successful implementation of a PES system. The level of adherence to quality documentation management practices, standards, and procedures is a measure of the quality of the final system. Proper documentation and adherence to procedures also provides sufficient information to define requirements and functionality that are used for quality assurance (QA).

19.2.2. Control Software Applications

Organizations will encounter many diverse applications of PES. These might include:

- use in a basic process control system (BPCS):
  —single process unit PLC;
  —single process unit DCS; or
  —plant-wide DCS.
- use in a safety interlock system (SIS).

As its name implies, the BPCS performs the moment-to-moment control function. While its capabilities are typically much more sophisticated, it is, in part, the functional equivalent of a traditional analog control loop. Typically, the most common application of a PES is to implement the BPCS function.

The SIS implements the safety interlock function and is the PES equivalent of the hard-wired, relay-based safety shutdown circuit. As with the BPCS, the SIS may offer sophisticated capabilities that go beyond those found in its traditional counterpart. In the past, the safety interlock function has not been commonly implemented via a PES, engineers and designers finding more comfort in the traditional hard-wired systems. However, as the reliability of the PES has increased, so has industry confidence, and PES-based SIS are becoming increasingly common.

Currently, it is common practice to separate the BPCS and SIS functions; an independent PLC or DCS would be required to implement each function. This is to ensure that the failure of a single device does not corrupt both the process control function and the interlock system provided to guard against loss of process control. Thus, in PES applications implementing both the BPCS and SIS functions, two sets of documentation will typically be generated. While it is up to each organization to establish its own documentation practices, it is anticipated that the documentation requirements for the SIS function would likely be more detailed and rigorous. This is in keeping with the more stringent performance standards (i.e., higher reliability) required due to the safety-critical function of the SIS.

In the balance of this chapter,the term *control system* will be used to indicate both the BPCS and SIS functions.

19.2.3. Software Life Cycle

Software development is best described by a *Life Cycle* model as defined in the *Life Cycle Stages* section that follows. This model covers the whole span of the software's existence from initial requirements analysis to final de-commissioning of the control system. *System* as used in this chapter means the complete entity comprising hardware, software, and documentation.

The life cycle is a sequence of managerial and technical activities that are grouped in stages. The suppliers of software and configuration databases manage the parts of the life cycle with which they are directly involved and follow procedures that permit the owner of the delivered system to support it through the remaining system life. Whether the software is a new project implemented by a supplier or an in-house operating plant modification by an end user, the same life cycle stages apply.

Whoever is responsible for the software or configuration database implementation part of the life cycle is also responsible for ensuring that acceptable QA procedures are available and utilized when providing documentation to satisfy these functions within the system.

19.2.4. Software Life Cycle Stages

Software development, particularly in the early stages of the life cycle is an iterative process where requirements are continually clarified, or changed, and alternative designs may be considered. The key system life cycle stages that require documentation are:

Definition Stage. This defines the user objectives that must be satisfied, determines the feasibility, develops the strategy required for meeting the user objectives and establishes the project resources and schedule.

Design Stage. During the design stage of the life cycle, the design documentation required is developed to achieve the objectives established during the definition stage.

Implementation Stage. During the implementation stage of the life cycle, the software is coded/configured and tested based on the design documentation.

Testing Stage. This stage follows implementation and ensures that the software meets the objectives set in the definition stage.

Installation and Checkout Stage. This stage follows module testing and ensures that the hardware and software, when installed together in the facility, meet the objectives set in the definition stage.

Operation and Maintenance Stage. This is typically the longest stage in the life cycle of the software, and includes its ongoing operation, maintenance, and overall management.

The documentation listed for the six (6) stages may or may not be developed formally on every project. On small projects, some parts may be unnecessary and others may be combined. Usually the documentation generated during the definition, design, testing, and installation and checkout stages is a one time effort that is kept for reference. Implementation stage documentation (module code) and operations and maintenance stage documentation comprise the long term documentation that is maintained and kept current.

Each stage should be fully documented to provide sufficient information for operation, software analysis, maintenance, and future modification.

19.3. Documentation of Control Software

19.3.1. Control Software Documentation Program

An organization's PSM program documentation should establish the activities required and assign related responsibilities for developing, documenting, and maintaining control software. The documentation challenges posed by this element are similar to those previously described for the process knowledge element, since both elements are broad in scope and involve a number of diverse activities. Additionally, the tasks requiring documentation are conducted by many people, over a staged life cycle, and the responsibilities for the resultant documentation must be transferred from one organization or individual to another at key milestones throughout that life cycle. Refer to Chapter 5, *Process Knowledge*.

The references listed in Section 19.7 provide guidance on creating such a control software documentation procedure. The issues that should be addressed by the documentation procedure include:

Responsibilities and Accountabilities for Documentation. Vendors, the system owner, programmers, and operators are all responsible within their areas of authority. Interfaces (especially with regard to transfer of documentation at life cycle milestones) and extent of responsibilities should be clearly defined.

Documentation Requirements. Documented procedures should define the documentation to be generated during the system life cycle and should identify who, by function or title, is authorized to approve the documentation.

Management of Change. The documentation procedure should include requirements for a management of change (MOC) system for software and supporting documentation. The mechanics of such a system are discussed in greater detail in Section 19.4 and in Chapter 10, *Management of Change*. It should be noted, however, the control software actually consists of two different types of software, only one of which is under the control of the system owner.

Application software is specific to the user application and is under the direct control of the system owner. In general, the term *software*, as used in this chapter, refers to *application software*.

Embedded software (also known as *firmware* or *system software*) is supplied by the vendor and is not accessible for modification by the system owner. Vendors may, however, periodically issue revisions to *embedded software*. While the system owner cannot control the vendor's modification of this software, the owner can control its implementation on the facility system. The MOC system should prevent the installation of such revised software until its potential impact on the operation and safety of the control system has been evaluated.

Auditing. The procedure should also establish requirements for the routine auditing of control software documentation and for promptly resolving any inadequacies that might be detected.

19.3.2. Definition Stage Records

The basic control software documentation for the *definition stage* falls into the following categories; descriptions of each follow:

- requirements specification;
- functional specification;

- system hardware specification; and
- test plan.

Requirements Specification. This document defines precisely how the system should function and highlights any special requirements, concerns, or constraints. The requirements specification is not intended to be a fully definitive statement of control detail, but should detail the scope requirements for the plant operations, control, automation, hardware constraints, management aspects, documentation requirements, etc. The requirements specification should include discussions of the following:

- introduction;
- references;
- process overview;
- operating philosophy;
- basic system constraints or preventive measures (what system will prevent from happening);
- start-up requirements;
- reliability requirements; and
- level of automation:
  —automatic control;
  —regulatory control; and
  —remote manual control.
- instrumentation:
  —signal levels;
  —mechanical packages inputs; and
  —instrumentation design requirements.
- control system:
  —standard loop configurations;
  —start-up sequence;
  —shutdown sequence;
  —interlocks (process actions);
  —engineering development system;
  —input/output (I/O) distribution;
  —I/O spares requirements;
  —redundancy;
  —uninterruptable power supply (UPS);
  —testing requirements;
  —failure mode analysis;
  —data highway communications; and
  —marshalling panels.
- operator interface:
  —operator terminal distribution;
  —graphics/displays; and
  —security and alarm management.

- supervisory control and management:
  —production reporting and logs;
  —batch reports;
  —laboratory interface;
  —inventory tracking system;
  —recipe management;
  —on-line help system; and
  —system security.
- project-specific issues:
  —system selection;
  —packaged equipment;
  —interface to existing distributed control system (DCS) or PLC; and
  —documentation.

Functional Specification. This document should be prepared once the requirements specification has been approved. During the definition phase the piping and instrument diagrams (P&IDs) are reviewed to develop the control strategy, interlock functions, user interfaces, process control database, etc., for the process. The functional specification should contain process control descriptions of the following in detailed text form:

- procedures:
  —system start-up requirements;
  —procedures to define the strategy for producing product; and
  —backup procedures.
- unit operations:
  —typically centered around a major piece of processing equipment; however, could also be defined to manage intensive manual activities or operator interactions within the system;
  —should be defined with logical, event driven steps; instructions and process parameters for the particular operation may be passed from outside the unit operation code (recipes);
  —defining unit operations is particularly important for a multipurpose plant or for a plant that has several similar or repetitive process operations; and
  —include procedures for situations where a system or part of a system is out-of-service.
- interlocks:
  —interlocks can be described by *if-then* structured text or by logic diagrams; and
  —the type should be the same type used to implement the system if logic diagrams are used.
- alarm and trip functions:
  —alarm definition, alarm prioritization, functionality, and grouping for the system are outlined in this part of the specification.

- matrix diagrams/failure mapping:
  —a batch sequence matrix can be used in conjunction with a process control description to completely describe a batch process. A key function of the matrix diagram is to illustrate the conditions that result in: a sequence failure; the action taken upon failure; and the recovery step. The matrix lists the sequence steps across the top, the analog and the digital devices on the left hand side of the matrix. The conditions that will cause a failure for the step and the actions that will be taken when a failure condition occurs are listed in the body.
- control module configuration:
  —requirements for complex control loops, i.e. overrides, loop initialization needs, ratio loops, cascade loops, etc., can be defined either by structured text or loop diagrams employing either Instrumentation

Society of America (ISA) symbols or Scientific Apparatus Makers Association (SAMA) symbols. Simple loops may be described by typical descriptions or diagrams.

- process controls database contains the information necessary to configure a control system. There is a record for each tag or device to be configured. The data base may include:
  —tag name;
  —description;
  —I/O addresses;
  —device information;
  —electrical requirements (120V, 24V, analog, digital, etc.);
  —process ranges and units;
  —operating setpoints; and
  —alarm trip points.
- ladder logic diagrams:
  —step processing is usually implemented utilizing a diagramming method of rungs and logic elements that sequence the execution of the steps. This type of logic development is well suited to programmable logic controllers (PLCs).
- data flow diagrams (supervisory systems):
  —an accepted structured functional analysis technique (such as Ward-Mellor) may be used to generate system requirements. High level *context* diagrams may be supported by data flow diagrams and a data dictionary to represent dynamic computer operations. This technique would be most applicable to host computer applications.
- man machine interface (MMI):
  —a detailed schematic sketch of each graphic display and configurable operator screen may be prepared to record design intentions before configuration is begun. Graphic color schemes should be stated.

- report functions:
  —report format definitions should be specified either by means of vendor-specific configuration sheets or graphical sketches.
- recipe management:
  —recipe management includes creating, editing, storing, and retrieving general, site, and master recipes and interfaces with the production scheduling and batch management activities. Statements of philosophy should be included and methods of automating these functions should be developed as required.
- batch management:
  —batch management of batch operations is the interaction of the user, recipe, production schedule, unit operations, and regulatory and discrete controls in performing their functions. The recipe selection and editing procedure, the initiation, supervision, and reporting of batch processes, and the management of batch resources all need to be described here.

A review of the functional specification should be made prior to the start of software development to ensure that all technical topics within the requirements specification have been addressed and satisfactorily resolved. The results should be recorded.

System Hardware Specification. The system hardware specification and the functional specification may be developed concurrently, and are complementary, but have separate functions. The process of developing the detailed functional specification helps to clarify and solidify the hardware requirements. The hardware specification is developed to ensure that the user's expectations of the control systems operation can be met by a vendor's hardware offering.

A hardware specification may address requirements for the following:

- process description;
- I/O types and quantities, and spares;
- I/O allocations;
- controller allocations;
- panel (front) layouts;
- marshalling;
- engineering development system;
- redundancy;
- uninterruptable power supply (UPS);
- data highway/communications;
- gateways/device drivers;
- operator interface; and
- vendor hardware/software training.

A preliminary factory acceptance test (FAT) specification can be produced and submitted with the hardware specification. The FAT specification is used as a basis for hardware acceptance at the vendor's shop. A FAT specification may include:

- visual inspection of vendor drawings;
- inventory check against the hardware bill of materials;
- check of inputs/outputs against I/O data base;
- vendor standard diagnostic checks;
- inspection of hardware component labeling; and
- vendor or user software testing and validation.

Test Plan. This test plan can be outlined at the end of the functional specification procedure. At this time, the system developer has an in-depth understanding of the philosophy, and equipment needed to test the software. A final specification of the acceptance test may be left until the installation stage. The test plan is a proposal for checking to ensure that the system contains the functionality defined in the functional specification (prior to final installation). This document addresses the methodology to be used for testing the system. The test plan includes:

- scope;
- test items;
- resources required; and
- schedule.

19.3.3. Design Stage Records

The basic control software documentation for the *design stage* falls into the following categories; descriptions of each follow:

- architectural design specification; and
- software module design specification.

Architectural Design Specification. The architectural design specification (ADS) is useful especially for software projects requiring more than one or two people to develop software. The goal of this document is to establish ground rules, conventions, and tools to make the whole software product consistent, even where several individuals are responsible for its development. This approach improves the efficiency of future updates and troubleshooting. The ADS should record how the selected system hardware and software will be engineered to meet the functional specification. The ADS includes, as appropriate:

- A baseline reference consisting of documents such as:
  —functional specification;
  —P&ID's, issued for design;
  —process equipment list, issued for design;
  —process design basis;
  —controls design basis; and
  —process control database.

- hardware block diagram;
- software block diagram showing a mapping of functions into hardware;
- details of file structures;
- program descriptions;
- source of programs;
- naming conventions;
- standard software templates developed during prototyping;
- standard software modules;
- device drivers; and
- software utilities or tools.

The architectural design specification should contain generic information about the software design. A working copy of the architectural design specification should be updated as software development progresses. A copy of changes/updates should be retained at least until the factory acceptance test (FAT) is complete, in case any related problems arise.

Software Module Design Specification. This document addresses specific design issues for each module. The extent of this documentation varies as required by the project. A module design specification should normally include the following:

- software design for each module in an appropriate form such as pseudocode or a functional representation of the logic built in to the code, flowchart, or state tables, or in the case of PLCs, the ladder logic tables;
- details specific to each software module not given as generic information in the architectural design specification;
- specific details of function block allocations;
- execution frequency if variable;
- specific timer and counter references and settings;
- program operating notes; and
- details of associated software modules.

The software design specification is expected to be an *evergreen* document, which develops as coding progresses and information is generated. It could contain all annotated code and listings thereby avoiding the need for separate files of listings. A review of the initial software design specification may be desirable before proceeding to the coding stage to ensure that the basic document structure has been created in a way that it can be used by the programmer to record details.

19.3.4. Implementation Stage Records

The basic control software documentation for the *implementation stage* falls into the following categories; descriptions of each follow:

- module software code; and
- module software test plan.

Module Software Code. Specific coding follows (using the hardware/software vendor's language) from the module design specifications developed previously. Basic documents include:

- source code;
- hard copy of program listings;
- calculations and formulas;
- editing permits; and
- revision authorizations.

Module Software Test Plan. Individual software modules may be tested using simulation software or a hardware simulator. The test plan should outline the method and procedure and criteria for success. Forms may be used during actual testing to organize the process and document results. APPENDIX 19A refers to the quality control of software. A test log should be maintained throughout the testing stages.

19.3.5. Testing Stage Records

The basic control software documentation for the *testing stage* falls into the following categories; descriptions of each follow:

- module test procedure; and
- module test results.

Module Test Procedure. A test procedure should be developed based on the testing requirements specified in the test plans. The test specification should address how each software module is to be checked. It is important to consider and identify success/failure criteria.

Module Test Results. The test results should identify and record the test parameters used during testing. Test results should be reported on standard forms. See Appendix 19A.

19.3.6. Installation and Checkout Stage Records

The basic control software documentation for the *installation and checkout stage* falls into the following categories; descriptions of each follow:

- site acceptance test plan;
- site acceptance test procedure; and
- site acceptance test results.

Site Acceptance Test Plan. A site acceptance test plan is a strategy for testing the software on the operating hardware after it is installed on the site. This may include:

- visual inspection of the equipment;
- start up, demonstration, and *run-in* of delivered hardware;

- selective tests of application software modules using the same I/O simulation as used for acceptance testing;
- demonstration of interfaces to equipment supplied by others; and
- loop checks.

Site Acceptance Test Procedure. This document defines the methodology to be used to test the system performance after installation.

Site Acceptance Test Results. A complete log of the system's performance during site testing should be maintained. This can be a valuable reference during the operation and maintenance stage.

19.3.7. Operations and Maintenance Stage Records

The basic control software documentation for the *operation and maintenance stage* falls into the following categories; descriptions of each follow:

- software source code
- reference and training manuals;
- additional system records;
- performance test records;
- change records (checklists); and
- preventive maintenance records.

Software Source Code. The focal record item during this stage would be the software itself. The source code should be periodically backed-up and a hard copy should be maintained.

Reference and Training Manuals. These manuals should be structured to be both a guide for training and as reference documents for actual operations. The manuals should address overall operating philosophy, as well as specific operator functions for operating the control system using the actual application software.

Additional System Records. These records, generated during earlier stages of the system life cycle, provide additional information describing the system. Some users may find it useful to compile this information to supplement the reference and training manuals. These records might include:

- block diagrams;
- program descriptions;
- source of programs;
- listing of I/O devices;
- loop descriptions (diagrams);
- interlock trip points;
- hardcopy program listings; and
- calculations and formulas used in control algorithms.

Performance Test Records. Performance test records are used to monitor the system over time. These records can be used to identify changes in the system response.

Change Records (Checklists). Change checklists are used to authorize and document and any hardware or software changes.

Preventive Maintenance Records. These records document routine maintenance procedures performed on the system.

19.4. Records Management

In addition to the description of the Control Software documentation program outlined in Section 19.3.1, the organization's PSM documentation should also define pertinent records management practices. The records management task for Control Software shares some issues in common with that of the Process Knowledge element. Refer to Chapter 5, *Process Knowledge*, and Chapter 4, *Records Management*, for additional guidance on records management issues.

19.4.1. Record Storage Locations

During the last stage of the software life cycle, *Operation and Maintenance*, system documentation storage practices will likely be dictated by the needs of two groups: those charged with the responsibility for maintaining the system and those actually using the system to control the process. Thus, records may be apportioned between the main computer room, the computer/instrumentation maintenance facilities, and the control room. This will likely be some overlap in the distribution of documentation; each organization must define records storage practices that meet their needs.

In earlier stages of the life cycle, responsibility for records storage will also be shared with the system vendor(s). The organization should clearly identify documentation that is to be provided at system turnover to ensure that the necessary information is available to support system operation and maintenance.

19.4.2. Media

Much of the control software documentation will be, or could be, stored on electronic media; however, most records systems will use a combination of electronic and hard copy media.

19.4.3. Management of Change

It is relatively easy to modify a control function through a software change in a PES; however, inappropriate modification might result in loss of process control and could cause serious injuries or significant damage to the facility. Changes in software may be proposed for a variety of reasons; e.g., as a result of an operational review, to clarify a technical issue, after discovery of a design limitation, to implement a scope change, or to correct a fault. Whatever the reason, control software documentation should be subject to MOC control procedures.

Changes to software or documentation (including design documentation such as a functional specification) should be authorized through the use of a system change form (SCF). Refer to Appendix 19A for an example form. This form is similar in intent to the change request forms discussed in Chapter 10, *Management of Change*. It provides a historical record of the nature of the change, the reason for the change, and the authorization for the change.

19.4.4. Configuration Management

Configuration management (CM) is the systematic control of all hardware, software, and related documents. The goal of CM is to ensure, throughout the life of the system, that there is traceability of engineering activities, that actual composition and pertinent characteristics of the system are documented and that only authorized changes are incorporated therein. Thus, CM helps document the MOC controls discussed in the previous section.

CM consists of identifying, controlling, and tracking the approved version of each document and/or software item. The following procedures outline the basics of CM:

- every item that is capable of alteration and change should have a unique title and filename;
- every page of a document including all appendixes and attachments should state the file name, file date, project number, plant name, issue, and revision;
- attachments, if any, should also state the source document;
- *issues* are the approval issue of a requirements specification, the design issue of a functional specification, or a full listing of the current system software and should be identified by whole numbers;
- *revisions* are minor corrections between issues and should be identified by two digits after a decimal point (thus a revision might appear as 1.09);
- each document should include a revision history recording the:
  —approval (when and by whom);
  —issue;
  —revision;
  —revised by;
  —date of revision; and
  —reason.
- changed items within a document should be shown by a revision marker in the margin next to the change;

- master copies of software and documents at each issue should be safely archived;
- preliminary documents should be stamped *Draft*;
- working copies of documents should be stamped *Copy* and dated;
- a reference list should be maintained for every specific document and should record the following information concerning the source of design input for the document:
  —title of source document; and
  —issue/revision of document.

19.4.5. *Replication, Storage, and Access Control*

Corruption of software and other electronic data is always a possibility and accurate records are a significant part of a speedy and successful recovery. Backing up of magnetic media is essential for the protection of documents and programs from either accidental deletion or media failure.

Copying should be carried out with copy verification as appropriate to the computer system used.

Disks should be kept in sleeves in a disk storage box and tapes in their plastic case, away from sources of light, heat, and magnetic fields, i.e., motors, loudspeakers, transformers, monitors, printers, etc.

Access to software databases should be limited to authorized personnel to guard against inadvertent or unauthorized modifications. Passwords, key-locks and administrative control procedures should be considered, as appropriate, for controlling access.

19.4.6. *Records Retention and Purge Schedules*

The control software documentation program should define requirements for retention and purging of documents. It should be particularly clear on the requirements pertaining to the records generated in the first five stages of the software life cycle. These records can tend to be voluminous; however, some (e.g., performance test results) may be irreplaceable, if lost.

19.5. Auditing

The PSM program documentation should outline the audit requirements for the control software documentation program, including frequency and procedures to be used. The first step in auditing control software documentation is to verify that a documented program exists for system development and maintenance. This program would include elements outlined in Section 19.3 that are pertinent to the needs of the organization or facility.

The second step is to verify that the results of implementing the program for control software have been documented in the manner prescribed by the program. Particular emphasis should be placed on software modification controls. One means of accomplishing this would be to audit the differences in successive versions of the software against the documented change authorizations.

Any deficiencies identified during the audit must be documented and resolved in a timely manner. Additional guidance on auditing can be found in Chapter 14, *Auditing*, and in the CCPS book, *Guidelines for Auditing Process Safety Management Systems*. The reader may also wish to refer to the two *Health & Safety Executive* publications listed in Section 19.7.

19.6. Examples

The events described below emphasize the importance of implementing adequate documentation controls for software programs that could affect safety-related systems and components. New and modified software should be properly documented and validated when safety-related functions are involved.

19.6.1. Improper Integration and Documentation of Software

In December 1992, technicians in a nuclear power plant performing routine logic testing of control software discovered that a time delay required to activate a safety-related feedwater pump was incorrectly implemented in the software. This situation resulted from attempts, made six months previously by a vendor technician, to customize the software. This activity was not documented and the system was not adequately retested (actual system logic not retested) afterwards. This failure to properly document software integration and test activities and to control the system configuration could have prevented the required automatic activation of pumps needed in an emergency and, thus, have led to very serious consequences.

19.6.2. Inadequate Test and Documentation of Control Software

In November 1991, routine systems testing by technicians in a nuclear power plant disclosed that an incorrect algorithm had been utilized in the software programming for a safety-related temperature monitoring system. The algorithm was to survey signals from a group of sensors and discard those values which deviated from the average by 100 percent or more. Instead, those deviating by only 10 percent or more were discarded. This error could have hidden excessively high temperatures in the system being controlled and could have caused a significant incident. This error resulted from a failure to adequately document, test, and audit the control software.

19.7. References

AIChE–CCPS, *Guidelines for Safe Automation of Chemical Processes,* New York, 1993.
ANSI/IEEE, Standard 610.12, *Standard Glossary of Software Engineering Terminology.*
ANSI/IEEE, Standard 828, *Standard for Software Configuration Management Plans.*
ANSI/IEEE, Standard 730, *Standard for Software Quality Assurance Plans.*
Health and Safety Executive, *Programmable Electronic Systems in Safety Related Applications*, 1 *an Introductory Guide*, 2 *General Technical Guidelines*, HMSO, London.
ISA SP88, *Batch Control Systems Models & Terminology*
ISA SP84, *Shutdown Systems.*
ISO 9000-3, *Guidelines for the Development, Supply and Maintenance of Software.*
Structured Development for Real Time Systems Yourdon Press.

Appendix 19A. Quality Control of Software Documentation

A documentation system is the preferred way of instituting and maintaining quality control of software. Chapter 19 has indicated how this could be performed and the type of documentation that could be used. This appendix provides some typical examples of the more important forms that could be used in its implementation. Other variants can be used, but care should be taken to ensure that they meet the basic objectives of the procedure.

SOFTWARE TEST RESULTS FORM

PROJECT NUMBER:	TEST REPORT No.
CLIENT NAME:	Sheet of
PROJECT NAME:	APPROVED BY:
PROJECT LOCATION:	DATE:
MODULE/PROGRAM TITLE:	TEST SPECIFICATION:

Functions tested:	Pass/ Fail

PERSONNEL PRESENT:

Name	Affiliation	Signature

TEST DATE:

SYSTEM CHANGE FORM

PROJECT NUMBER:	SCF NUMBER:
CLIENT NAME:	ISSUED BY:
PROJECT NAME:	DATE:
PROJECT LOCATION:	DCR NUMBER:

MODIFICATION:

REASON:

AFFECTED DOCUMENTS	Y/N	Completion Date	By
Requirements Specification			
Functional Specification			
Arch. Design Specification			
Module Design Specification			
Code / Configuration			
Hardware Test Specification			
Hardware Specification			

CATEGORY OF CHANGE			
Operational Need		Simplification	
Personal Hazard		Process Change	
Equipment Hazard		Design Correction	
Operational Benefit			

SOFTWARE REGISTRATION FORM

<table>
<tr><td colspan="3">PROJECT NUMBER:</td><td colspan="3">PROGRAM NAME:</td></tr>
<tr><td colspan="3">CLIENT NAME:</td><td colspan="3">ISSUE:</td></tr>
<tr><td colspan="3">PROJECT NAME:</td><td colspan="3">VERSION No.</td></tr>
<tr><td colspan="3">PROJECT LOCATION:</td><td colspan="3">DATE:</td></tr>
<tr><td colspan="6">USER DOCUMENTATION:</td></tr>
<tr><td colspan="6">MAINTENANCE DOCUMENTATION:</td></tr>
<tr><td colspan="6">TEST DOCUMENTATION:</td></tr>
<tr><td colspan="2">PREPARED BY:

DATE:</td><td colspan="2">CHECKED BY:

DATE:</td><td colspan="2">APPROVED BY:

DATE:</td></tr>
</table>

SYSTEM FAULT REPORT FORM

DATE:

PROJECT NUMBER:
CLIENT NAME:
PROJECT NAME:
PROJECT LOCATION:

No.	Date	Description	Rep. By	Corrective Action	By	Date	Retest	By

20

Document Life Cycle

20.1. Overview

20.1.1. Introduction

Chapters 5 through 19 of this *Guidelines* book provide specific documentation suggestions for individual process safety management (PSM) elements and PSM-related topics. This chapter is intended to put the information presented in these chapters into perspective and to illustrate to the reader the evolution of the documentation task for a typical chemical process. The illustrations and descriptions in this chapter generally match the five phases of Plant Life Cycle previously defined in Chapter 8, which are:

- Design;
- Fabrication and Inspection;
- Installation;
- Start-up and Initial Operation;
- Continued Operation and Maintenance.

This division is somewhat arbitrary since the phases overlap and some organizations may divide their project cycle based on other project milestones. While the chapter describes the documentation activities for a grassroots facility, many of the activities described here would be applicable to facility modifications, or the procurement of a major piece of process equipment. Thus, owners and operators of existing facilities can use pertinent parts of the information to review their documentation flow and status.

General categories of documents and their relationships are defined in this chapter and then presented as Figure 20-1, Overall Document Flow and in Figure 20-2, Document Quantities.

The narrative in Section 20.3, is a description of the Document Life Cycle Timeline (Figure 20-3). For the purpose of illustration, this timeline begins with a chart of the five phases of the Plant Life Cycle. The example life cycle illustration shows a two-year time frame from initial concept through completion of construction, plus the first two years of an extended period of operation. The remainder of Figure 20-3 addresses various document categories generated or used during the five phases.

In addition, a detailed example of the *Document Life Cycle* for a single piece of equipment is presented in Section 20.4 and Figure 20-4.

20.1.2. Goals and Benefits

The primary goal of providing information on *Document Life Cycle* is to provide an insight into the relationships between the various types of documentation that are needed to support Chemical Process Industry (CPI) projects and facilities, as well as the relationships between the various parties that are responsible for creating, using, and maintaining this documentation. Careful planning to address the documentation needs for a facility can permit:

- assigning responsibilities;
- establishing accountability;
- acquiring appropriate resources;
- establishing sound records management practices; and
- realistically determining document retention and purge schedules.

There is a risk that an organization may be overly ambitious in their documentation program by attempting to maintain too large a quantity of documents, many of which might be of little value to the organization. By not being sufficiently discriminating, an organization can burden itself with a document management task that detracts from the truly important records. Careful planning can help minimize this risk.

Benefits derived from long range planning based on an understanding of the *Document Life Cycle* include:

- efficiently acquiring and maintaining the information required to support facility operations;
- creating an optimal records management system that ensures authorized users prompt, easy access to information;
- efficient use of capital and human resources; and
- as a consequence of the above, improved facility safety.

20.2. Description of Documentation

20.2.1. What Is a Document?

The following definition of the word *document* was taken from the 1993 electronic version of the American Heritage Dictionary:

> **a.** A written or printed paper that bears the original, official, or legal form of something and can be used to furnish decisive evidence or information. **b.** Something, such as a

recording or a photograph, that can be used to furnish evidence or information. c. A writing that contains information.

As Chapter 21 will discuss in detail, the emerging multimedia work environment requires that the concept of *document* be expanded beyond the traditional paper, or hard copy, context used in the past. While documents will increasingly be found on many diverse media, the underlying theme of the document being a means of recording and conveying information will remain.

20.2.2. Overall Document Flow

The *Introduction* defined five phases for a typical project/facility life cycle. There is a pattern of information flow (and, therefore, document flow) through this life cycle, and the pattern is generally the same whether the facility is an oil refinery, a chemical plant, or a pharmaceutical plant. This flow pattern is illustrated in Figure 20-1.

Each organization needs to identify and categorize the documents it must handle during the life cycle of a facility. Categorization can assist in making determinations such as:

- who has responsibility for the document with regard to:
  —creation;
  —use; and
  —maintenance.

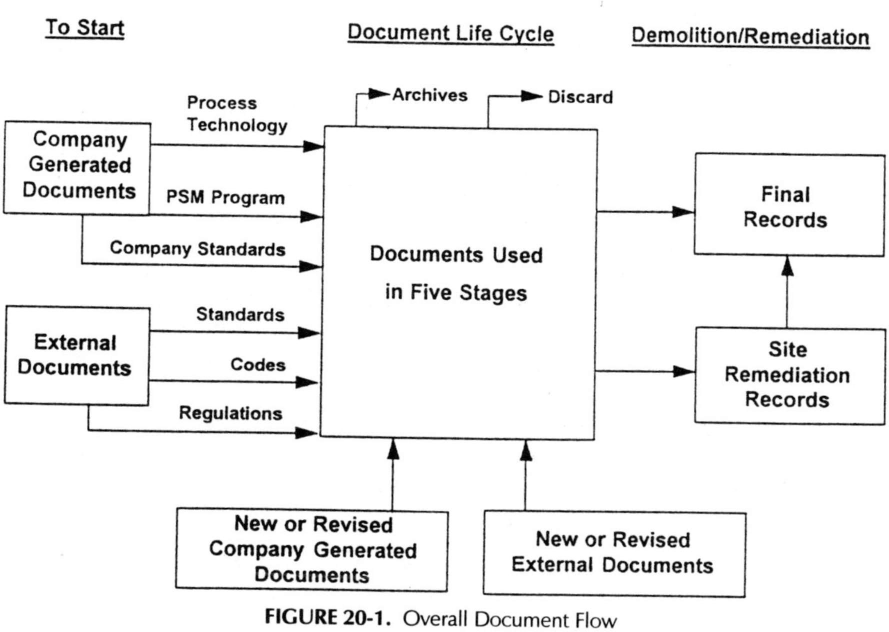

FIGURE 20-1. Overall Document Flow

- which documents must be retained, and which can be destroyed:
  —where are documents retained; and
  —what is the schedule for document destruction.
- what are the appropriate document media.

At he inception of a project there are two classes and a total of six categories of documents to consider. These are:

- Company Generated Documents:
  —process technology;
  —PSM program;
  —company standards; and
- External Documents:
  —standards;
  —codes; and
  —regulations.

These six types of documents are updated and used throughout each of the five phases of the plant life cycle. These documents, in turn, provide the information required to create other documents needed to design, build, operate, and maintain the facility.

It is helpful to further categorize company generated documents and external documents into three main groups of documents, reflecting their typical usage during the plant life cycle. These are:

- Working Documents;
- Reference Documents; and
- Archival Documents/Records.

Each of these categories of documents are subsequently discussed in greater detail.

20.2.3. Company Generated Documents

Company Generated Documents consist of PSM program documents, company standards, and process technology documents.

PSM Program Documents. When building a new facility, it is necessary to create many PSM plans and procedures required for use during the various phases of the project life cycle. PSM program documents are *reference documents* and serve as the basis for more detailed, project or facility-specific procedures. The more detailed procedures are *working documents*, and might include:

- employee participation plan;
- audit plan;
- incident investigation plan;
- emergency response plan;
- management of change procedure; and
- schedule for conducting PHA.

Company Standards. Many companies have their own design standards, and require their use by the engineering contractor or design department. Safe work practices and construction standards are often provided prior to beginning construction operations. For example, a company design standard on insulation materials and installation requirements will be used during detailed design for a variety of purposes, including piping design, and preparing the mechanical subcontract. Such standards are *reference documents* intended to ensure that the facility is designed, built, and maintained according to established best practices.

Process Technology Documents. These would normally include items such as R&D results, Process Flow Diagrams (PFDs) from another facility, a licensed Process Package, or pilot plant operations data.

Enough information must be gathered to define the process, to size and design equipment, and to identify safety considerations that will be further reviewed during subsequent PHA activity.

Process technology documents serve as the basis for many of the other *working documents* created during the design, and subsequent, phases of the facility life cycle.

20.2.4. External Documents

One type of external documents includes technical standards and codes, regulations from federal, state, and local governmental agencies, patents, licensing documents, and similar input material affecting the project in technical, legal or economic areas. This category, which becomes *reference documents*, is kept on file permanently, and updated periodically, so that any changes in the process or operation will be subject to the latest guidance or requirements.

Another type of external documents includes documents provided by equipment vendors, such as vendor drawings, parts lists and maintenance procedures. These documents can be either *working documents* or *reference documents* depending on their nature.

20.2.5. Working Documents

Working documents are those documents used on a frequent, perhaps daily, basis. These are generally documents that define objects, that prescribe or describe specific actions, or that record or show results of current activities. Examples of working documents are cited below for each of the facility life cycle phases.

Typically, working documents are *vital*, in both senses of the word: they are *important* to the task they support and they are *alive* in that there is activity associated with them. Working documents are usually in the process of being created, revised, used, filled out (as with a form), etc. Often, but not always, the user of a working document has a role in its creation or maintenance.

Some working documents have a limited service life. Such temporary working documents may become *reference documents* or *archive documents* after their short-term utility has been fulfilled.

While most working documents are company or organization generated, some can be externally generated; e.g., by equipment vendors, contract design agencies, etc.

Design Phase. During the Design phase, many documents are created which become working documents that will be maintained for the life of a facility. Others may be destroyed after being used for a short time. Some examples of working documents that might be retained are:

- P&IDs (final issue);
- detailed design drawings:
  —architectural;
  —civil;
  —mechanical;
  —heating, ventilation, and air conditioning;
  —electrical.
- detailed process knowledge documentation; and
- detailed procedures for implementing PSM program requirements, such as:
  —conducting and documenting process hazard analysis (PHA);
  —implementing a design management of change procedure.

Some examples of documents that might be destroyed are:

- P&IDs (interim issue);
- vessel drawings from unsuccessful bidders; or
- engineering evaluations of design alternatives.

Fabrication and Inspection Phase. Design Phase equipment specifications along with codes and standards from the reference documents are used by fabricators and inspectors to provide equipment that meets the design requirements. Fabrication and inspection documents are created to record the results and communicate what is being delivered to the facility for installation. Many documents created during this phase are temporary working documents that are gradually transferred to the archives as their need for accessibility diminishes.

Installation Phase. Generally, working documents created during the Design Phase are used for installation (construction). New working documents are also created during installation. These include documents such as:

- Receiving Reports/Arrival Inspections
- Calibration Records
- Welding Procedures
- Audits results

- As-built drawings of:
  —P&IDs
  —Underground Pipe drawings
  —Electrical Instrument Loop drawings
  —Master Electrical One-Line drawings
- Work Schedules, and
- Contractor performance records

Some of these (e.g., calibration records or as-built drawings) are maintained as reference documents. Others (e.g., audit results or arrival inspections) are transferred to the archives after their short-term need has been satisfied.

Start-up and Initial Operation Phase. Start-up and initial operation follows the installation phase. This activity is sometimes known as commissioning. Additional testing and performance records are created during this phase. Some examples are:

- pressure testing results for equipment and pipe;
- line cleaning records; and
- initial operating logs.

These types of records are initially part of working documents because the operators are comparing results with intentions established during the Design Phase. The results from this activity are ultimately transferred to the archives and/or used to update certain working Documents (e.g., P&IDs, Operating Instructions).

Continued Operation and Maintenance. Working documents generated in this phase could include training documents and revisions, operating procedures, maintenance procedures, details on system design revisions, plus related supporting records and authorizations.

Records of audit schedules, reports, and results are generated as required either by organizational or regulatory requirements. Incident investigations are conducted as needed and generate reports and follow-up actions that must be tracked and resolved. Many of these documents are temporary working documents that are eventually transferred to the archives.

20.2.6. Reference Documents

Reference documents, like working documents, can also describe objects or actions; however, there are important differences that distinguish them from working documents. The first distinction pertains to their content and the manner in which the are used. As their name implies, reference documents are not used as frequently as working documents; nor do they contain the same level of detail. For example, reference documents describing or prescribing actions might include standards, codes, and regulations. Such documents would describe actions in terms of policies or broad requirements that would serve as the basis for more specific actions embodied in working documents such as procedures.

Thus, as a second distinction, reference documents are often issued by organizations working at a strategic level, as contrasted with the tactical or applications focus of groups that use working documents. Such reference documents are often externally generated (e.g., by regulatory bodies or industry standards organizations). If company generated, the reference documents often are standards or policies with company-wide application. In either case, the user often has no direct role or input in the content of these documents.

Additionally, there is a second type of reference document, consisting of records of recent activities. These typically are, or are derived from, working documents and are maintained to provide a readily accessible historical record required in support of current activities. This function can be contrasted with that described below for archival documents or records.

External Documents such as standards, codes, and regulations as well as some company generated documents (e.g., company standards and PSM program documents) become reference documents that are typically maintained for the life of the facility. Changes in these documents often lead to changes in working documents. Refer to Chapter 16 for further information on standards, codes, and regulations.

20.2.7. Archival Documents or Records

Some working documents will have a relatively brief existence. These documents, after serving their immediate function, will have little, or no, residual utility and can be destroyed. Examples might include: a request-for-bid issued to a vendor who was not selected as the successful bidder; or a training examination that need not be retained after it has been graded, shown to the student, and the score recorded.

Other documents, however, may provide an important long-term historical record of an activity. Even though frequent reference to the record may not be required, there is a continuing need for the organization to retain the information supplied in the documentation. In some instances, the document itself may be as important as the information it contains (e.g., legal documents for which the signed originals must be maintained).

Such working documents (and, occasionally, reference documents) can be placed in long-term, or archival, storage. In this situation, the primary goal is not necessarily the speed or convenience with which the document can be retrieved but, rather, the security with which the document is protected.

Archive storage does not necessarily constitute the final disposition of all documents placed there. While some documents may have long-term reference value, few documents are maintained indefinitely. Most organizations will establish retention and purge schedules requiring that documents be destroyed after a certain time. These schedules are established to meet the logistical and potential legal issues related to the storage of large volumes of obsolete documentation. Refer to Chapter 4, *Records Management*, for information on records storage practices.

Archival documents or records usually include original permitting, contractual, regulatory, and related legal documents. Also included might be copies of incident investigation reports and audits in which deficiencies or discrepant items or situations were reported. The purpose for keeping these types of documents is to provide a solid historical basis for any legal, compliance or technical knowledge needs in the future.

20.2.8. Quantities of Documents

Figure 20-2, Document Quantities, illustrates the increases and decrease in the relative amounts of documentation in the three groups: *Working Documents*, *Reference Documents*, and *Archival Documents/Records*.

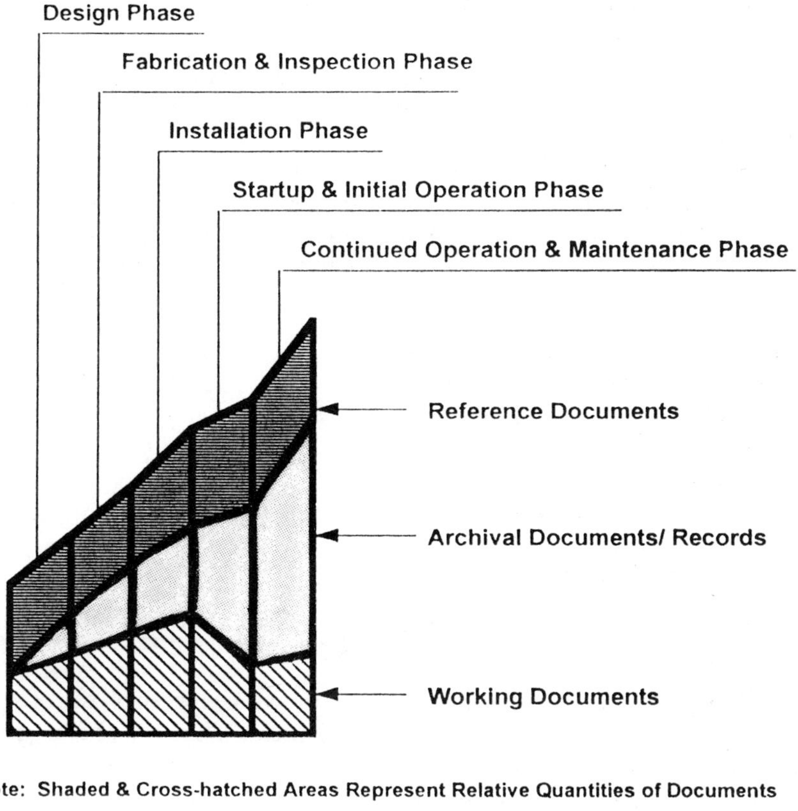

FIGURE 20-2. Document Quantities

The first group, *Working Documents* starts off with the Process Technology Documents, grows until Start-up and Initial Operation phase, then reduces somewhat to a steady-state for the Continued Operation and Maintenance phase. A deliberate effort to constrain the inventory of *Working Documents* to a manageable size is sometimes necessary.

The second group, *Reference Documents,* starts off initially as a collection of External Documents, Company Standards, and PSM Program Documents. This group grows throughout the life cycle as new standards, codes, policies, and regulations are issued; although, the rate of growth is typically not very great during the Continued Operation and Maintenance phase.

The third group, *Archival Documents/Records* starts off at zero and grows as *Working Documents* are put aside when their importance diminishes. While Figure 20-2 shows this category growing continually in size, this trend would not be anticipated to continue indefinitely. After a point, records purge schedules would typically tend to stabilize the size of the archival inventory.

20.3. Description of Document Life Cycle

The objective of analyzing *Document Life Cycle* is to show the existence of and interaction between documentation requirements on typical CPI projects and facilities throughout their life.

Document Life Cycle is depicted in detail in Figure 20-3 in a timeline which, for purposes of illustration, only covers four years. The period shown as the later two years (1996, 1997) can be extended indefinitely for as long as the facility is operational. Figure 20-3, is divided into the following parts and subparts:

- Plant Life Cycle;
- Company Generated Documents:
  —PSM Program Documents;
  —company standards; and
  —process technology.
- External Documents:
  —standards and codes;
  —regulations;
- Working Documents.

The information that follows on Plant Life Cycle Activities supplements the descriptions of documents covered in Section 20.2.

Plant Life Cycle. This section of Figure 20-3 shows the typical phases of a project and assigns some hypothetical durations to them. The five phases generally proceed in the following manner.

The Design, Fabrication, and Inspection, and Installation Phases, which overlap, typically take from fifteen to twenty four months to complete. Early

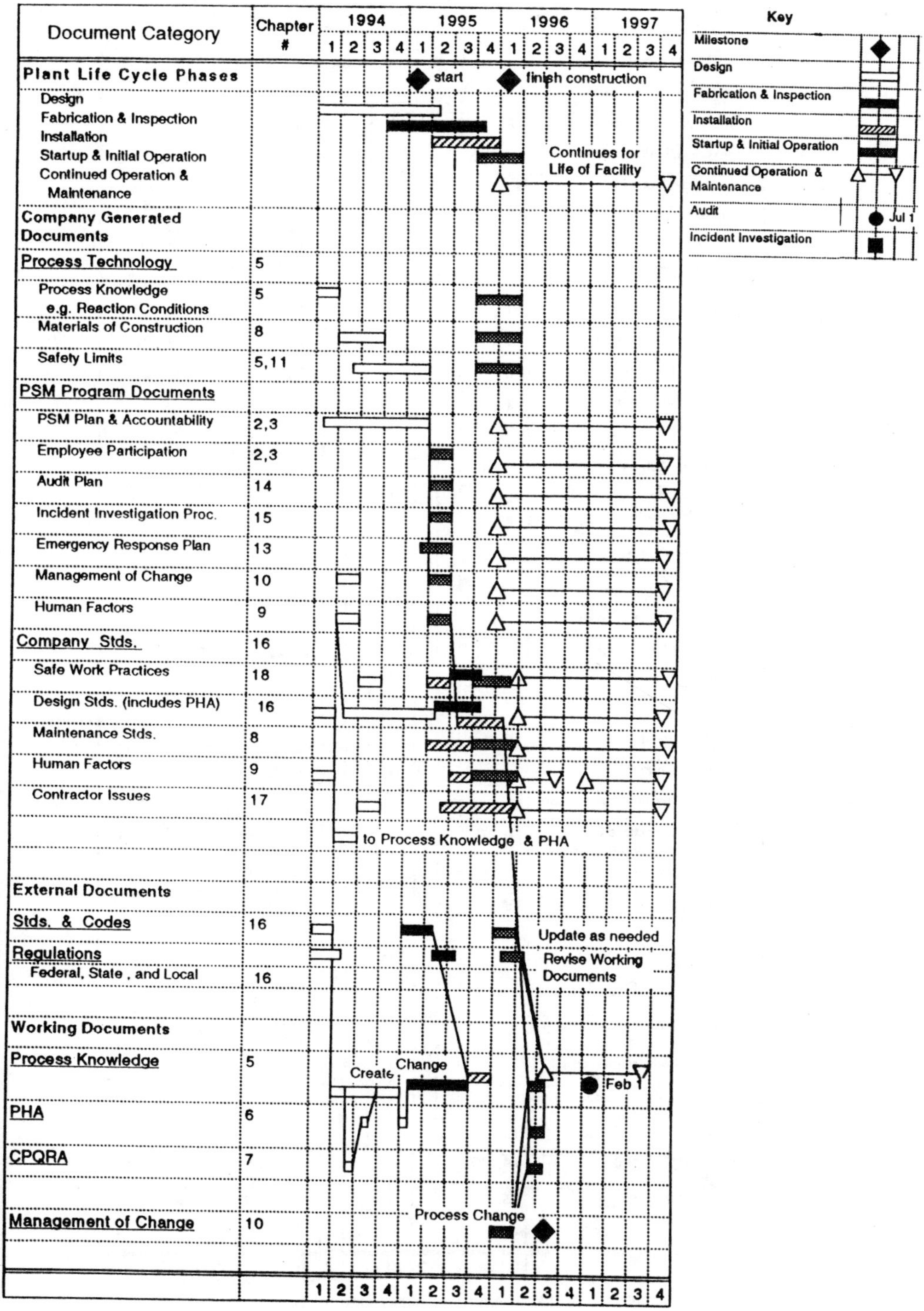

FIGURE 20-3. Document Life Cycle Timeline (Sheet 1 of 2)

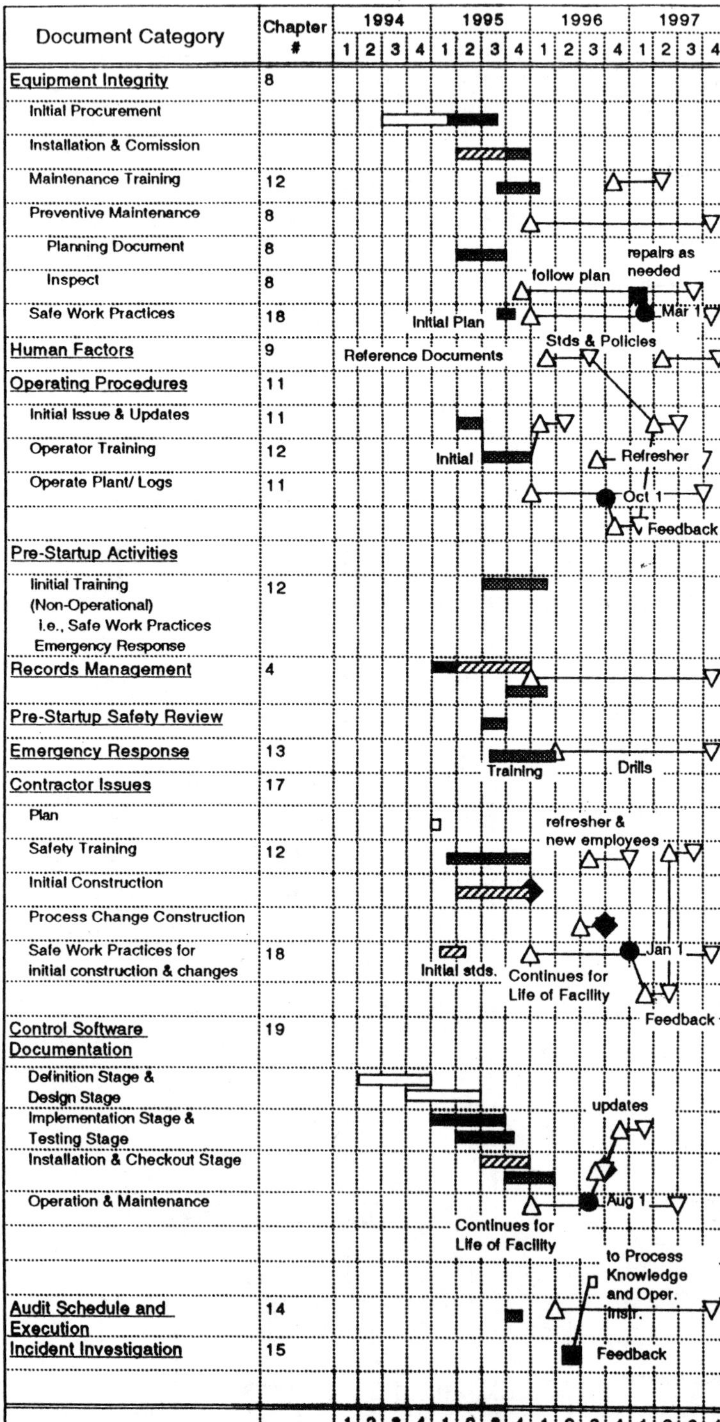

FIGURE 20-3. Document Life Cycle Timeline (Sheet 2 of 2)

activities, which include process concept development and creation of the initial process knowledge documents, typically take from three to six months. Initial analysis of the process functionality, economics, and safety takes place during the Design Phase. Preliminary PHAs may take place at this stage to decide whether to continue, cancel, or modify the process or location.

Detailed design is next. This is primarily the further development of Process Knowledge and generation of related documents that are necessary to build, operate, and maintain the facility. Once the organization commits itself to detailed design, the organization begins preparing the planning documents and the detailed procedures required to implement them in the operation and maintenance of the facility.

Procurement is next. Equipment suppliers are selected who provide documentation on how to maintain and operate each item of purchased equipment.

The Construction (Installation) Phase follows and more documentation is accumulated. Documents move from an engineering office to the actual facility, creating the need for on-site records management. A collection of working documents is gathered that are maintained for the life of the facility. These documents will ensure that management of change can be accomplished, that training activities are based on accurate information, and that operations and maintenance personnel have correct records upon which they can plan and execute their work safely.

The Start-up and Initial Operation Phase is next. At this point, planning and policy documents are close to completion. Testing and early operations records are being generated. Operations and maintenance training could be done either in this phase or could have begun during the construction phase. This time frame includes Pre-Start-up Activities, a special part of the Start-up and Initial Operation which could include generation of documents for training, creation of access controls and operating plans, planning for subcontractors' safe work practices, contractor and owner implementation of these plans, definition of records management requirements and required programs, plus final PHAs, and follow-up reports. Final verification of design drawings, procedures, and completion of training take place during this time.

The Continued Operation and Maintenance Phase takes place next for an extended period, typically for years, depending on technological and/or economic developments. Documents created will be dominated by *records*, that is, logs of actual results that historically track what goes on in a facility. Feedback from actual operations and maintenance activities will be used to update the working documents which include: PSM procedures, operating instructions, emergency response plans, maintenance instructions, process knowledge documentation, and similar documents. Incident investigations and audits will also generate feedback. Records Management will be used to handle, store, and make accessible all the various categories of documents.

Company Generated Documents. *Company Generated Documents* are depicted separately on Figure 20-3 to show relative needs and timing requirements for this hypothetical example. Later in the figure, the documents are listed wherever relevant as part of *Working Documents.*

External Documents. *External Documents* (standards, codes, and regulations) are shown at several phases of the life cycle. Changes in these documents often lead to changes in *Working Documents.* A typical link is shown.

Working Documents. *Working Documents* are listed to generally match the document categories covered by Chapters 4–19 in this book. The second column of Figure 20-3 lists the corresponding chapter in this book that describes the suggested documentation in detail. Selected examples are shown to indicate typical time frames and interrelationships between the categories at various points in the life cycle. This is a hypothetical example showing just a few of the links; many more exist. Owners/operators of facilities need to analyze their own situation in order to fully understand all the possible needs for document creation, use, and sharing.

20.4. Detailed Example

The documentation associated with each piece of equipment has its own life cycle. For example, a shop-fabricated storage tank goes through various phases, each with its own documentation, such as:

- Design Phase:
  —process specification;
  —mechanical specification;
  —procurement;
  —bid submittals;
  —selection process;
  —technical selection; and
  —commercial selection,
- Fabrication & Inspection Phase:
  —shipping;
  —delivery; and
  —inspection.
- Installation Phase:
  —final inspection.
- Start-up and Initial Operation Phase
- Continued Operation and Maintenance Phase
  —periodic inspection;
  —maintenance;
  —on-site modifications; and
- Demolition.

Prior to start-up, equipment files are usually established to include items such as:

- design basis information;
- current vessel drawing;
- inspection requirements;
- inspection records;
- maintenance history; and a
- summary of modifications.

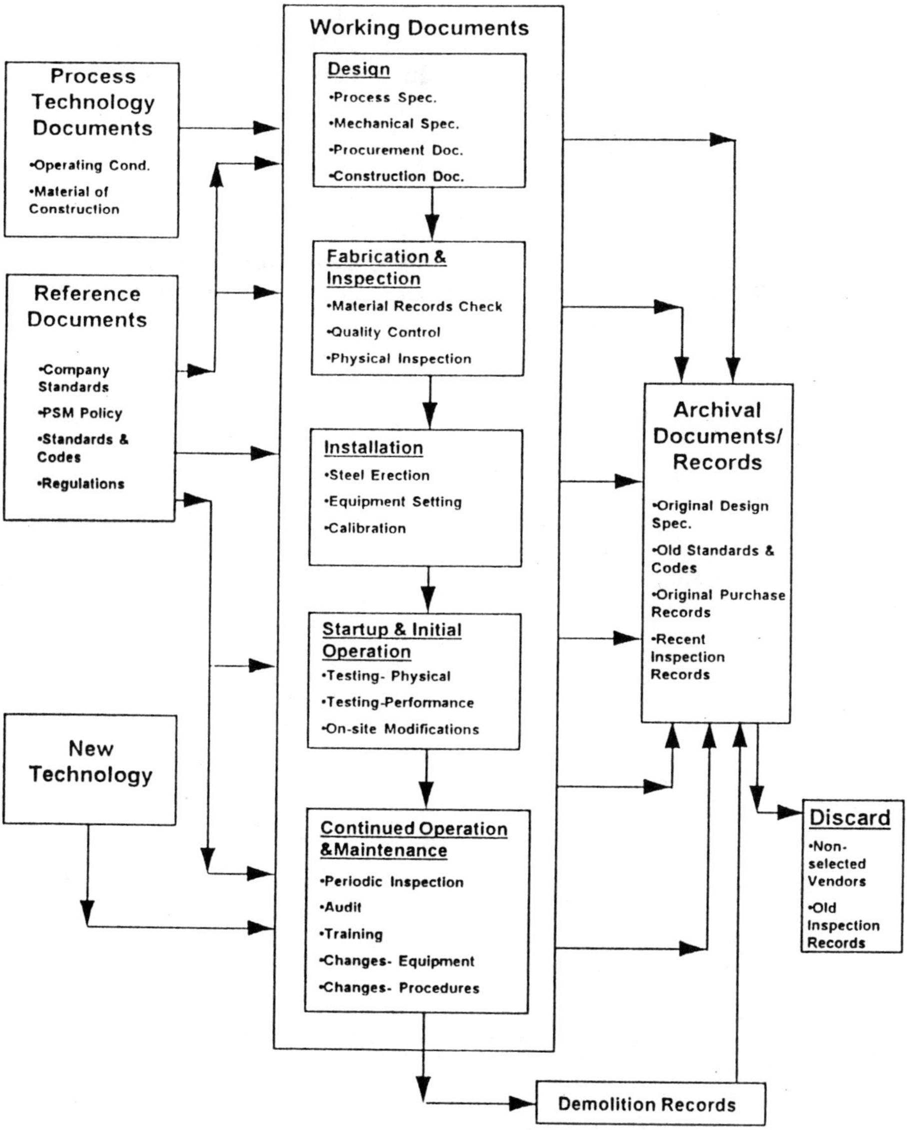

FIGURE 20-4. Equipment Life Cycle Example

This equipment file constitutes the working documents for this item. These are generally kept where they will be readily accessible to operating and maintenance personnel having a genuine need. The remainder of the documents for this storage tank will generally be kept in the archives if the records provide a valuable historical record or, if not, they will likely be destroyed. For example, bid documentation submitted by unsuccessful bidders would typically not be retained.

Each piece of equipment in a facility will have an equipment file of working documents. The contents will vary but the purpose remains the same. As company standards, codes, and regulations are updated, it maybe necessary to modify the equipment file. Figure 20-4 illustrates the equipment life cycle and documentation requirements for an individual piece of equipment.

20.5. References

AIChE–CCPS, *Guidelines for Auditing Process Safety Management Systems*, 1993.

AIChE–CCPS, *Guidelines for Technical Management of Chemical Process Safety*, 1989.

21

Emerging Technologies, Research, and Development

21.1. Overview

21.1.1. Introduction

Increasingly complex process plant operations and burgeoning governmental regulations are two prime factors that have increased the scope and the magnitude of the documentation task within the chemical process industry (CPI) over the last fifteen years. It is anticipated that monitoring, record keeping, and reporting requirements will continue to increase, driven by new needs and, in part, by technological advances in the ways that information can be generated.

Keeping up with evolving requirements, most likely with fewer employees, will be a continuing challenge for plant owners and operators. Fortunately, industry has a variety of new tools to assist in meeting this challenge, especially in the area of computerized technology. The trend has been, and will likely continue to be, away from paper documents and toward the electronic generation, handling, and storage of information.

Emerging electronic capabilities require new ways to think, act, and talk. In the past, tribal history was passed down from generation to generation by word of mouth, and it worked. Kings could not read, but they had scribes to tell them what was on the papyrus, and it worked. Copies of documents were limited to the clarity obtainable through layers of carbon paper, and it worked. We must adjust, as did our ancestors, to emerging technologies, which present new opportunities and new vocabularies. For example:

- Instead of sheets of paper we have screens;
- Instead of drawings we have views in windows;
- Instead of static information we have dynamic interaction;
- Instead of physical location we have a data path;
- Instead of footnotes we have hypertext; and
- Instead of reference symbols we have objects with data links.

As documentation gradually transitions from paper to electronic form, organizations must look at work flow and the various stages in the life cycle of documentation. In a world historically dominated by paper, the same medium has been used for document creation, storage, access, distribution, and use. There has been no compelling need to consider these as separate entities. There has also been no compelling need to distinguish between the format of a document and the medium in which it is embodied.

Digital technologies open a world that is both more complex and, in some ways, simpler. It is more complex because now the same document or document format may be intrinsically represented in different media for different purposes, creating the need to distinguish between the format and the medium. Furthermore, different media can be used interchangeably for different stages of document handling.

Upgrading to new computerized technology is costly. Timing, choice of hardware and software, and connectivity to and upgrading of existing systems are important issues. Owners and operators of existing facilities are faced with the decision of keeping existing systems in place, converting to electronic systems or combining existing with new technologies in a workable hybrid system. The decision is not often simple. This chapter defines some of the emerging technologies that are now, or soon will be, available and explains some important concepts and issues relevant to their implementation. The intent is to relate *Emerging Technologies* to specific process safety management (PSM) elements to indicate how these new tools can be used to create and maintain effective process safety document management systems.

21.1.2. Goals and Benefits

One of the goals of applying new technology to chemical process documentation is to improve the management of information required for the safe operation of facilities that handle highly hazardous materials. Proper application of computer technology can produce many benefits, including:

- improved ability to handle larger amounts of information, ensuring that it is reliably and readily available;
- improved analysis during the consideration and implementation of change;
- more effective incident investigations;
- prompt retrieval of correct and concise information for use in emergency situations; and
- more effective training documentation.

Collectively, these, and similar, benefits can enhance the overall PSM program and contribute to the creation of a safer workplace.

21.1.3. Impact of Regulatory Changes

Documentation is becoming increasingly complex, in part, due to the different types and levels of regulations that governmental agencies impose on the chemical industry. For example federal PSM regulations required by the Clean Air Act Amendments have been implemented by OSHA and proposed by EPA. In addition, PSM regulations have been issued by the states of Delaware, New Jersey, California, and Nevada. All of these federal and state regulations contain lists of covered substances. The number of specific chemicals covered by these regulations varies from 90 to about 300; in addition, some entire classes of chemicals are covered (e.g., flammables).

Thus, some facilities may have to comply with one state and two federal regulations and three different lists of covered chemicals. Additionally, each regulation has its own PSM requirements, many of them documentation related. The poses a challenge to avoid duplication of work and, thereby, minimize the costs of compliance.

Proper use of new information management technologies may help ease the burden of regulatory compliance. Some organizations in industry, and even some regulatory agencies, have already found this to be true. In 1988, a federal mandate that each state monitor every underground storage tank within its borders threatened to overwhelm Alabama's Department of Environmental Management (ADEM). The department faced the burden of filing, storing, and retrieving the documents necessary to track 41,000 tanks located at more than 14,000 sites throughout the state. ADEM used document imaging technology to dramatically simplify the document management task. A database of document images and map images was integrated with the department's information database for each tank. With the two databases linked, the imaging system enhanced the staff's ability to retrieve documentation and handle inquiries about any tank in its database.

21.2. Description of Emerging Technologies

21.2.1. Electronic Information Management

Electronic Information Management (EIM) is, in effect, the process of turning paper documents and the information on those documents into electronic data.

There are three basic operations involved in electronic documentation: establishing the system, getting information into the system, and getting information out of the system. The system involves databases, networks, telecommunications facilities, and storage devices. Getting information into the system includes both hardware (devices) and software (user or sensor interface) and includes both human and machine input. Getting information out of the system involves hardware (devices) and software (user interface and navigational techniques).

There are a number of factors that the information specialist should consider when establishing the system, including:

- information needs;
- distribution needs;
- data and document record formats;
- external information resources; and
- general information technology tools.

A systematic, analytical approach provides the best chance for the successful implementation of EIM. The following activities are important:

- needs analysis
  —work flow evaluations provide important input to the system design;
- fit analysis
  —system integration studies may need to be incorporated in the final system design to assure compatibility with other internal systems;
- implementation management
  —an implementation plan should be prepared and coupled with a staff sufficient to manage the plan;
- installation;
- integration; and
- training
  —requires time and a custom plan based on the complexity of the organization's system, current skill levels, and operational requirements.

It is important that the users of an EIM system have a sound understanding of how the system will work and what it will, and will not, do. Some important considerations to recognize when implementing EIM systems are:

- when committing information on paper documents to electronic media the work-flow of the department or organization may be restructured (e.g., documents may be circulated, reviewed, and approved without ever being printed as a hard copy);
- the information on the paper documents is often the information used to actually manage process safety (i.e., most organizations will never become totally free of paper); and
- the paper documents themselves will frequently be assets of the corporation (i.e., some documents, such as signed legal papers, may have to be retained, even though the information they contain has been entered into the EIM system).

With regard to the last item, paper, microfilm, and microfiche storage are all generally acceptable for legal documents. However, magnetic and optical disk formats at this time may not meet existing legal standards. Because data contained in these technologies can be altered electronically, the courts have been slow to recognize these formats as legal documents. In the future, employee numbers, passwords, and software keys will make it possible to electronically record document access and revisions along with time, date, and authority information; hopefully such document security features will be sufficient to make these media acceptable to the legal system.

21.2.2. New Information Management Tools

Plant owners and operators can select from a great variety of recently developed software and hardware to meet PSM documentation challenges. These include:

- touch-screen technology;
- interactive multimedia systems;
- photo compact disks (CD);
- digital photography;
- bar code readers;
- data loggers (hand-held);
- voice-activated systems;
- personal digital assistants;
- laptop or notebook computers;
- LCD projection systems; and
- fully integrated or linked software.

Each of these tools can become part of an EIM solution. Emerging technologies, which include some of the tools just listed above, are discussed in the next section.

21.2.3. Types of Emerging Technologies

This section reviews some emerging technologies that are anticipated to play an increasingly important role in EIM.

Multimedia. Multimedia is a term generally describing a group of user-interface forms that are used simultaneously. The basic forms are text, graphics, video, and sound. These terms do not just refer to the type of storage *media* (such as diskettes, hard disks, CD-ROM, etc.), but are broad concepts that include a variety of technologies. After first being used in games and entertainment, multimedia has been increasingly utilized in business as more powerful computers and software have become available.

Multimedia will expand the use of computers in business by combining audio, video, text, graphics, still images, and animation into a unified approach to presenting information. Multimedia presentations will allow businesses to convey messages more effectively.

The basic strategy when introducing multimedia technology is to combine multimedia products with an organization's existing networks, thus moving multimedia beyond the standalone environment of desktop computers. The ability to combine text, voice, graphics, and other data types, such as document imaging, in a multi-vendor networked environment is important in application building.

Close Range Photogrammetry. Photogrammetry is a technique that combines photography and computer technology to obtain physical survey results with accuracy possible to within several thousandths of an inch. Close range photogrammetry (CRP) could be used to document as-is conditions in existing plants for future reference; e.g., in establishing the basis for management of change. CAD drawings could be created and data attached to individual drawing objects in a more manageable way. CRP techniques could also be used to accurately document the aftermath of an explosion for use during incident investigation. The need for teams of people to hand measure facilities could be eliminated.

FAX Technology. Facsimile technology is used to increase office efficiency and reduce communication costs. Fax machines can be used to send or receive document copies within minutes to or from any part of the world. Because of recent digital innovations and a competitive marketplace, fax prices have fallen within the budget range of most companies. Fax machines have multiple applications, such as obtaining corporate records from a centralized source, improving the accuracy of communications (e.g., in contrast to verbal communications) and sending photographs. Emerging industry trends include using personal computers to perform fax functions and to provide increased portability.

Imaging. This is a general term for various methods of inputing and storing electronic images of paper-based documents for subsequent retrieval and usage on video display terminals. Imaging technology offers the opportunity for the business community to electronically capture, store, process, display, and retrieve information on-line, with significant savings in time and effort.

Image processing technology can make PSM activities that currently depend on paper-based information more efficient and more responsive. The benefits of imaging technology include:

- significant reductions in the manual handling of documents and files;
- faster and better in-house response time;
- improved control of documents and work flows;
- reduced clerical labor costs;
- reduced document storage costs; and
- as a consequence of the above, an improved, more efficient work environment.

Although the potential benefits of this technology are many, only a few companies have implemented imaging systems, partly because the technology is still in development. The key to tapping the potential of imaging lies with fully integrating imaging applications with existing office procedures and systems in a cost effective manner. Maximum flexibility should be designed into imaging products through the use of open systems and system components that comply with industry standards. It is anticipated that the information imaging industry will be provide advancing technologies capable of accommodating additional document types and media.

Optical Character Recognition. If one wishes to search the text of scanned documents, the scanned images must be converted to text, a process known as optical character recognition (OCR). As this technology improves, it will become an integral part of information transfer. For example, incoming faxes could be automatically received, converted into ASCII code, and compressed for efficient electronic storage.

Text-based Management Systems. Text-based management systems (TBMS), a new and still evolving technology, can help organizations keep track of the ever increasing mass of information that must be processed each day. A TBMS is similar to a database, but also has the ability to locate entire documents based on context as well as content. The search and retrieval features of a TBMS allow users to collect and evaluate information according to their needs. The seven types of text-based management systems are:

- authoring;
- text scanning and optical character recognition;
- indexing, search, and retrieval;
- hypertext;
- document assembly;
- electronic and optical publishing; and
- knowledge-based.

Magnetic Media. Low cost and versatile, magnetic media such as floppy disks, floptical disks, hard disks, and magnetic tape comprise the bulk of electronic storage for information today.

CD-ROM, WORM, and Rewritable Magneto-optical. Compact disk read-only memory (CD-ROM), write-once, read-many (WORM), and rewritable magneto-optical are the three current optical technologies. CD-ROM is used primarily for software distribution. WORM products are ideal for applications requiring permanence or an unalterable audit trail, but adhere least to established standards. Rewritable optical storage, best suited to unattended backup, archival storage, and document management, is an emerging technology where users expect to take advantage of the portability of the disks to transport data to various systems.

Computer Network Systems. The first systems for electronic information exchange used main frame computers with terminals located at remote locations. This gave organizations the ability to input and extract data from one central source. As central processing units became physically smaller and more powerful, main frames were replaced with local area networks (LANs) where personal computers communicate with other personal computers as well as the server, which functions as the electronic switcher for all information. Devices such as personal computers, printers, hard drives and other storage devices are all typically part of a network. A network can also function to connect to on-line information services such as Internet and Telnet.

Hand Held Devices. Improvements in chip technology have made possible many hand-held information management devices. Notebook computers, cellular phones, personal digital assistants, pagers, and portable barcode readers are some of the devices available today. These more powerful, and thus more useful, devices will gradually replace manual recording of information and improve access to information.

21.3. Emerging Technologies Applied to Documentation

Maintenance of accurate information is a dynamic process whereby documents flow from one stage to another. The four stages are:

- Creation or revision;
- Storage;
- Retrieval and Distribution; and
- Destruction.

Many organizations currently handle documents without the use of computers, others have automated some parts of this information flow, and a few have computerized much of the process. Those organizations applying emerging document processing technologies have typically found that it helps streamline many of the cumbersome tasks previously faced in creating, maintaining, and communicating information throughout their organizations.

Among the more basis approaches to document processing is database management, which enables users to store, sort, and search for data. Although this approach to document management may not reduce the amount of paper stored, the technology aids users in accessing the large amounts of information within their work environments. Some organizations are building productive document management systems based simply upon computerized techniques to help them keep track of their paper documents.

Other organizations are integrating more sophisticated technologies into their information management systems. For example, one company is instituting a document management system for product documentation created over the last 15 years. The system is designed to rapidly get images of hard-copy documents in front of customer representatives. This type of system (integrating stored graphic images with indexing based on keywords and references) is another simple form of document management that can be used by organizations that formerly used paper files and microfilm.

Newer document management system software can accommodate text files (word processor output), CAD files and associated data in one system. One industry trend has been to provide easily recognizable graphical displays linked to data as the documentation management and retrieval solution. By using such a system, plant organizations can make various types of information (e.g., training materials, operating instructions, maintenance procedures) available to employees without extensive training in information management practices.

The four document management stages require equipment and trained personnel to make them function. Equipment and facility requirements for a computerized system might include:

- Creation or Revision;
  —word processing
  —CAD system with database
- Storage;
  —file server
  —document locator terminal
  —drawing plotter
  —document printer
  —reproduction equipment
  —hard copy storage area
- Retrieval and Distribution;
  —remote graphic terminal(s)
  —local printer
- Destruction;
  —document shredder.

In order for a documentation management system to be truly successful, the two stages of the system that users most frequently see must effectively function in a way that supports the users' needs. These two stages are: the way the information enters the system (creating and updating); and the way the information is presented (retrieving and distributing).

Entering complex and large amounts of process safety information into a system is a challenge. For many organizations, especially the smaller ones, this must be achievable without the support of large Information Technology (IT) resources. Proper design of the system to accommodate the unique attributes of the major categories of process safety documentation would be a first step. Requiring contractors, vendors, and consultants to provide as much of the required information for new installations, expansions or alterations of a facility in the prescribed multimedia format is the next step. Actual entry of existing documentation may still fall under the current IT resource framework, however, information management consulting services could be contracted to assist. For large, multisite companies, a greater emphasis on wide area networks (WANs) would not only ensure a higher degree of standardization of process safety documentation, but also help to keep overall costs and the size of IT staffs reasonable. Striving to use newer integrated systems at the working level can also minimize IT specialist costs.

When planning for the retrieval of information from the system, it should be remembered that a document is more than just an information carrier. Experts in document management today recognize that the value of information often hinges as much on presentation as content. Presentation of information will continue to improve in effectiveness as technologies such as multimedia presentation become more prevalent. With such emerging technologies, the definition of a document will expand to include elements such as sound, animation, and video.

The multimedia presentation supplies the user the requested information in media that is close to the real-life subject. If "a picture is worth a thousand words", then an on-line, interactive, narrated, multiscreen video presentation on "How To Change a Pump Seal" is worth countless written procedures, safety warnings, and vendor drawings on the same subject. What the industry utilizes as end-data today will exist strictly as script/production source information in the future.

21.4. Implementation of Emerging technologies

21.4.1. Considerations on the Use of Emerging Technologies

When an organization changes its way of doing business, there is always the question of how and/or when to integrate new systems with existing systems.

There is no perfect time to adopt a new technology, upgrade an existing one, or abandon an outmoded system. New technologies usually hold great potential for improved efficiency, ease of use, and less chance of error; however, new technologies often carry high implementation cost and a risk that the technology will not achieve its stated goal. As a technology ages it becomes accepted, the price drops and the technology becomes proven and commonplace and the decisions become easier to make.

When new technology becomes available at reasonable cost it is readily assimilated. Copiers, printers, facsimile machines, telephone voice mail systems, word processing, and computer invoicing and accounting are a few examples of technologies that have been adopted and, essentially, are required to conduct business today. Laser printers have become an accepted, even preferred, means of hard copy output because they are faster, quieter, and produce better output than dot matrix printers. Typewriters are still in use, however, for filling out preprinted forms. As these forms become a part of the electronic technology, typewriters will become obsolete.

Emerging technology is changing at a rapid pace and an important concern is the threat of hardware and system obsolescence. For example, will the long-term storage formats currently in use be readable with the equipment commonly available decades from now? Owners of 8-track audio tapes and Beta format video tapes can appreciate the problems this presents.

When considering investment in an emerging technology, the consumer should:

- give preference to systems based upon open standards, as opposed to proprietary systems;
- look at the company behind the product as well as the technology and the product;
- avoid the expensive, but incremental improvement (fancy $500 turntables for stereo systems are worthless with today's audio CD); and
- insist on flexibility, expandability, and portability of data.

The new technology must fit the way the organization works or the organization must be willing to change the way it works (in reality a compromise often achieved between an organizations procedures and system capabilities).

21.4.2. General Issues for Implementation

Emerging technologies are complex, diverse and, by definition, always changing. Selecting hardware, software, and other solutions utilizing emerging technology is a formidable task.

Doing nothing to a fairly good system is a proper approach in some instances. Postponing installation of systems utilizing new technology (enlightened procrastination) is another approach. For example, a university in the northeastern United States waited to install fiber optic cable while other universities were tying all buildings together. When the university did proceed, the cost of fiber optics had decreased markedly, allowing the installation of a more sophisticated fiber optic cable system, direct to each dorm room. The university is now ready for high definition television, interactive real-time communications and other technologies yet to emerge.

The world of emerging technologies is complex and potentially costly and intimidating. Yet there is no doubt that industry is moving toward electronic information interface. To realistically approach utilization of emerging technology an organization must:

- know what is available;
- understand the technology concepts;
- set priorities on media, database information and conversion; and
- be willing to change employee interaction with documentation.

The primary goal of applying emerging technology to the PSM documentation task is achieve enhanced safety of facility operations. Continued use of existing workable systems while planning carefully for new technologies may be the best approach for many existing facilities. On the other hand, if existing systems are inadequate for facility needs, prompt adoption of new technologies may be the best approach. The type and age of the existing system within an organization, along with the available budget will also be factors in determining whether an organization will:

- continue with the existing system;

- scrap the existing system in favor of new technology; or
- integrate new technology with the existing system.

New systems can either be purchased as a total system or acquired by the organization, component by component. Designing in-house software is compelling for many organizations. However, this approach can lead to an extensive effort, which sometimes may result in eventual replacement by commercial software that does a better job. Implementing an extensive search for available commercial software to meet facility needs may ultimately be a time and money saving exercise. A key concept is to get information into *nonproprietary* electronic formats. Whatever system or combination of hardware and software chosen, newer, better, faster systems are coming and portability of data from one system to another is essential. Closed proprietary systems may seem attractive in the short-term, but can eventually create significant difficulties for the organization. Consequently, many software developers are avoiding such closed-end solutions.

The following example shows one way plant owners and operators can implement new technology that will improve information handling and simultaneously address several requirements.

Chemical Database. It is helpful if a plant establishes a chemical database for its facility. This can be used to store information on inventories, location, toxicity, physical properties, hazards, fire protection requirements, shipping requirements, etc. Each piece of information fulfills a particular need.

If a plant produces a variety of blends from a batch process, each blend may require a unique MSDS, and similar product-specific documentation. A computer-based system for the generation of each MSDS, DOT shipping label, etc. could be utilized to facilitate compliance with OSHA, EPA, and DOT requirements. When new product blends were created, the safety, engineering, or shipping departments would update information within the portion of the database under their responsibility. However, the end user would have only to request from the database those forms needed for a particular blend.

Use of fully integrated software that includes CAD, a relational database, and a report writer can accomplish this task. This example has already been implemented by a major U.S. pharmaceutical company resulting in a multimillion dollar savings.

21.4.3. Implementation for Specific Elements of PSM

The following paragraphs present some ideas on the application of emerging technologies to the individual elements of PSM:

Records Management (Chapter 4). Retention, access, and tracking of documents is vitally important in PSM. Emerging technologies can already greatly enhance records management. Costs of implementation may be, in some cases, a problem but this situation will improve. Indexes, cross-referencing of records that relate to more than one of the twelve elements, and duplicate records at remote locations for fire and security are integral parts of electronic document management systems. New methods of finding documents, impossible with manual systems, are available. These include full text searches using boolean methods (e.g., find: *chemical* and *vessel* not *pipe*) and searches by document name, subject or keywords.

Process Knowledge (Chapter 5). Documentation of process knowledge is descriptive and definitive in nature. Information about chemicals, process technology, and equipment are used for the design, fabrication, installation, and testing of equipment. It is also used to support a comprehensive process safety management (PSM) program. Assembling, managing, and maintaining process knowledge throughout the life of the facility is essential to safe facility operations. Documents for process knowledge are predominantly text-based, with a second component of diagrams or drawings.

At the most basic level, computer indexing of process knowledge documents can be a low-cost, high return investment. A database of all process knowledge documents showing name, number, type, and location can provide quick access to these documents, whether they be in paper or electronic form. At a higher level, computer text-based management systems can provide immediate access, security, and control of text-based documents. Drawings, diagrams, and images could be part of a computer drawing-based management system. The fully integrated solution would be to combine database, text-based and drawing-based systems into one seamless operation with navigational tools to retrieve information in various ways according to the user's preference. For example, linking equipment records to P&ID symbology is possible with off-the-shelf software.

Process Hazard Analysis (Chapter 6). The PHA process and the recording of PHA findings can be enhanced with emerging technologies. Presentation, recording, and interactivity are the key issues. There are many good software programs currently available to assist the PHA team; refer Chapter 6 for some examples of available software specific to the various techniques. Most of these programs offer guidance in terms of a checklist or series of screens to be filled in and provide a report generating capability. Using new hardware such as active matrix displays or direct color projection devices attached to a computer in the conference room where the analysis is taking place can greatly enhance the analysis. The analysis program screens can then be viewed and commented on by all team members. When the session is completed, the results are already recorded. Records can then be printed and/or made available over the computer network.

If all drawings and text-based documents were made available on-screen by networking the computer and using a view program with the ability to redline, then the entire analysis process could become electronic. The team could interact with the drawings, PHA software and each other.

Plant designers and safety experts can use 3D CAD walk-through capability during PHA work to review operability and worker access to equipment. This approach will become more common, especially for new facilities.

Chemical Process Quantitative Risk Analysis (Chapter 7). Plant and site layout, community impact and risk reduction alternatives are key concepts in risk analysis. The nature of risk analysis is computational and statistical, based on a large volume of data. Using computers to store data, compute results and automatically display information in meaningful ways is customary.

One concern with CPQRA is that the data upon which analyses were based can continually and quickly change. The use of emerging technologies to hand large quantities of information, and to facilitate the tracking and updating of this information could facilitate and enhance CPQRA. Outside information could be entered in the system through CD-ROM readers or information services. In-house information could be available on a local area network. With a complete and current computer database, facilities could share common data and comparisons could be automated. A computer system designed to retrieve population statistics, calculate, and report on changes, for example, could provide plant owners with trends in risk.

Process Equipment Integrity (Chapter 8). Process equipment must perform as expected and adequate documents must be available to allow it to be operated in a reliable, safe manner. Maintaining up-to-date, accurate, and easily accessed data for equipment plays an important role in process safety. Improved user interfaces for input of data (e.g., inspection reports) and for viewing information (e.g., display of an exploded view of a piece of equipment that could be rotated to provide various perspectives) will become more important as organizations adopt computerized information systems. Acceptance of such systems by the employees is essential to successful system implementation. Acceptance will be based on ease of use and will, with time, engender trust in electronic rather than paper-based systems. Eventually, plant floor multimedia presentations (e.g., of how to change a pump seal) may offer means of communicating maintenance details that are clearer, and less error prone than reference to an oily, dog-eared copy of a vendor manual.

Specialized software can also improve equipment integrity over the life cycle of a plant. For example, a company specializing in nondestructive testing created a software product that is used to create reports from field measurements of wall thicknesses of pipe and equipment. By using a set of software tools, which includes CAD, a relational database, and a built-in report writer, this company has been able to deliver the work product in a very efficient manner. Successive wall thickness measurements and the use of the relational database has allowed prediction of replacement dates, thus avoiding costly incidents. Including a link to the actual vessel drawing or a piping isometric has enhanced communication and helped with problem analysis.

Human Factors (Chapter 9). The study of human factors related to process and equipment design and operation involves human reliability and ergonomics (consideration of the man/machine interface). Technologies that provide clear, easily assimilated information will reduce the potential for human error. For example, as computer terminals and electronic displays replace mechanical controls, clear graphical design is essential. Replacing clearly labelled mechanical buttons with a visually confusing monitor screen increases the potential for error. On the other hand, real-time video, sound, animation, simulation, and other multimedia techniques can improve human access to and understanding of information.

Management of Change (Chapter 10). When a facility is first completed, documentation should correctly describe the facility as it then exists. As change occurs, documents must be revised so as to continue to represent the conditions of the facility. Management of change (MOC) is often made more difficult because reference documents are not up-to-date, have been relocated or misplaced, or are in a state of deterioration. MOC will become easier with the use of emerging technologies that relate and link drawings with text-based and database information. Search engines could be specifically tailored to retrieve all documentation related to a proposed change. Document revision would become an integral part of the change process if MOC was implemented via a full-function electronic document management system. Users would soon learn that such a system was the easiest method to implement a change.

Operating Procedures and Training (Chapters 11 and 12). Operating procedures provide clear, sequenced steps to safely perform a task. Training provides employees with skills, knowledge, and understanding of these instructions. Using multimedia technologies, computer-based operating procedures could be more easily understood and followed. By combining video, sound, drawings, and text, the procedure could be conveyed with less opportunity for misunderstanding. Using multimedia, computer-based training could improve employee skill and understanding in more efficient ways. This technique may combine classroom and plant floor instruction. If an employee can view a training presentation utilizing video, sound, text, drawings, and interactive response testing, acquiring the knowledge and skill becomes easier and the results are potentially longer lasting. Eventually, virtual reality systems may be used for training. Meanwhile, making good use of training videos from equipment manufacturers and professional organizations can be effective. Preparing video tapes of operators or maintenance employees performing their normal job can be used for new employee or refresher training courses.

Emergency Response and Incident Investigations (Chapters 13 and 15). Good process safety management will reduce the need for emergency response, but if an incident occurs, a system must be in place to effectively minimize the impact of the incident. Emerging technologies can improve the speed and efficiency of discovery, evaluation, notification, and action. These technologies include: networked communications on and off the plant site; software that can clearly indicate location, type, and potential consequences of an incident; and prompt electronic retrieval of documents (drawings, specifications, procedures) to assist in decision making. Technologies such as cellular phones, continuous video cameras, and automatic contact systems that enhance a smooth and rapid response, should be considered. Technologies that provide automatic recording of events prior to and as a part of the incident, as well as those providing automatic logging of response efforts will streamline the process of response reporting and incident investigation.

Auditing (Chapter 14). An audit is a systematic review to verify conformance with established guidelines. Emerging technologies can make the audit team more systematic, more independent and more effective . For example, hand-held notebook computers can provide checklists and provide data input capabilities for the audit team. Audit software programs can organize this input and produce audit reports.

When emerging technologies are used in process safety procedures, operations, and documentation, the audit team can confirm and verify a variety of information by direct access to networked documentation. A systems audit, which confirms that procedures and practices exist, could thus be facilitated. Process safety audits, which provide detailed verification that the actual performance matches procedure, could be accomplished with reduced effort with more concise reports, providing incentive for more frequent audits. Compliance reviews, which confirm that the documentation and procedures meet regulatory requirements could be conducted whenever regulatory requirements change, again with less effort.

21.5. Vision for the Future

One way to anticipate what is ahead for plant owners and operators is to look at new technologies in the general business and entertainment marketplace. New devices and new technology introduced in high volume and high profit sectors will later find application in manufacturing plant operations. There are thousands of CD-ROM titles available today, with the majority in games, education, and interactive applications. Some plant equipment suppliers are beginning to provide catalogs, manuals, and operating instructions on CD-ROMs. Increasingly information will be supplied in interactive multimedia format.

Notebook computers and personal digital assistants are in widespread use today in business. This type of devices will become more prevalent in day-to-day plant operations for input and recording of information, as well as retrieval of information on the plant floor. Cellular phones with voice activation, fax, and voice mail capabilities will continue to improve for the huge communications market. These devices will also improve intraplant communications. Access to operating or maintenance instructions will become instantaneous at any location in the facility.

Techniques developed for libraries and on-line information services will be available for process safety management (PSM) as well. Search and retrieval operations will become more streamlined. Expert systems will help to solve specific problems.

Virtual reality is used for games and flight simulation training today. In the future, operator training will include virtual reality. Imagine training for emergencies by simulating the emergency in a virtual reality environment. The individual or group will learn and be tested in, what to them, is a real situation.

People will have information at their finger tips to handle any challenge. Geography and time will increasingly become nonissues for information delivery. Powerful chip development, innovative software, multimedia interfaces, virtual environments, expert systems and artificial intelligence are all currently in their infancy. Advancements in these areas will lead to a safer workplace because owners and operators will gain more control through improved access to necessary information.

As the creation, storage, and retrieval of document images enters the digital world, office productivity is expected to soar. Work flow will be streamlined, with significant reductions in paper and storage costs. As emerging technologies improve, we should be able to:

- drastically reduce amount of paper handled;
- access information without any specialized training;
- have the system handle coordination of information; and
- present information in a more understandable and usable way.

The future is exciting. Emerging technologies will continually evolve from dreams to every day use. As a result, the chemical process industries will be able to more safely operate existing processes and safely operate new, more complex processes.

21.6. References

Corbitt, Walter, *IMC Journal,* IMC, Jan/Feb 1992

Shen, Grace, and Early, William F., *Workflow Management of Technical Documents*, 1994 Process Plant Safety Symposium, AIChE—South Texas Section, 1994.

Index

Q

R

U

V

W

Geoffrey W. Kinison, P.E.
19626 Sweet Forest Lane
Humble, Texas 77346-2174
Phone: 281-852-7895